# Ferroelectric Materials: Science and Technology

# Ferroelectric Materials: Science and Technology

Edited by
Charlotte Murray

STATES
ACADEMIC PRESS
www.statesacademicpress.com

Published by States Academic Press,
109 South 5th Street,
Brooklyn, NY 11249, USA



ISBN: 978-1-63989-198-6

**Cataloging-in-Publication Data**

Ferroelectric materials : science and technology / edited by Charlotte Murray.
p. cm.
Includes bibliographical references and index.
ISBN 978-1-63989-198-6
1. Ferroelectric devices. 2. Electronic apparatus and appliances. 3. Piezoelectric devices.
4. Electronics. I. Murray, Charlotte.
TK7872.F44 F47 2022
621.381--dc23

For information on all States Academic Press publications
visit our website at www.statesacademicpress.com

STATES
ACADEMIC PRESS

# Contents

# Preface

Ferroelectricity is the property of some materials that have an electric polarization which can be reversed by applying an external electric field. The materials that exhibit ferroelectricity are known as ferroelectric materials. All the ferroelectric materials are pyroelectric. The permittivity of ferroelectric materials is very high especially when close to the phase transition temperature. They are used to make capacitors with tunable capacitance. Ferroelectric capacitors are made up of a pair of electrodes sandwiching a layer of ferroelectric material. This book attempts to understand the multiple branches that fall under the discipline of ferroelectric materials and how such concepts have practical applications. It covers in detail some existent theories and innovative concepts revolving around ferroelectric materials. This book will prove to be immensely beneficial to students and researchers in this field.

This book unites the global concepts and researches in an organized manner for a comprehensive understanding of the subject. It is a ripe text for all researchers, students, scientists or anyone else who is interested in acquiring a better knowledge of this dynamic field.

I extend my sincere thanks to the contributors for such eloquent research chapters. Finally, I thank my family for being a source of support and help.

**Editor**

1

# Polarization and Thermally Stimulated Processes in Lead-Free Ferroelectric Ceramics

Aimé Peláiz-Barranco, Yuslín González Abreu,
José de los Santos Guerra, Jinfei Wang,
Tongqing Yang and Pierre Saint-Grégoire

**Abstract**

Over the last 20 years there has been an enormous effort in developing lead-free ferroelectric ceramics in order to obtain good dielectric, ferroelectric, piezoelectric and pyroelectric properties than those the conventional ferroelectric ceramics based on lead, such as lead zirconate titanate. An important group of lead free ferroelectric materials belong to the Aurivillius family, compounds which are layered bismuth $[Bi_2O_2]^{2+}[A_{n-1}B_nO_{3n+1}]^{2-}$. $SrBi_2Nb_2O_9$ is a member of this family where the ferroelectric properties can be affected by the crystallographic orientation due to their anisotropic crystal structure. The divalent Sr cation located between the corner-sharing octahedra can be totally or partially replaced by other cations, most commonly barium. The chapter presents the analysis of the thermally stimulated current in $Sr_{1-x}Ba_xBi_2Nb_2O_9$ ferroelectric ceramic system with x = 0, 15, 30, 50, 70, 85, 100 at%. A numerical method is used to separate the real pyroelectric current from the other thermally stimulated processes. The remanent polarization is evaluated considering the hysteresis ferroelectric loops; the pyroelectric coefficient and the merit figure are evaluated too. $Sr_{0.70}Ba_{0.30}Bi_2Nb_2O_9$ shows better ferroelectric and pyroelectric properties.

**Keywords:** Ferroelectrics, Aurivillius, Ceramics, Polarization, Pyroelectricity

## 1. Introduction

It is known that lead-based ferroelectric systems exhibit very good properties for different kinds of applications [1-6] and also that high-level ferroelectric and piezoelectric activities have remained confined to these materials. The only drawback in the technology as a whole has been the environmentalist's nightmare of its dependence on a high lead-containing family of materials [7]. Therefore, over the last 20 years there has been an enormous effort made in developing lead-free ferroelectric systems in order to obtain better dielectric, ferroelectric, piezoelectric and pyroelectric properties than those of conventional ferroelectric ceramics based on lead, such as lead zirconate titanate (PZT) [8-10].

An important group of lead-free ferroelectric materials belong to the Aurivillius family $\{[Bi_2O_2]^{2+}[A_{m-1}B_mO_{3m+1}]^{2-}\}$, which was discovered by Bent Aurivillius in 1949 [11]. These compounds have a complex structure, which is composed of perovskite blocks ($[A_{m-1}B_mO_{3m+1}]^{2-}$) interleaved between bismuth and oxygen layers ($[Bi_2O_2]^{2+}$), where *m* is the number of perovskite blocks in the structure. The *A* sites of the structure are typically occupied by elements such as $Sr^{2+}$, $Ba^{2+}$, $Ca^{2+}$ and $Bi^{3+}$, with low valence; the *B* sites are typically occupied by elements with high valence such as $Ti^{4+}$, $Nb^{5+}$ and $W^{6+}$ [12-16]. These materials have received great attention due to their large remanent polarization, low real dielectric permittivity at room temperature, lead-free nature, relatively low processing temperatures, high Curie temperatures, high electromechanical anisotropy and coercive fields, and excellent piezoelectric properties [8-9], which have suggested them as good candidates for high-temperature piezoelectric applications and memory storage. The bismuth layers $[Bi_2O_2]^{2+}$ constrain the size of perovskite blocks establishing a limit for the incorporation of elements into them and providing the mixing of different elements between *A* sites and bismuth sites in the layered structure [17]. The ferroelectricity depends strongly on the crystallographic orientation of these materials, which is the subject of continuing researches. The main contribution to its spontaneous polarization comes from the displacement of the *A* cation in the perovskite block, which is quite different for the perovskite structure. It is well known that these have the majority polarization vector along the a-axis in a unit cell and that the oxygen vacancies prefer to stay in the $Bi_2O_2$ layers, where their effect upon the polarization is thought to be small, and not in the octahedral site that controls polarization [18].

Figure 1 shows the structure for some Aurivillius systems with m=1, m=2 and m=3, at the paraelectric phase, as examples. The structural studies on these materials have shown a relation between the number of perovskite blocks and the symmetry of the cell, i.e., the number of perovskite blocks is related to the crystallographic orientation and to the plane of polarization in these materials [12-16, 18-19]. The polarization vector has also shown a relation to the number of perovskite blocks [18]. For even-layered systems, it has been reported to be a restriction on the polarization to the *a-b* plane of the cell and an orthorhombic symmetry with $A2_1am$ space group [18]. For odd-layered systems, the polarization has shown a component in c and orthorhombic phase with *B2cb* space group [18]. Other results have shown a strong relation between the elements in *A* sites of the structure and the symmetry of the cell [13-14, 17].

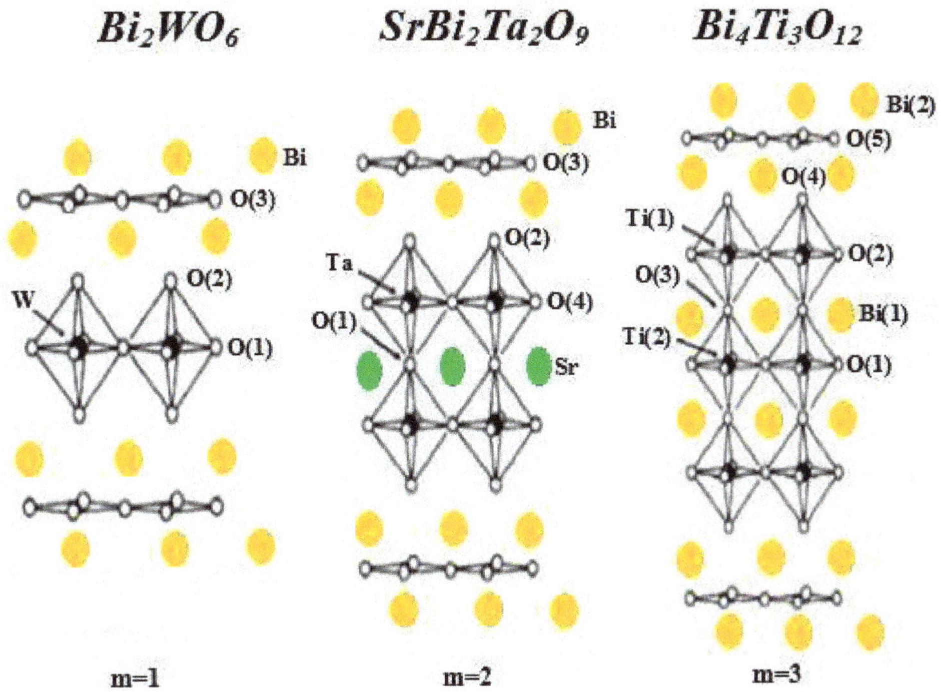


**Figure 1.** Structure of some Aurivillius materials with m=1, m=2 and m=3, at the paraelectric phase.

$SrBi_2Nb_2O_9$ (SBN) is a member of the Aurivillius family in which the ferroelectric properties can be affected by the crystallographic orientation due to their anisotropic crystal structure [13-14]. This system has received particular attention due to its large fatigue resistance, which has been associated with the migration of oxygen vacancies in the material [20]. The $Sr^{2+}$cation, which is located between the corner-sharing octahedral, can be totally or partially replaced by other cations, as barium is an important element for improving fatigue resistance [20]. The studies on the barium-modified $SrBi_2Nb_2O_9$ system have shown interesting results from the structural and dielectric point of view [13-14, 20-25]. Structural studies have shown an orthorhombic symmetry with $A2_1am$ space group for pure and doped SBN samples [22]. The mixing of different elements between *A* sites and bismuth sites, which occurs to equilibrate the lattice dimensions between the $(Bi_2O_2)^{2+}$ layers and the perovskite blocks, has been also analysed [22]. The oxygen vacancies, which are the results of $Bi^{3+}$ for $Ba^{2+}/Sr^{2+}$ substitution, could have an important influence in the properties of these compositions [22].

For the $Sr_{1-x}Ba_xBi_2Nb_2O_9$ system (x= 0, 15, 30, 50, 70, 85, 100 at%), the barium concentration dependence of $T_m$, as well as the temperature of the corresponding maximum for the real part of the dielectric permittivity, has suggested a cation site mixing among atomic positions, which has been supported by structural analysis [22]. For compositions with $x \leq 30$ at%, $T_m$ increased with the barium concentration; for $x \geq 50$ at%, a decrease of $T_m$ and a widening of the curves was observed with the increase of the barium concentration. The structural studies have shown the mixing of $Sr^{2+}$, $Ba^{2+}$ and $Bi^{2+}$ into *A* sites and the bismuth sites of the structure [22]. For lower barium concentrations ($x \leq 30$ at%), the presence of bismuth into *A* sites and the increasing of the strontium concentration into this site, has been discussed as the principal reason for the increase of the $T_m$ value. The higher barium concentration into *A* sites was obtained for the compositions with x ≥ 50 at% [22], supporting the decreasing of $T_m$ [22].

On the other hand, a change from normal ferroelectric-paraelectric phase transition to relaxor behaviour has been observed when the barium concentration is increased [22]. For the compositions showing relaxor behaviour, an increase of the frequency dispersion degree was also observed with the increase of barium concentration. The relaxor behaviour is typical of materials with a disorder distribution of different ions in equivalent sites of the structure, which is called compositional disorder. For the studied materials, the relaxor behaviour has been explained with reference to the positional disordering of cations at *A* sites of the structure, which delays the evolution of long-range polar ordering [23, 26].

## 1.1. Ferroelectric behaviour and pyroelectricity in ferroelectric materials

It is known that ferroelectric materials present a spontaneous polarization in the absence of an electrical field (*E*), for temperatures below the temperature of the phase transition from the ferroelectric to the paraelectric phase [1]. These materials have regions with uniform polarization, which are called ferroelectric domains. If an electrical field is applied to the material, the structure of domains changes due to the reorientation of the dipoles with E. In ferroelectrics with normal ferroelectric-paraelectric phase transitions, if the electrical field is strong enough the system can reach a saturated state, showing a high percentage of oriented domains in the E direction, which depends on the structure of the system. When the electrical field is removed, the system exhibits a remanent polarization ($P_r$), which corresponds to the configuration of the minimal energy. On the other hand, for relaxor ferroelectrics, typical slim loops suggest that most of the aligned dipole moments switch back to a randomly oriented state upon removal of the field.

Ferroelectric materials, good isolators by their nature, exhibit temperature-dependent polarization, i.e., when the sample is heated the polarization changes and an electrical current is produced (pyroelectric current) which disappears at a certain temperature [1]. For normal ferroelectrics, the pyroelectric current ($i_P$) achieves a maximum value when the temperature ($T$) increases, and then decreases until zero at the ferroelectric-paraelectric phase transition temperature. For relaxor ferroelectrics, the pyroelectric current is different from zero even at higher temperatures than $T_m$, as well as the temperature of the corresponding maximum for the real part of the dielectric permittivity [27].

However, the study of the pyroelectric behaviour and its corresponding physical parameters may be quite difficult in many ferroelectric systems because, apart from the localized dipolar species, free charges can also exist in the material. The decay of the electrical polarization could be due to dipolar reorientation, the motion of the real charges stored in the material and its ohmic conductivity. The first of these is induced by thermal excitation, which leads to decay of the resultant dipole polarization, and the second is related to the drift of the charges stored in the internal field of the system and their thermal diffusion. During the temperature rise, the dipoles tend to be disordered gradually owing to the increasing thermal motion, and the space charges trapped at different depths are gradually set free. Therefore the pyroelectric behaviour is usually overlapped by other thermally stimulated processes, and a detailed analysis of this phenomenon is very important in order to separate the different components of the electrical current during the heating of the material (*i-T* dependences), to then make a real pyroelectric characterization of any system.

The thermally stimulated discharge current method is a typical technique for this analysis, which has been applied with very good results to ferroelectric materials [28-30]. By using this method, the pyroelectric current can be separated from other stimulated processes (including the electrical conductivity mechanisms), providing better knowledge of the material response in a wide temperature range.

Several analytical methods have been developed to analyse the thermally stimulated processes [30-33]. Among these can be mentioned the initial rise method, the peak shape method and the numerical method using Gaussians [30-33].

For the initial rise method, it is considered that measurements do not depend on the heating rate in the initial rise region [33]. Then, a slow heating rate can be used, reducing the problems related to the difference of temperature between the samples and the thermocouple or gradients of temperature in the sample. The peak shape methods [34] depend on the constant heating rate, but consider more experimental points concerning the initial rise method. However, these methods do not consider the overlapping of several peaks in the material response, as the Gaussian method does. This method considers the overlapping of several peaks in *i-T* dependences, which is very useful for a better understanding of the material response.

The Gaussian method was proposed by Faubert and Sánchez [32]. It consists of fitting the rightmost part (highest temperatures) of the curve with a single time relaxation theoretical curve (Gaussians), and then a new spectrum is obtained by subtracting the theoretical curve from the experimental one (Figure 2). The operation is repeated until the resulting spectrum is smaller than a fixed limit. The final test is carried out summing all the theoretical curves, which may offer the experimental spectrum.

From the so-called area method given by equation (1), where $T_F$ is the final temperature and $\beta$ is the constant heating rate, which is constant, the relaxation times ($\tau$) can be calculated for each theoretical curve (single curves). On the other hand, it is known that the relaxation times usually show a temperature dependence which can be expressed by the Arrhenius law (equation 2, where $k_B$ and $\tau_0$ are constants). Then, the activation energy value (U) for each process can be obtained from the $\ln \tau$ vs $1/T$ dependence.

$$\tau(T) = \frac{\int_T^{T_f} i(T)dT}{\beta . i(T)} \qquad (1)$$

$$\tau(T) = \tau_0 e^{\frac{U}{k_B T}} \qquad (2)$$

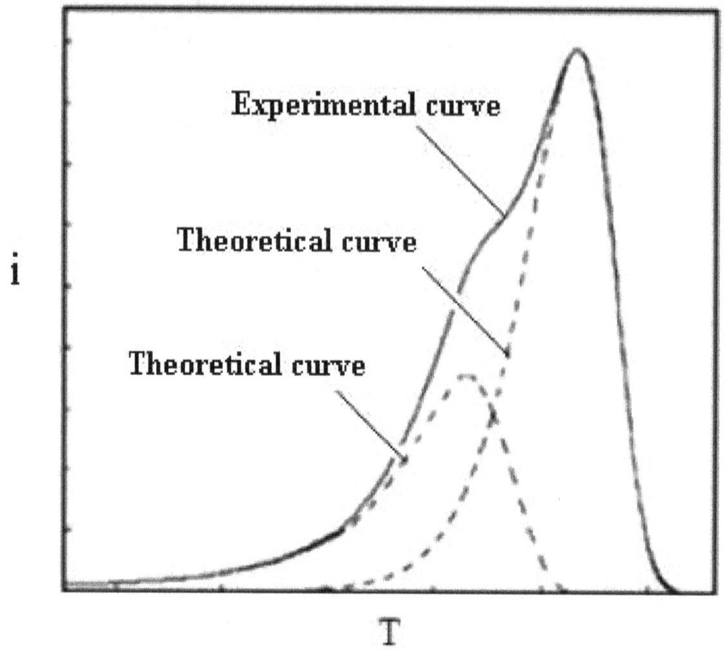


**Figure 2.** Theoretical decomposition of the *i-T* dependence using the Gaussian method.

The remanent polarization ($P_r$) can be obtained from the pyroelectric current $i_p(T)$ using equation 3, where $A$ is the area of the samples. The integration is made from the operation temperature $T$ (usually room temperature) until $T_m$ (or a higher temperature in the case of relaxor ferroelectrics).

$$P_r = -\frac{1}{A\beta}\int_{T}^{T_m} i_p dT \tag{3}$$

Other parameters can be evaluated from the $i_p(T)$ dependence, such as the pyroelectric coefficient ($p$) and several merit figures. The pyroelectric coefficient is related to the variation of $P_r$ (equation 4). The current response parameter ($R_v$) is one of the important merit figures which are associated with pyroelectric behaviour, and can be obtained using equation 5.

$$p = \frac{dP_r}{dT} \tag{4}$$

$$R_v = \frac{p}{\varepsilon'} \tag{5}$$

There are not many reports concerning the pyroelectric behaviour of ferroelectric systems from the Aurivillius family. Most of the studies have been carried out on pure and modified bismuth titanate [35-36]. For niobium- and thallium-modified bismuth titanate, it has been reported that there is a pyroelectric coefficient of 12 μC/m$^2$K at room temperature [36], which is better than that for pure bismuth titanate ceramics [35]. The *P-E* hysteresis loops have showed a remanent polarization of 3.49 μC/cm$^2$ at room temperature [36].

The chapter presents studies on ferroelectric properties and thermally stimulated processes which have been carried out on the $Sr_{1-x}Ba_xBi_2Nb_2O_9$ ferroelectric ceramic system with x = 0, 15, 30, 50, 70, 85 and 100 at%. The dependence of the polarization on the applied electric field is discussed at room temperature, for normal and relaxor ferroelectrics compositions. For the thermally stimulated current, the Gaussian method is used to separate the pyroelectric contribution from the other contributions to the total *i(T)* response in the studied samples. The remanent polarization is evaluated, at room temperature, considering the hysteresis ferroelectric loops and the pyroelectric current dependence $i_p(T)$. The pyroelectric coefficient and the current response merit figure are also evaluated.

## 2. Experimental Procedure

### 2.1. Sample preparation

$Sr_{1-x}Ba_xBi_2Nb_2O_9$ (x = 0, 15, 30, 50, 70, 85, 100 at%) ferroelectric ceramic samples were prepared by solid-state reaction method (Figure 3).

The powders of the starting materials SrO, BaO, $Bi_2O_3$ and $Nb_2O_5$ were mixed with a desired weight ratio. The mixture of oxides was milled with alcohol for two hours, dried and pressed by applying 100 MPa. The pressed samples were calcined in air atmosphere at 950 °C for two hours. After calcination the samples were milled again for one hour, dried and pressed by applying 200 MPa. The sintering process was made in a sealed alumina crucible at 1100 °C for one hour. Samples with density values higher than 90 % of the theoretical density values were obtained. Silver electrodes were deposited on the opposite faces of the disk-like samples by using a heat treatment at 590 °C. The samples were named SBN (x=0), SBBN-x (x=15-85 at%) and BBN (x=100 at%), respectively.

### 2.2. Ferroelectric measurements and thermally stimulated discharge current experiments

Polarization-electric field (*P–E*) loops were obtained at room temperature for 10 Hz by using a precision ferroelectric analyser (Premier II, Radiant Technologies Inc.), which is combined with a high-voltage power supply (TReK Model 663A). The highest applied electric field was 90 kV/cm for the studied samples.

The study of thermally stimulated depolarization currents was carried out in sequential thermal cycles as follows: (*i*) zero-field heating - heating from room temperature to 60 °C under zero electrical field; (*ii*) field cooling – cooling to room temperature while a polarizing electrical field is applied ($E_P$ = 2 kV/mm); (*iii*) zero-field heating – heating from room temperature to temperatures higher than $T_m$ under zero electrical field. The thermal discharge current was

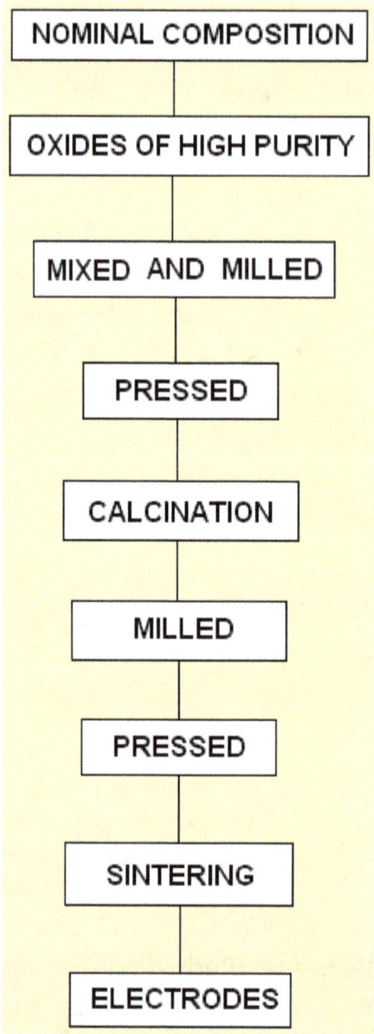


**Figure 3.** Solid-state reaction method for the sample preparation.

measured (during the third step) using a Keithley 6485 Electrometer, while keeping a temperature rate of about 5 K/min.

## 3. Results and discussion

### 3.1. Ferroelectric properties

The hysteresis loops at room temperature are shown in Figure 4 and Figure 5 for the studied samples. The compositions with $x \leq 30$ at% show polarization-electrical field (*P–E*) loops typical of normal ferroelectric materials. The compositions of SBN and SBBN-15 show wide loops at room temperature. This behaviour could be associated with high dielectric losses in these samples. The composition with 30 at% of barium shows the better response with a clear tendency to saturation with the applied electric field. The samples with $x \geq 50$ at% show thin hysteresis loops, which are typical of relaxor ferroelectric systems. These compositions have shown relaxor behaviour in the corresponding dielectric analysis [22]; relaxor ferroelectrics do not show a tendency to saturation in the *P–E* dependence even in a very high electric field.

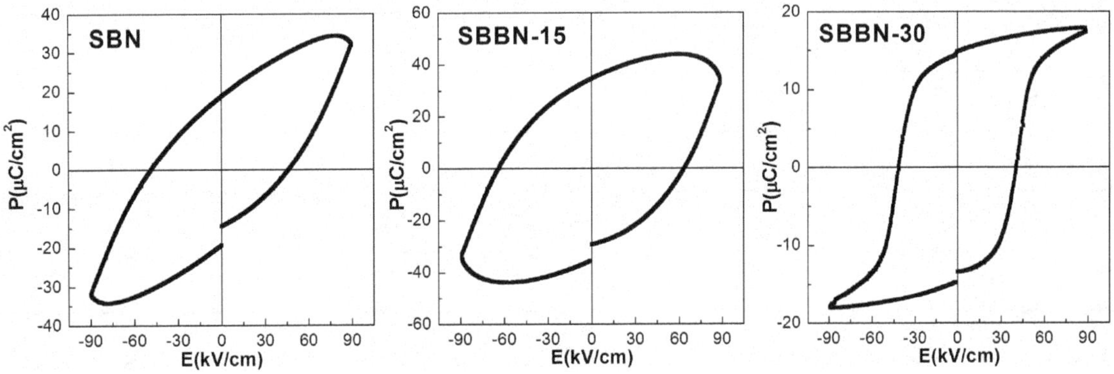


**Figure 4.** Polarization (*P*) dependence on the applied electric field (*E*), at room temperature, for samples with x ≤ 30 at %.

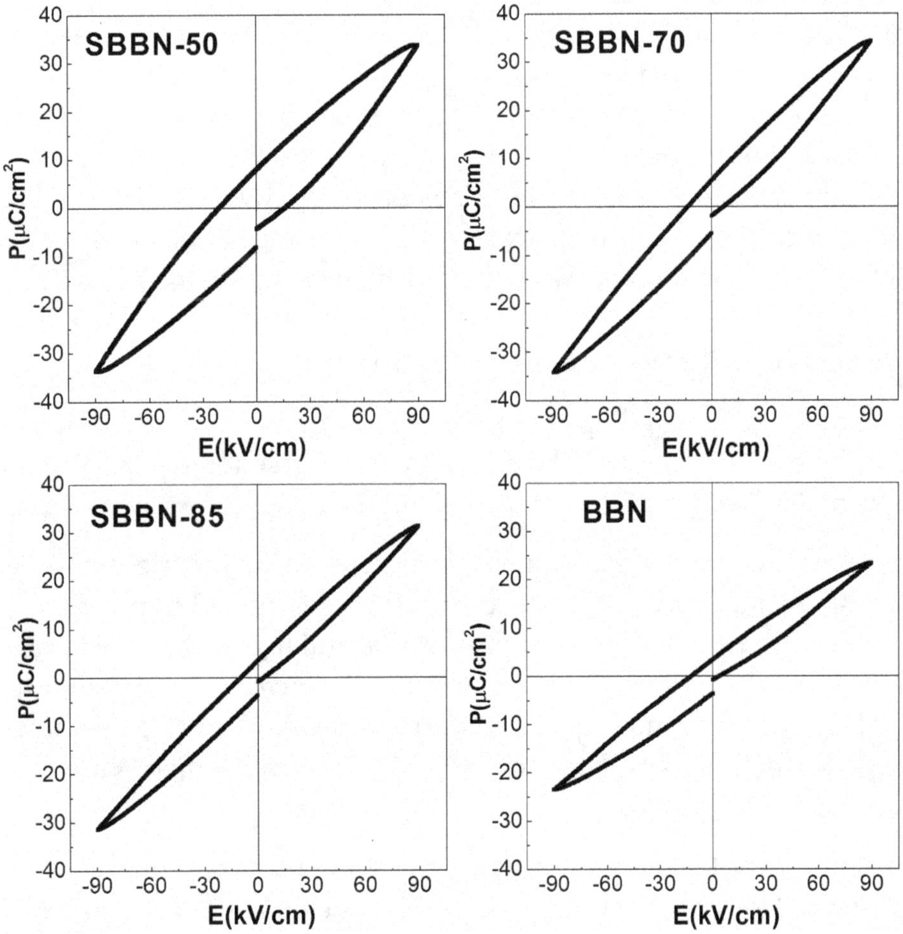


**Figure 5.** Polarization (*P*) dependence with the applied electric field (*E*), at room temperature, for samples with x ≥ 50 at%.

| Composition | $P_r$ ($\mu C/cm^2$) | $P_r/P_{max}$ | $E_c$ ($kV/cm$) |
|---|---|---|---|
| SBN | 18.96 | 0.55 | 45 |
| SBBN-15 | 34.48 | 0.80 | 63 |
| SBBN-30 | 13.45 | 0.75 | 42 |
| SBBN-50 | 8.10 | 0.24 | 22 |
| SBBN-70 | 5.44 | 0.16 | 14 |
| SBBN-85 | 3.68 | 0.11 | 12 |
| BBN | 3.55 | 0.14 | 13 |

**Table 1.** Values of the remanent polarization ($P_r$), $P_r/P_{max}$ relationship and the coercive field ($E_c$) for the studied samples at room temperature.

Table 1 shows the values of the remanent polarization ($P_r$), the $P_r/P_{max}$ relationship and the coercive electric field ($E_c$), at room temperature, for the studied compositions. $P_{max}$ is the polarization at the highest applied electric field. The SBN and SBBN-15 samples show the highest $P_r$ values. This is associated with higher dielectric losses for theses samples. The compositions with x ≥ 50 at% show the lower $P_r$ values. These ceramics have also presented lower piezoelectric activity [37]. The sample with 30 at% of barium shows the better ferroelectric response with a high $P_r$ value and a $P_r/P_{max}$ relation, showing a good saturation. For this composition, a better piezoelectric response has been reported [37]. The $E_c$ values tend to decrease with the increase of barium concentration. For compositions with x ≤ 50 at%, the $E_c$ values are higher than those for other ferroelectrics materials from the Aurivillius family [35].

### 3.2. Thermally stimulated processes and pyroelectricity

Figure 6 and Figure 7 show the dependence of the thermally stimulated current (*i*) on the temperature in the studied samples. The black points represent the experimental curve and the lines represent the fitting, which was carried out using the Gaussian method.

For the compositions with x ≤ 30 at% (Figure 6), three different contributions were observed below the transition temperature ($T_m$). The contribution at higher temperatures (blue line) is observed from the increase of *i* at temperatures near to and higher than the transition temperature. The pyroelectric contribution is characterized by an increase to a maximum value, when the temperature (*T*) increases, and then a decrease to zero at the ferroelectric-paraelectric phase transition temperature. From this point of view, the third contribution is not the pyroelectric contribution.

The dielectric analysis of the studied samples has shown a strong influence of the electric conductivity on the dielectric parameters at the higher temperature range [22]. The third contribution could be associated with the electric conductivity processes in this temperature range. The influence of this contribution on the second (black line) is remarkable; this second contribution must be associated with the pyroelectric response. The first contribution (red line)

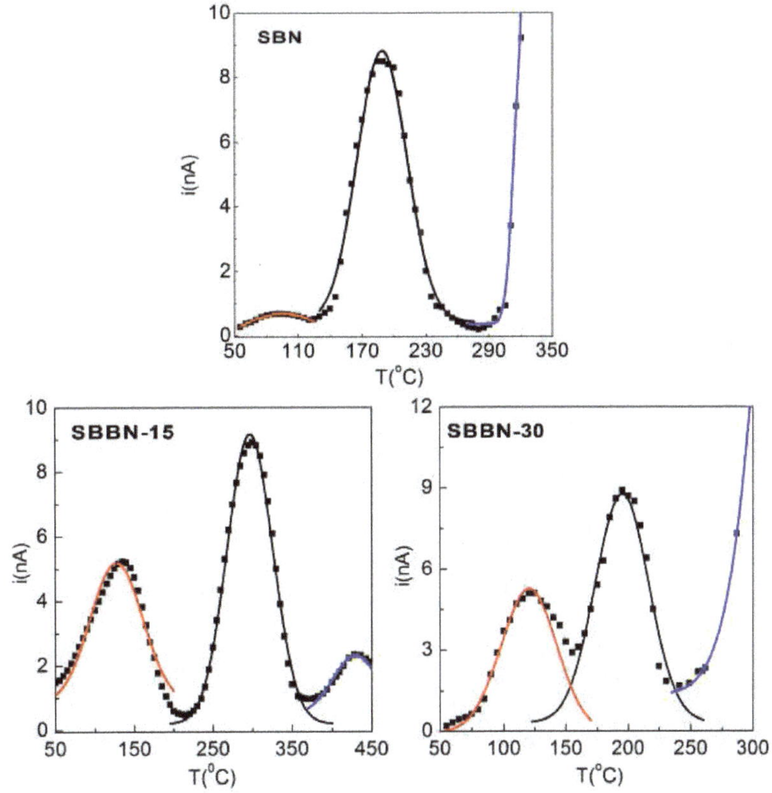


**Figure 6.** Thermally stimulated current curves ($i$) in a wide temperature range for the SBN, SBBN-15 and SBBN-30 samples. The black points show the experimental data and the red line (first contribution), black line (second contribution) and blue line (third contribution) represent the fitting using the Gaussian method.

is observed at the lower temperature range; it could not be associated with the pyroelectric response or electrical conductivity processes.

The compositions with x > 30 at% show the three contributions for temperatures lower than $T_m$ as well (Figure 7).

From the theoretical curves $i(T)$, which were obtained by using the Gaussian method, the values of the relaxation time ($\tau$) were calculated (equation 1). The temperature dependence for $\tau$ (Figures 8 and 9) was obtained for the first and second contributions, showing a typical Arrhenius dependence (equation 2). The values of $\ln \tau$ are represented by points and the lines represent the fitting using equation 2. From the fitting, the corresponding activation energy values (U) for each contribution were obtained, and are shown in Table 2.

The activation energy values for the first contribution are between 0.40 and 0.60 eV. This contribution is observed in the lower temperature range, showing lower current values than those obtained for the second contribution. The first contribution could be related to space charge, which is injected during the polarization process. For the second contribution, which is associated with the pyroelectric current, the activation energy values tend to increase with the increase of the barium concentration until 30 at%; above that concentration, this parameter decreases.

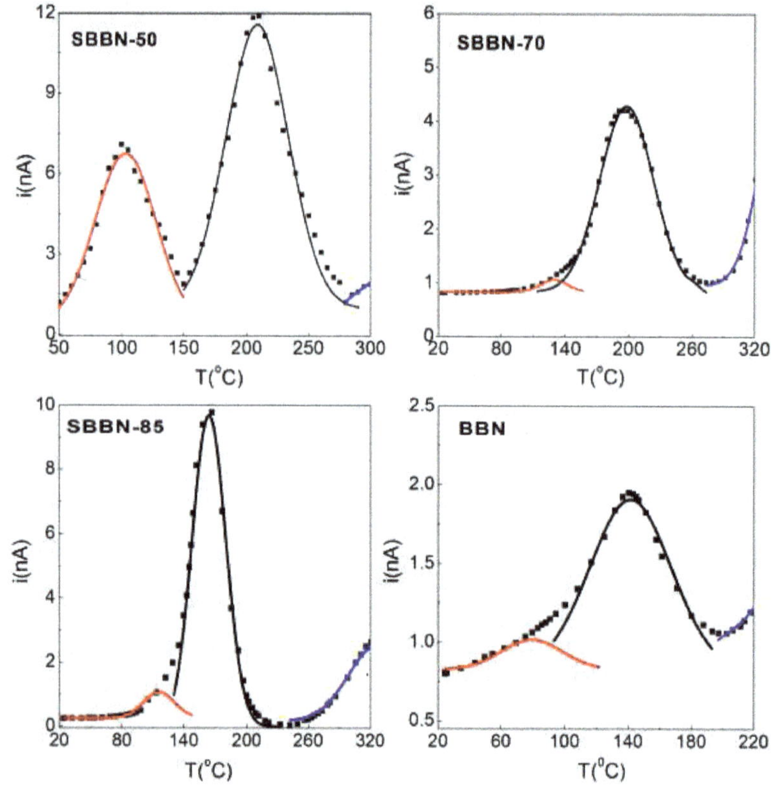


**Figure 7.** Thermally stimulated current curves (*i*) in a wide temperature range for the compositions with x > 30 at%. The black points show the experimental data and the red line (first contribution), black line (second contribution) and blue line (third contribution) represent the fitting using the Gaussian method.

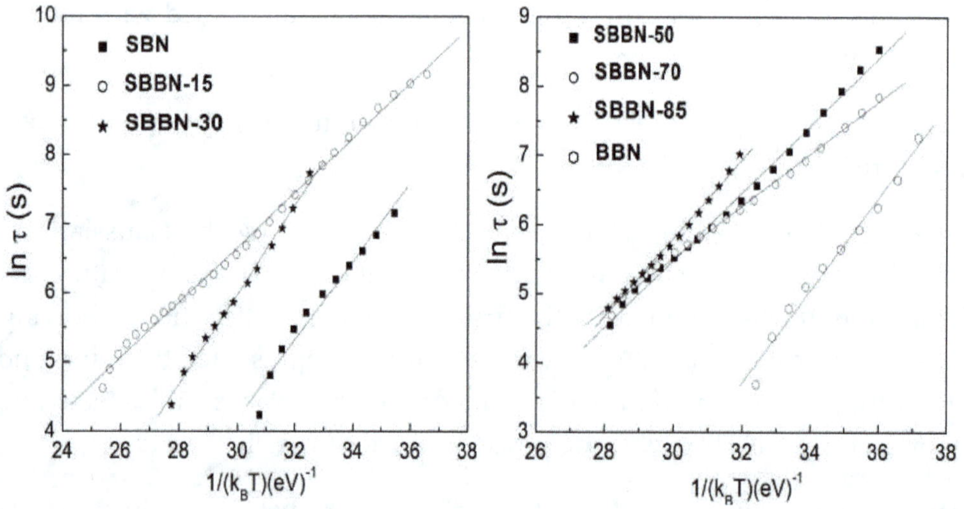


**Figure 8.** Arrhenius dependence of the first contribution on the *i-T* dependence. The solid points correspond to the relaxation time values, which were obtained using equation 1; the solid lines correspond to the fitting using equation 2.

For materials from the Aurivillius family, the major contribution to the spontaneous polarization comes from the motion of the *A* cation in the perovskite blocks [20, 23, 26]. The analysis of the dielectric behaviour for the studied samples has shown a lower ferroelectric-paraelectric

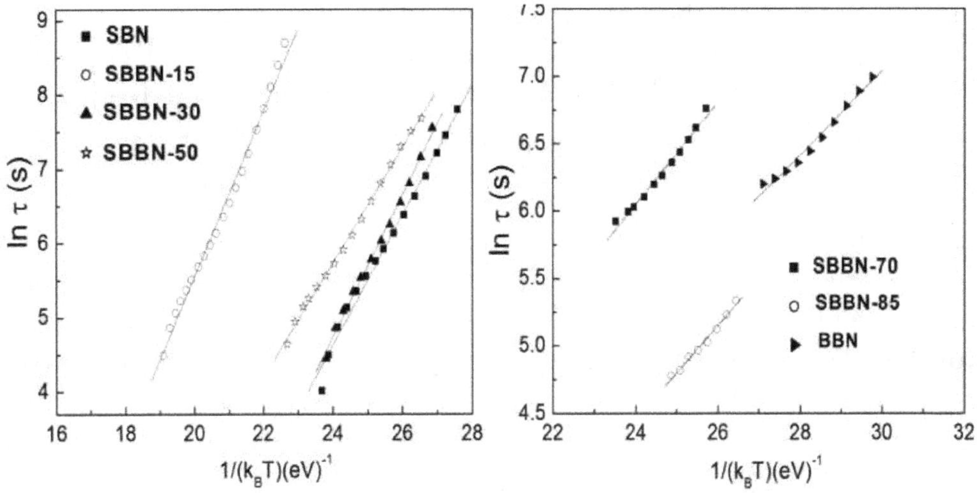


**Figure 9.** Arrhenius dependence of the second contribution on the *i-T* dependence. The solid points correspond to the relaxation time values, which were obtained using equation 1; the solid lines correspond to the fitting using equation 2.

transition temperature for the SBN sample [22] than that of the previous report [20]. This result suggests a decrease of the thermal energy, which is necessary to transition from a ferroelectric phase to a paraelectric phase. The structural study for this composition has also shown a higher occupancy of $Bi^{3+}$ in *A* sites of the structure [22] than previous reports [14], which can explain the lower $T_m$ value considering the lower radii ionic of the $Bi^{3+}$ than that of the $Sr^{2+}$. Therefore, a lower activation energy value is necessary for the thermal depoling (pyroelectric contribution) of the studied SBN sample compared to previous reports. The SBBN-15 and SBBN-30 samples show an increase of the activation energy value with respect to the SBN composition, which is in agreement with the $T_m$ behaviour from 0 to 30 at% [22].

| Composition | First contribution U (eV) | Second contribution U (eV) |
|---|---|---|
| SBN | 0.57 | 0.88 |
| SBBN-15 | 0.40 | 1.11 |
| SBBN-30 | 0.51 | 0.98 |
| SBBN-50 | 0.49 | 0.82 |
| SBBN-70 | 0.37 | 0.38 |
| SBBN-85 | 0.57 | 0.35 |
| BBN | 0.60 | 0.33 |

**Table 2.** Activation energy values, which were obtained from the fitting shown in Figures 8 and 9.

For compositions with $x > 30$ at%, the activation energy values for the pyroelectric contribution have shown a decrease with the increase of the barium concentration in the structure. These results are in agreement with the dielectric behaviour of these compositions, which is shown in a decrease of the $T_m$ values when the barium concentration increases [22]. The SBBN-50 and SBBN-70 samples have shown a greater increase in the occupancy of $Ba^{2+}$ and $Bi^{3+}$ in *A* sites than was observed in the SBBN-30, since the concentration of $Ba^{2+}$ is higher than that of $Bi^{3+}$, which explains the decrease of $T_m$ from 30 to 70 at% of barium and then by extension the lower activation energy values for the pyroelectric contribution.

For SBBN-85 and BBN, a greater decrease of the $Ba^{2+}$ occupancy in *A* sites is observed than in SBBN-70 [22], but the $T_m$ values are lower. For both compositions, it has also been reported that there is a higher $Ba^{2+}$ occupancy in $Bi^{3+}$ sites than in the other compositions [22]. A higher $Ba^{2+}$ occupancy in $Bi^{3+}$ sites and the corresponding generated oxygen vacancies would distort the ionic dipoles due to the *A* sites' ions. Then, the decay of the spontaneous polarization could be affected, providing a decrease of the $T_m$ values and the activation energy values for the thermal depoling process (pyroelectric response).

For the third contribution, there were not enough experimental points in some compositions. Thus, the activation energy was only estimated for the studied samples, showing values between 0.7 and 1.50 eV. These values are related to electrical conductivity processes, which are governed by double ionized oxygen vacancies [21, 24]. The oxygen vacancies in the structure of the studied samples are generated to compensate the electrical charge unbalance, which is caused by the substitution of trivalent $Bi^{3+}$ ion for divalent $Ba^{2+}$ and $Sr^{2+}$ ions.

### 3.3. Ferroelectric and pyroelectric parameters

Figure 10 shows the temperature dependence of the remanent polarization ($P_r$), which has been obtained from the pyroelectric current dependence $i_P(T)$ using equation 3. It can be noted that there is an important influence of the barium concentration on the $P_r$ values. At room temperature, an increase of $P_r$ is observed for the lower barium concentration ($x \leq 30$ at%); above 30 at%, $P_r$ decreases. These results are in agreement with the $P_r$ behaviour, which has previously been discussed in relation to the hysteresis loops (*P-E* dependence).

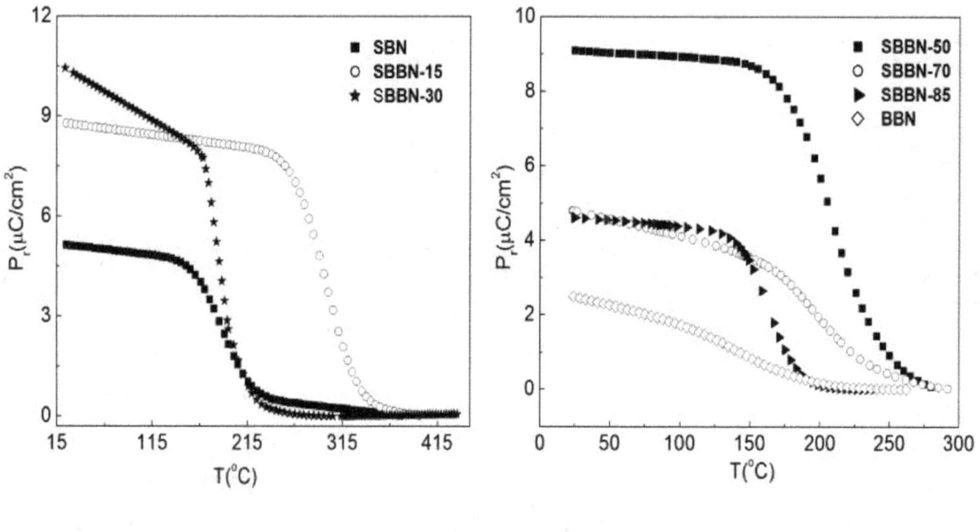


**Figure 10.** Temperature dependence of the remanent polarization ($P_r$) for the studied compositions.

| Composition | $P_r$ ($\mu C/cm^2$) | $p$ ($\mu C/m^2$ °C) | $R_V$ ($\mu C/m^2$ °C) |
|---|---|---|---|
| SBN | 5.03 | 34 | 0.12 |
| SBBN-15 | 8.63 | 38 | 0.16 |
| SBBN-30 | 10.4 | 73 | 0.62 |
| SBBN-50 | 9.02 | 20 | 0.07 |
| SBBN-70 | 4.79 | 4 | 0.01 |
| SBBN-85 | 4.61 | 10 | 0.04 |
| BBN | 2.49 | 4 | 0.02 |

**Table 3.** Values of the remanent polarization ($P_r$), the pyroelectric coefficient ($p$) and the current response parameter ($R_V$).

Table 3 shows the values for the remanent polarization ($P_r$), the pyroelectric coefficient ($p$) and the current response parameter ($R_V$), at room temperature, for the studied samples. The last two parameters were obtained using equations 4 and 5, respectively. The SBBN-30 sample shows the better pyroelectric parameter. The values for the pyroelectric parameters are analogous to previous reports in other Aurivillius materials [24]. However, these are lower than those for conventional lead-based ferroelectric systems [38-39]. Further researches are in progress.

## 4. Conclusions

The ferroelectric properties and thermally stimulated processes were studied in the $Sr_{1-x}Ba_xBi_2Nb_2O_9$ ferroelectric ceramic system with x = 0, 15, 30, 50, 70, 85, 100 at%. The dependence of the polarization on the applied electric field was discussed, at room temperature, for normal and relaxor ferroelectric compositions. The Gaussian method was used to separate the pyroelectric contribution from the other contributions to the total i(T) response in the studied samples. Three different contributions were obtained in the studied temperature range. The first contribution was associated with space charge, the second with the pyroelectric current and the third with the electric conductivity processes. The remanent polarization, the pyroelectric coefficient and the current response merit figure were evaluated at room temperature. The SBBN-30 showed better ferroelectric and pyroelectric properties.

## Acknowledgements

The authors would like to acknowledge the Third World Academy of Sciences (RG/PHYS/LA Nos. 99-050, 02-225 and 05-043), and the ICTP, Trieste-Italy, for financial support of the Latin-American Network of Ferroelectric Materials (NET-43). Thanks to CNPq and FAPEMIG agencies of Brazil. Dr Aimé Peláiz-Barranco acknowledges Le Conseil Régional de Languedoc-

Roussillon for her invitation to the University of Nîmes, France. Thanks to the Embassy of France in Havana, Cuba, for financial support for the scientific cooperation between the University of Nîmes and Havana University. Dr Aimé Peláiz-Barranco would like to thank Tongji University, Shanghai, China.

## Author details

Aimé Peláiz-Barranco[1*], Yuslín González Abreu[1], José de los Santos Guerra[2], Jinfei Wang[3,4], Tongqing Yang[3] and Pierre Saint-Grégoire[5]

*Address all correspondence to: pelaiz@fisica.uh.cu

1 Facultad de Física - Instituto de Ciencia y Tecnología de Materiales, Universidad de La Habana. San Lázaro y L, Vedado. La Habana, Cuba

2 Grupo de Ferroelétricos e Materiais Multifuncionais, Instituto de Física, Universidade Federal de Uberlândia. Uberlândia – M G, Brazil

3 Functional Materials Research Laboratory, College for Materials Science and Engineering, Tongji University, Caoan, Shanghai, China

4 Dwo Chemical (China) Investment, LTD. Pudong district, Shanghai, China

5 University of Nîmes, Department of Sciences and Arts, Nimes cedex, France

## References

[1] Xu Y. Ferroelectric Materials and Their Applications. The Netherlands: Elsevier Science Publishers B.V.; 1991.

[2] Haertling GH. Ferroelectric ceramics: History and technology. Journal of the American Ceramic Society 1999; 82(4) 797-818.

[3] Ivan IA, Rakotondrabe M, Agnus J, Bourquin R, Chaillet N, Lutz P, Poncot J, Duffait R, Bauer O. Comparative material study between PZT ceramic and newer crystalline PMN-PT and PZN-PT materials for composite bimorph actuators. Advanced Materials Science 2010; 24(15-16) 1-9.

[4] Rauls MB, Dong W, Huber JE, Lynch CH. The effect of temperature on the large field electromechanical response of relaxor ferroelectric 8/65/35 PLZT. Acta Materialia 2011; 59(7) 2713-2722.

[5] Li F, Zhang S, Xu Z, Wei X, Luo J, Shrout TR. Temperature independent shear piezoelectric response in relaxor-$PbTiO_3$ based crystals. Applied Physics Letters 2010; 97(25) 252903.

[6] Li F, Zhang S, Lin D, Luo J, Xu Z, Wei X, Shrout TR. Electromechanical properties of $PbIn_{1/2}Nb_{1/2}O_3$–$PbMg_{1/3}Nb_{2/3}O_3$–$PbTiO_3$ single crystals. Journal of Applied Physics 2011; 109(1) 014108.

[7] Wang H, Ren MF. Characteristics of Ag/$Bi_{3.25}La_{0.75}Ti_3O_{12}$/p-Si heterostructure prepared by sol-gel processing. Journal of Sol-Gel Science and Technology 2007; 42(3) 247-250.

[8] Zhang H, Yan H, Reece MJ. Microstructure and electrical properties of Aurivillius phase $(CaBi_2Nb_2O_9)_{1-x}(BaBi_2Nb_2O_9)_x$ solid solution. Journal of Applied Physics 2010; 108(1) 014109.

[9] Yi ZG, Li YX, Liu Y. Ferroelectric and piezoelectric properties of Aurivillius phase intergrowth ferroelectrics and the underlying materials design. Physica Status Solidi A 2011; 208(5) 1035-1040.

[10] Cui Y, Fu X, Yan K. Effects of Mn-doping on the properties of $BaBi_4Ti_4O_{15}$ bismuth layer structured ceramics. Journal of Inorganic and Organometallic Polymers and Materials 2012; 22(1) 82-85.

[11] Aurivillius B. Mixed bismuth oxides with layered lattice. I. Ark. Kemi 1949; 1(54) 463-480.

[12] Wachsmuth B, Zschech E, Thomas N, Brodie S, Gurman S, Baker S, Bayliss S. Structure model of Aurivillius compounds. Physica Status Solidi A 1993; 135(1) 59-71.

[13] Ismunandar, KB. Structure of $ABi_2Nb_2O_9$ (A = Sr, Ba): Refinement of powder neutron diffraction data. Journal of Solid State Chemistry 1996; 126 136-141.

[14] Blake S, Falconer M, McCreedy M, Lightfoot P. Cation disorder in ferroelectric Aurivillius phases of the type $Bi_2ANb_2O_9$ (A=Ba, Sr, Ca). Journal of Materials Chemistry 1997; 7(8) 1609-1613.

[15] Mercurio D, Trolliarda G, Hansenb T, Mercurio J. Crystal structure of the ferroelectric mixed Aurivillius phase $Bi_7Ti_4NbO_{21}$. International Journal of Inorganic Materials 2000; 2(5) 397-406.

[16] Perez-Mato J, Aroyo M, García A, Blaha P, Schwarz K, Schweifer J, Parlinski K. Competing structural instabilities in the ferroelectric Aurivillius compound $SrBi_2Ta_2O_9$. Physics Review B 2004; 70(21) 214111.

[17] Haluska M, Misture S. Crystal structure refinements of the three-layer Aurivillius ceramics $Bi_2Sr_{2-x}A_xNb_2TiO_{12}$(A=Ca, Ba; x= 0, 0.5, 1) using combined x-ray and neutron powder diffraction. Journal of Solid State Chemistry 2004; 177(6) 1965-1975.

[18] Newnham R, Wolfe R, Dorrian J. Structural basis of ferroelectricity in the bismuth titanate family. Materials Research Bulletin 1971; 6 1029-1040.

[19] Newnham R, Wolfe R, Horsey R, Diaz-Colon F. Crystal structure of (Sr,Ba)$Bi_2Ta_2O_9$. Materials Research Bulletin 1973; 8(10) 1183-1195.

[20] Huang S, Feng Ch, Chen L, Wang Q. Relaxor behavior of $Sr_{1-x}Ba_xBi_2Nb_2O_9$ ceramics. Journal of the American Ceramic Society 2006; 89(1) 328-331.

[21] Wu Y, Forbess MJ, Seraji S, Limmer SJ, Chou TP, Nguyen C, Cao GZ. Doping effect in layer structured $SrBi_2Nb_2O_9$ ferroelectric. Journal of Applied Physics 2001; 90 5296-6002.

[22] González-Abreu Y, Peláiz-Barranco A, Guerra JDS, Gagou Y, Saint-Grégoire P. From normal ferroelectric transition to relaxor behavior in Aurivillius ferroelectric ceramic. Journal of Materials Science 2014; 49(21) 7437-7444.

[23] Peláiz-Barranco A, González-Abreu Y. Dielectric relaxation mechanisms in relaxor bi-layered perovskites. Ferroelectrics 2012; 426(1) 122-131.

[24] Peláiz-Barranco A, González-Abreu Y. Ferroelectric ceramic materials of the Aurivillius family. Journal of Advanced Dielectrics 2013; 3(4) 1330003.

[25] Ismunandar, Kennedy BJ. Effect of temperature on cation disorder in $ABi_2Nb_2O_9$ (A=Sr, Ba). Journal of Materials Chemistry 1999; 9(2) 541-544.

[26] González-Abreu Y, Peláiz-Barranco A, Aráujo EB, Franco Júnior A. Dielectric relaxation and relaxor behavior in bilayered perovskites. Applied Physics Letters 2009; 94(26) 262903.

[27] Cross LE. Relaxor ferroelectrics: An overview. Ferroelectrics 1994; 151(1) 305-320.

[28] Liu W, Randall CA. Thermally stimulated relaxation in Fe-doped $SrTiO_3$ systems: II. Degradation of $SrTiO_3$ dielectrics. Journal of the American Ceramic Society 2008; 91(10) 3251-3257.

[29] Almeida A, Correira TM, Chaves MR, Vilarinho PM, Kholkin AL, Costa AM. Study of polar relaxation processes in $Sr_{(1-1.5x)}La_xTiO_3$ ceramics by using field-induced thermally stimulated currents. Journal of the European Ceramic Society 2008; 27(13) 3701-3703.

[30] Peláiz Barranco A, Calderón Piñar F, Pérez Martínez O. Pyroelectricity and mechanisms of conductivity in $PbZr_{0.53}Ti_{0.47}O_3$ + 2.5 mol% $La_2O_3$ ferroelectric ceramics. Journal of Material Science Letters 2001; 20(15) 1439-1441.

[31] Chen R, Krish Y. Analysis of Thermally Stimulated Processes. Oxford: Pergamon Press; 1981.

[32] Faubert F, Sánchez M. Numerical decomposition of a complex thermostimulated depolarization current spectrum in single time relaxation peaks. Journal of Applied Physics 1998; 84(3) 1541-1545.

[33] Chen R, Haber G. Calculation of glow curves activation energies by numerical initial rise method. Chemistry Physics Letters 1968; 2(7) 483-485.

[34] Grossweiner LI. A note on the analysis of first-order glow curves. Journal of Applied Physics 1953; 24(10) 1306-1307.

[35] Kong LB, Ma J, Zhu W, Tan OK. Preparation of $Bi_4Ti_3O_{12}$ ceramics via a high-energy ball milling process. Material Letters 2001; 51(2) 108-114.

[36] Hou J, Vaish R, Ou Y, Krsmanovic D, Varma KBR, Kumar RV. Dielectric, pyroelectric and ferroelectric properties of $Bi_4Ti_{2.98}Nb_{0.01}Ta_{0.01}O_{12}$ Ceramics. Materials Chemistry Physics 2010; 121(1-2) 32-36.

[37] González-Abreu Y, Peláiz-Barranco A, Guerra JDS, Saint-Grégoire P. Piezoelectric behavior in $Sr_{1-x}Ba_xBi_2Nb_2O_9$ Aurivillius-type structure ferroelectric ceramics. Physica Status Solidi B 2013; 250(8) 1-5.

[38] Zhang MF, Wang Y, Wang KF, Zhu JS, Liu J-M. Characterization of oxygen vacancies and their migration in Ba-doped $P(Zr_{0.52}Ti_{0.48})O_3$ ferroelectrics. Journal of Applied Physics 2009; 105(6) 061639.

[39] Peláiz-Barranco A, García-Wong AC, González-Abreu Y, Gagou T, Saint-Grégoire P. Thermally stimulated processes in samarium-modified lead titanate ferroelectric ceramics. Applied Physics A 2013; 112(2) 419-423.

2

# Dynamic Amplification of Optical Signals by Photorefractive Ferroelectric Liquid Crystals

Takeo Sasaki

**Abstract**

In this chapter, the photorefractive effect in photoconductive ferroelectric liquid crystal blends is described. A blend of liquid crystals that form chiral smectic C phase and photoconductive compounds shows the photorefractive effect. The ferroelectric liquid crystal blends containing a photoconductive chiral compound were found to exhibit a gain coefficient of over 1200 $cm^{-1}$ and the response time of sub-millisecond. Using the blend, real-time dynamic amplification of an optical image signal of over 30 fps was demonstrated.

**Keywords:** Ferroelectric liquid crystals, photorefractive effect, semiconductor, photoconductivity, hologram

## 1. Introduction

Ferroelectric liquid crystals (FLCs) have attracted significant interest from both fundamental and practical perspectives [1, 2]. Even though liquid crystals (LCs) are liquid, some of them exhibit ferroelectricity. The response of FLCs to an applied electric field is very fast, in the range of a few tens of microseconds to a few milliseconds. The fast response of FLCs is advantageous for display applications; however, the treatment of FLCs requires sophisticated techniques because FLCs are very viscous. Displays that employ FLCs are not commercially available nowadays, due to the difficulty of fabrication into wide area displays. Recently, the photorefractive effect in FLCs was investigated, and FLCs were found to exhibit very high performance. The photorefractive effect is a phenomenon in which a dynamic hologram is formed in a material. When two laser beams interfere in a photorefractive material, a refractive index grating is formed. This phenomenon is applicable to devices related to diffraction optics, including 3D displays, optical amplification, optical tomography, novelty filters, and phase-conjugate wave generators [3]. The unique feature of the photorefractive effect is the asymmetric energy exchange in two-beam coupling; when two laser beams interfere in a photorefractive material, the energy of one of the interfering beams is transferred to the other beam. The asymmetric energy exchange can be used to coherently amplify signal beams [4]; therefore, it has the potential to be used as a transistor for electrical circuits in a wide range of optical technologies (Figure 1). Optically transparent materials that possess both photovoltaic and electro-optic properties also exhibit the photorefractive effect. There have been a lot of studies to develop photorefractive materials, such as inorganic photoconductive ferroelectric crystals, organic photoconductive ferroelectric crystals, organic photoconductive polymers, amorphous organic photoconductive materials, photoconductive amphiphilic compounds, and photoconductive LCs [5-8]. Since 1991, organic materials have attracted significant interest because they exhibit large photorefractivity and a shorter response time compared with that of inorganic crystals [3, 7, 8]. The photorefractive effect induces a change in the refractive index by a mechanism that involves both photovoltaic and electro-optic effects (Figure 2). An organic photorefractive material is composed of a photoconductive compound, an electro-optic compound, and an electron-trap reagent. Thus, the active wavelength can be determined by the selection of the photoconductive compound. With interfering two laser beams in an organic photorefractive material, charge generation occurs at the bright positions of the interference fringe and the generated charges diffuse within the material. The distances of the diffusions of positive and negative charges are different in an organic material because the mobilities of the positive and negative charges are different. That leads to a formation of charge-separated state. The higher mobility charge diffuses over a longer distance than the lower mobility charge. While the charges with low mobility stay in the bright areas, the charges with high mobility diffuse to the dark areas. Thus, the bright area and dark area are charged with opposite polarities, and an internal electric field (space charge field) is formed in the area between the brightest and the darkest positions. The refractive indices of the areas are changed by the electro-optic effect. In such a way, a refractive index grating is produced. A high electric field of 10–50 V/μm is typically applied to a polymer film, aside from the internal electric field, to obtain photorefractivity. The polymer material is typically 100 μm thick, so that a voltage of 1–5 kV is required to apply on the film to achieve photorefractivity, which is almost comparable to the breakdown voltage of the polymer film. This electric field is necessary for increasing the charge generation efficiency.

It is expected that a multiplex hologram display devices can be developed using photorefractive polymers [9, 10]. It was shown that clear 3D images were recorded in the photorefractive polymer film. However, if polymer photorefractive materials are put into practical uses, a high voltage (3–5 kV) required to activate the photorefractive effect in polymer materials must be improved. There have been reports on the photorefractive effect of LCs [11]. LCs can easily be driven by a low electric field because LCs are liquid in nature. The most well-known LC phases are nematic and smectic phases (Figure 3). LCs that show nematic phase are used in liquid crystal displays (LCDs). However, LCs that show smectic phases are very viscous so that they are seldom utilized in practical applications. The LCs of which the photorefractivity was firstly investigated were nematic LCs [11]. It was found that a large photorefractivity was obtained with the application of low activation voltage (a few volts). The photorefractivity of ferroelectric liquid crystals doped with photoconductive compounds has been reported [12-14]. FLCs are considered to be candidates for practical photorefractive materials. A sub-millisecond refractive index grating formation time and a large gain coefficient are easily obtained in photorefractive FLCs.

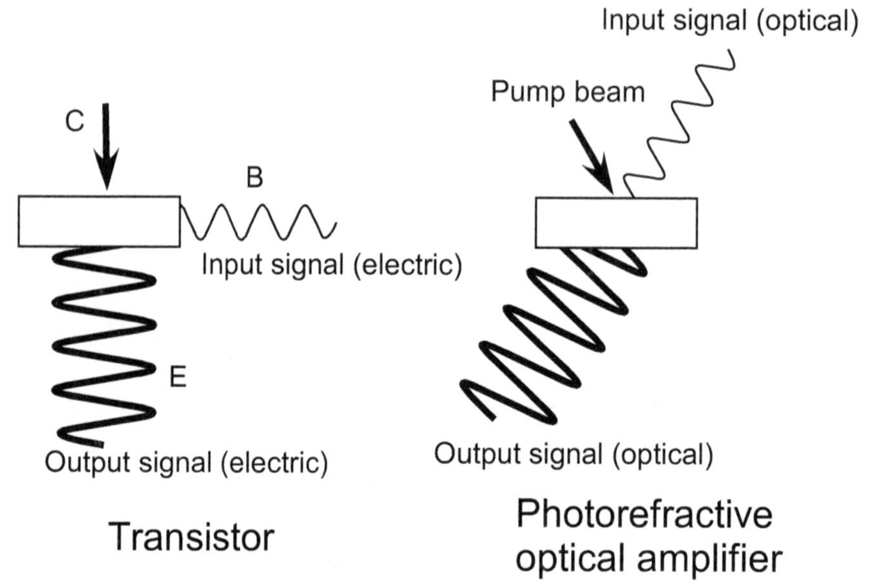


**Figure 1.** Transistor in electric circuit and photorefractive amplifier.

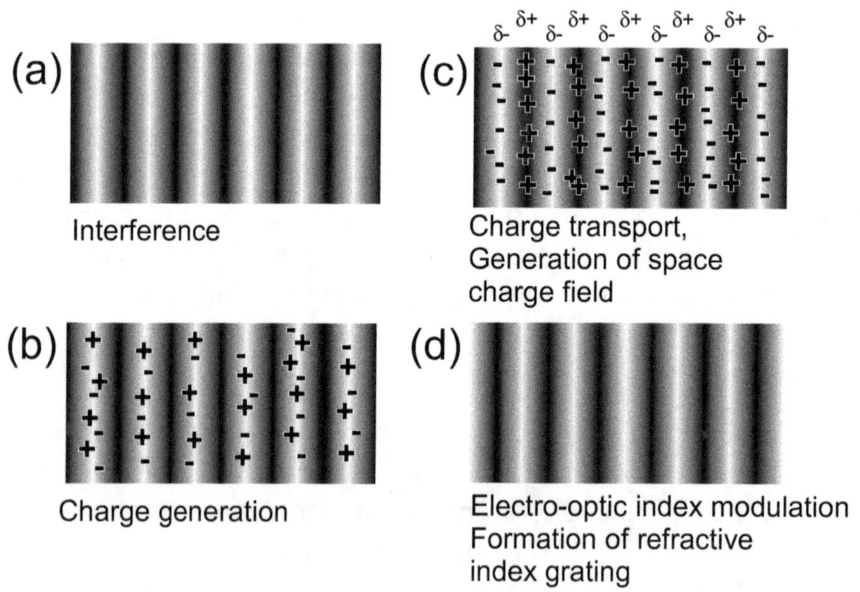


**Figure 2.** Schematic illustration of the mechanism for the photorefractive effect. (a) Two laser beams interfere in the photorefractive material; (b) charge generation occurs at the light areas of the interference fringes; (c) electrons are trapped at the trap sites in the light areas, holes migrate by diffusion or drift in the presence of an external electric field, and generate an internal electric field between the light and dark positions; (d) the refractive index of the corresponding area is altered by the internal electric field that is generated.

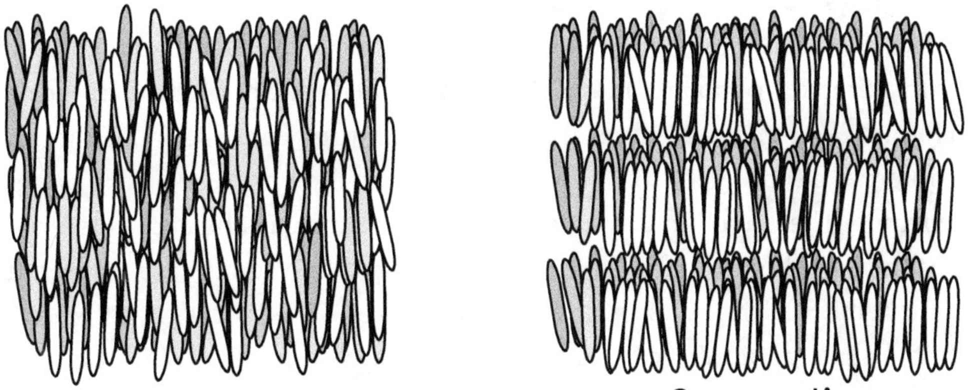


**Figure 3.** Structures of the nematic and smectic phases.

## 2. Ferroelectric liquid crystals

In 1974, Meyer suggested theoretically that LCs that form an LC phase with low symmetry should exhibit ferroelectricity [15]. Meyer instructed the synthesis of a liquid crystalline molecule that exhibits the smectic C (SmC) phase and also possesses chiral structure and permanent dipole moment. The structure of the LC molecule (DOBANBC) is shown in Figure 4. As Meyer had expected, DOBANBC was found to exhibit ferroelectricity. After the discovery of ferroelectricity in LCs, an intensive search for FLCs was conducted by many researchers, including LCD developers. As a result, it was determined that not only pure liquid crystalline compounds, but also mixtures of smectic LC compounds and chiral compounds exhibit ferroelectric phases. This was very advantageous for practical applications because clear, defect-free large-area panels could be realized using such mixtures and the properties of the mixtures could be adjusted by the selection of appropriate component compounds. In 1997, a fast FLC display panel was commercially released by Canon.

FLCs belong to the class of smectic LCs that have a layered structure [1, 2, 15]. A typical FLC molecule consists of a central core, a carbonyl group, and a chiral unit (Figure 4). The dipole moment of an FLC molecule is perpendicular to the long molecular axis. FLCs exhibit a chiral smectic C phase (SmC*) that possesses a helical structure. It should be noted here that to observe ferroelectricity in these materials, the FLCs must be formed into thin films [1] with a thickness of a few micrometers. When an FLC is sandwiched between glass plates to form a film with a thickness of a few micrometers, the helical structure of the SmC* phase uncoils and a surface-stabilized state (SS-state) is formed in which ferroelectricity appears (Figure 5). The thickness of the FLC film is typically 2 μm when it is used in display applications. The FLC molecules are restricted to only two directions in such thin films. The state is termed as SS-state. The direction of the alignment of FLC molecules is changed by the change in direction of the spontaneous polarization (Figure 6). When an alternating electric field is applied to the SS-FLC, the FLC molecules perform a continuous switching motion. The response time of the electrical switching in the FLCs is typically shorter than 1 ms. The direction of spontaneous polarization is governed by the applied electric field, which causes a change in the properties according to the direction of polarization. The mechanism for the photorefractive effect in FLCs is shown in Figure 7. When laser beams interfere in a mixture of an FLC and a photoconductive compound, internal electric fields are produced between the bright and dark positions of the interference fringe. The direction of spontaneous polarization in the area between the bright and dark positions of the interference fringes is changed by the internal electric field and a striped pattern of the FLC molecule orientations is induced. The process is different from those occur in other organic photorefractive materials. In a photorefractive FLC, the bulk polarization responds to the internal electric field. That is the base of the fast switching of FLC molecules.

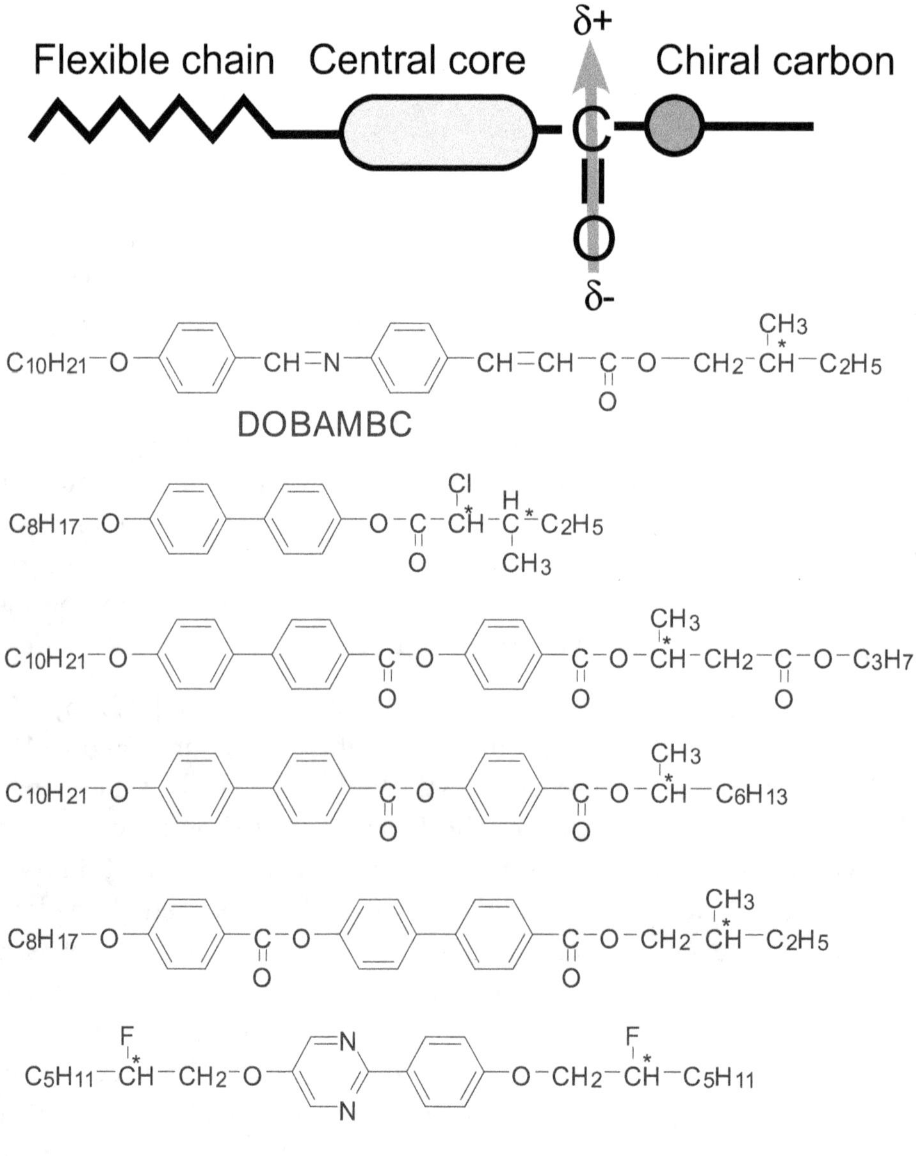


**Figure 4.** Molecular structures of FLCs.

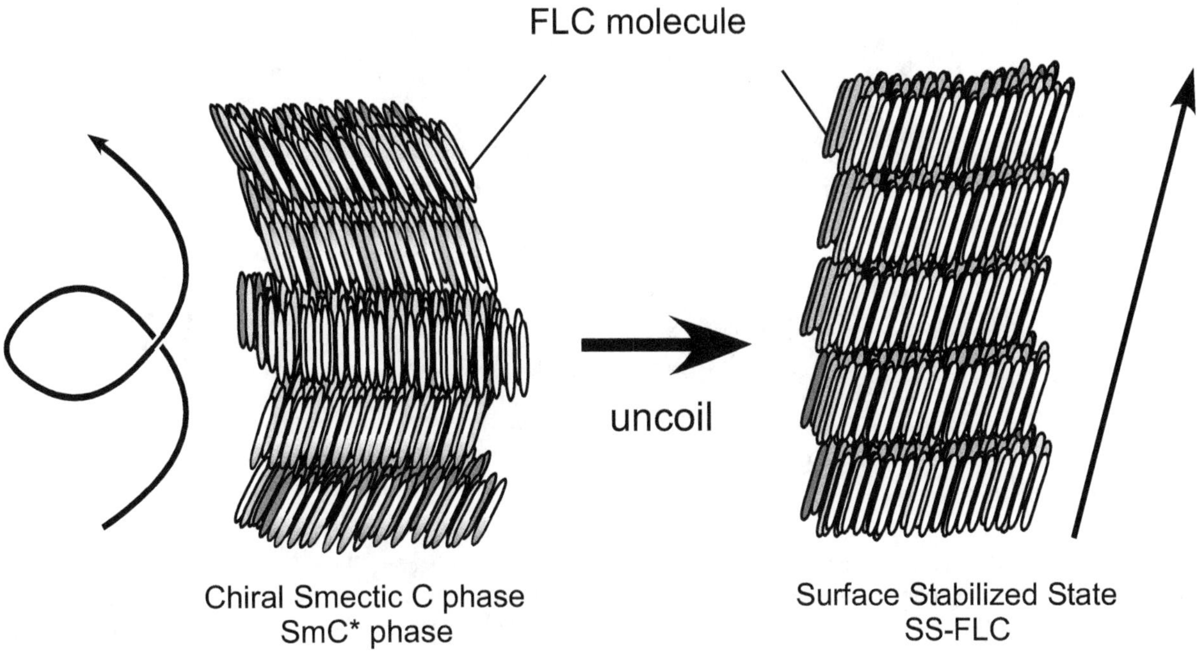


**Figure 5.** Structures of the SmC phase and the SS-state of the SmC phase (SS-FLC).

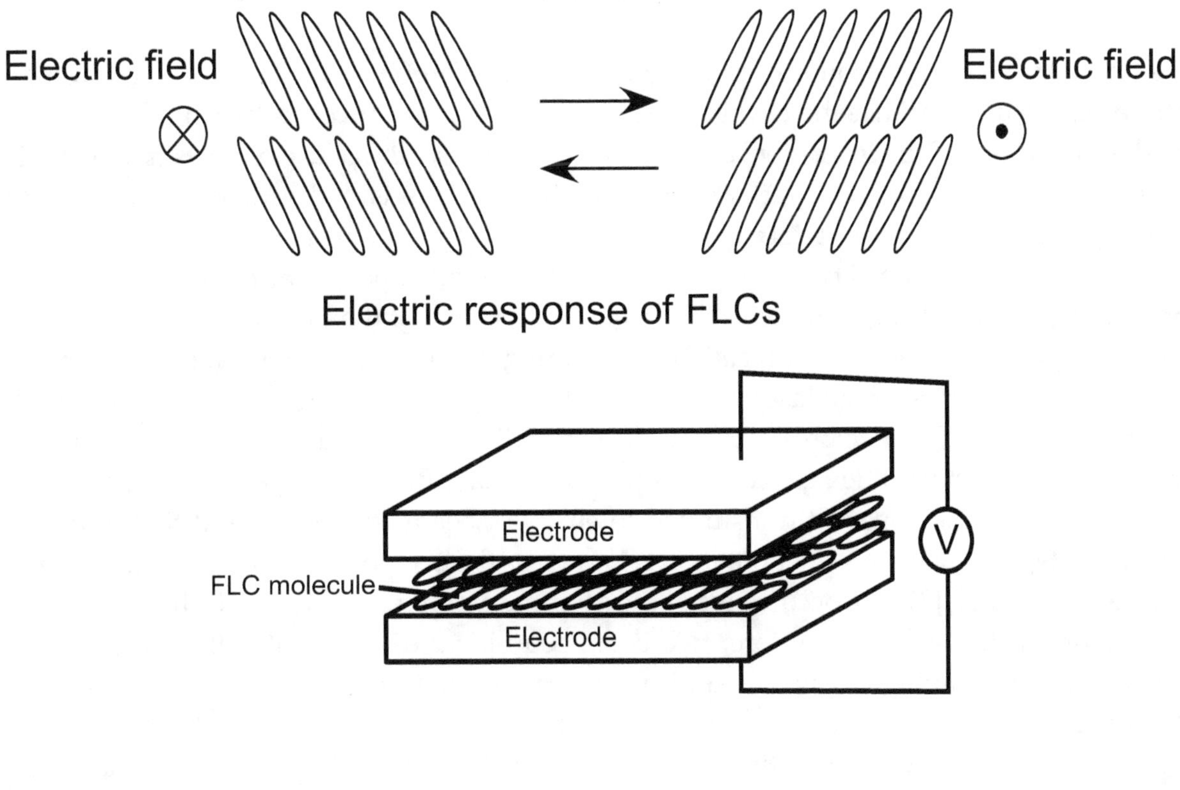


**Figure 6.** Electro-optical switching in the SS-state of FLCs.

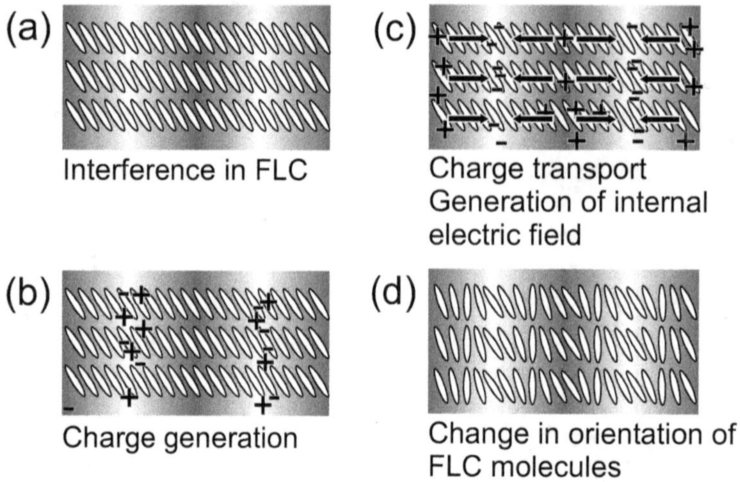


**Figure 7.** Schematic illustration of the mechanism for the photorefractive effect in FLCs. (a) Two laser beams interfere in the SS-state of the FLC/photoconductive compound mixture; (b) charge generation occurs at the light areas of the interference fringes; (c) electrons are trapped at the trap sites in the light areas, and holes migrate by diffusion or drift in the presence of an external electric field to generate an internal electric field between the light and dark positions; (d) the orientation of the spontaneous polarization vector (i.e., orientation of mesogens in the FLCs) is altered by the internal electric field.

## 3. Asymmetric energy exchange in photorefractive materials

The phase of the refractive index grating formed by photorefractive effect is shifted from the interference fringe because the change in refractive index is induced between the bright and the dark positions of the interference fringe. In an ideal case, the phase of the refractive index grating is shifted from the interference fringe by $\pi/2$.This is a distinctive feature of the photorefractive effect. When a refractive index grating is formed via a photochemical reaction or thermal change in density, the change in refractive index occurs at the bright areas and forms a refractive index grating with the same phase as that of the interference fringe (Figure 8(a)). The induced grating diffracts the interfering laser beams; however, the apparent transmitted intensities of the laser beams do not change because beam 1 is diffracted to the direction of beam 2, and beam 2 is diffracted in the direction of beam 1. However, when the phase of the refractive index grating is shifted from that of the interference fringe (photorefractive grating), the apparent transmitted intensity of beam 1 increases and that of beam 2 decreases (Figure 8(b)). This phenomenon is known as the asymmetric energy exchange in the two-beam coupling [3]. The occurrence of the asymmetric energy exchange is the evidence for the photorefractive effect. Therefore, two-beam coupling measurement is the most straightforward way to unambiguously distinguish between the photorefractive effect and other types of grating. In LCs and low-glass-transition temperature polymers, the sign of the electro-optic coefficient is determined by the direction of the applied electric field. A change in the electric field polarity reverses the sign of the gain coefficient due to the change in sign of the electro-optic coefficient. Thus, the amplification and attenuation of beam 1 and beam 2 switches when the polarity of the applied electric field is reversed.

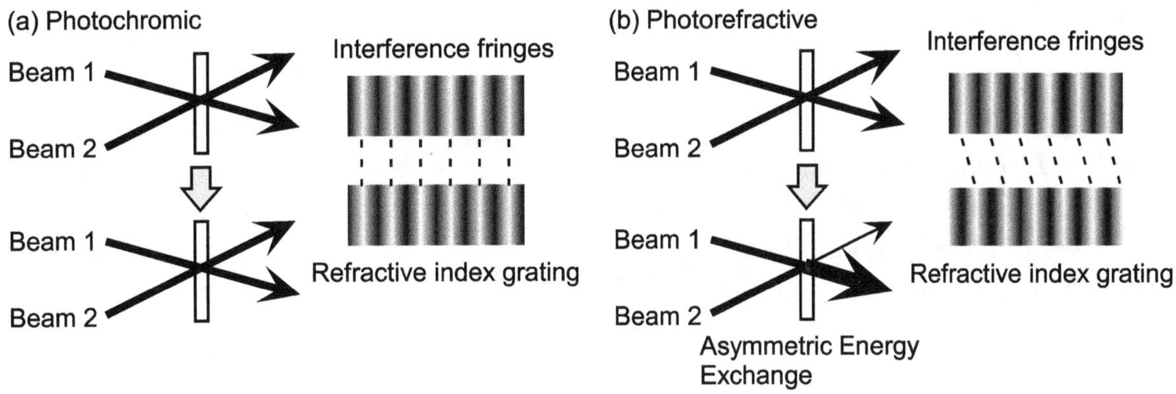


**Figure 8.** Schematic illustrations of (a) photochromic and (b) photorefractive gratings.

A schematic illustration of the setup used for the two-beam coupling experiment is shown in Figure 9(a). Laser beams (p-polarized) are interfered in the sample. An electric field (external electric field) is applied to the sample to increase the efficiency of charge generation in the film. The transmitted intensities of the laser beams through the sample are monitored. If a material exhibits photorefractive effect, an asymmetric energy exchange is observed. The magnitude of photorefractive effect is evaluated by the magnitude of gain coefficient, which is obtained from the two-beam coupling experiment [3]. According to the standard theory of the photorefractive effect with the limit of the ratio of beam intensities (pump/signal) >>1, the intensity of the transmitted signal beam is given by:

$$I = I_0 \exp(\Gamma l), \tag{1}$$

where $I_0$ is the signal beam intensity, $l$ is the interaction length in the sample, and $\Gamma$ is the gain coefficient. In order to obtain the two-beam coupling gain coefficient, the diffraction condition must be clarified whether it is in the Bragg diffraction regime or in the Raman-Nath diffraction regime. The diffraction conditions are distinguished by a parameter $Q$ [3]:

$$Q = 2\pi\lambda L / n\Lambda^2, \tag{2}$$

where $\lambda$ is the wavelength of the laser, $L$ is the interaction path length, $n$ is the refractive index, and $\Lambda$ is the grating spacing. If the $Q$ value is larger than 1, the condition is classified to the Bragg regime of optical diffraction in which only one order of diffraction is allowed. If $Q$ value is smaller than 1, the condition is classified to the Raman-Nath regime of optical diffraction in which many orders of diffraction are allowed. In order to guarantee the entire Bragg diffraction regime, $Q > 10$ is often required.

The two-beam coupling gain coefficient $\Gamma$ ($cm^{-1}$) for the Bragg diffraction condition is calculated according to the following equation [5]:

$$\Gamma = \frac{1}{D}\ln\left(\frac{gm}{1+m-g}\right), \tag{3}$$

where $D = L/\cos(\theta)$ is the interaction path for the signal beam ($L$ = sample thickness, $\theta$ = propagation angle of the signal beam in the sample), $g$ is the ratio of the signal beam intensities behind the sample with and without a pump beam, and $m$ is the ratio of the beam intensities (pump/signal) in front of the sample.

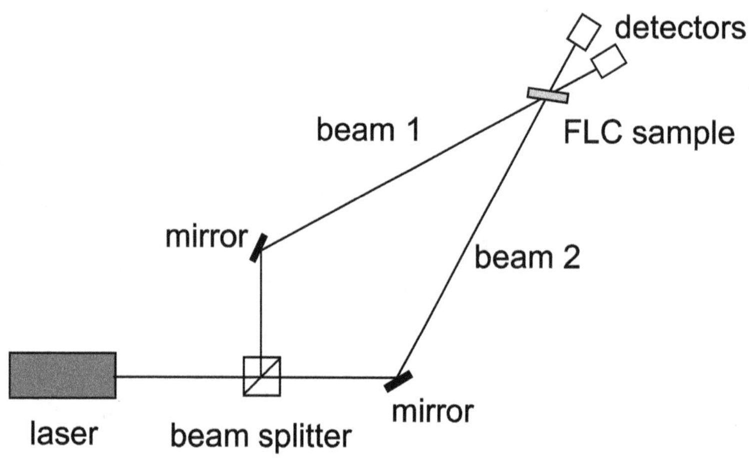


**Figure 9.** Schematic illustration of the experimental setup for the two-beam coupling experiment.

## 4. Photorefractive effect of FLCs

### 4.1. Two-beam coupling experiments on FLCs

Investigations into the photorefractive effect of FLCs started in the year 2000 [12, 13]; the photorefractive FLC is a mixture of FLC and photoconductive compounds. Further details of the photorefractivity in FLC materials have since been investigated by Sasaki et al. and Talarico et al. [14-19]. The structures of the photoconductive compounds used are shown in Figure 10. A commercially available FLC, SCE8 (Clariant, SmC* 60 °C SmA 80 °C N* 104 °C I, spontaneous polarization = 4.5 nC/cm$^2$), was used in preliminary investigations. SCE8 is a mixture of LC compounds and chiral compounds. Carbazole diphenylhydrazone (CDH) as a photoconductive compound and trinitrofluorenone (TNF) as a sensitizer were used at concentrations of 2 wt.% and 0.1 wt.%, respectively. The samples were injected into a 10 μm gap glass cell equipped with 1 cm$^2$ indium tin oxide electrodes and a polyimide alignment layer (Figure 11). A typical example of asymmetric energy exchange observed in the FLC (SCE8)/CDH/TNF sample under an applied DC electric field of 0.1 V/μm [17] is shown in Figure 12. The grating formation was within the Bragg diffraction regime. Interference of the laser beams in the FLC medium resulted in increased transmittance of one beam and decreased transmittance of the other. The transmitted intensities of the two beams changed symmetrically, indicating that the phase of the refractive index grating is shifted from that of the interference fringe as shown in Figure 12.

The temperature dependence of the gain coefficient of the FLC (SCE8) doped with 2 wt.% CDH and 0.1 wt.% TNF is shown in Figure 13(a). In this sample, asymmetric energy exchange was observed only at temperatures below 46 °C. Figure 13(b) shows the temperature dependence of the spontaneous polarization of the identical sample. When the temperature was raised above 46 °C, the magnitude of the spontaneous polarization dropped to zero. Thus, the photorefractive effect of the FLC was observed only at temperatures where the sample exhibits ferroelectric phase. The reorientation associated with spontaneous polarization is induced by the internal electric field in the ferroelectric phase. The change in the direction of spontaneous polarization causes a change at the orientation of FLC molecules in the corresponding area. A maximum resolution of 0.8 μm was obtained for this sample [16].

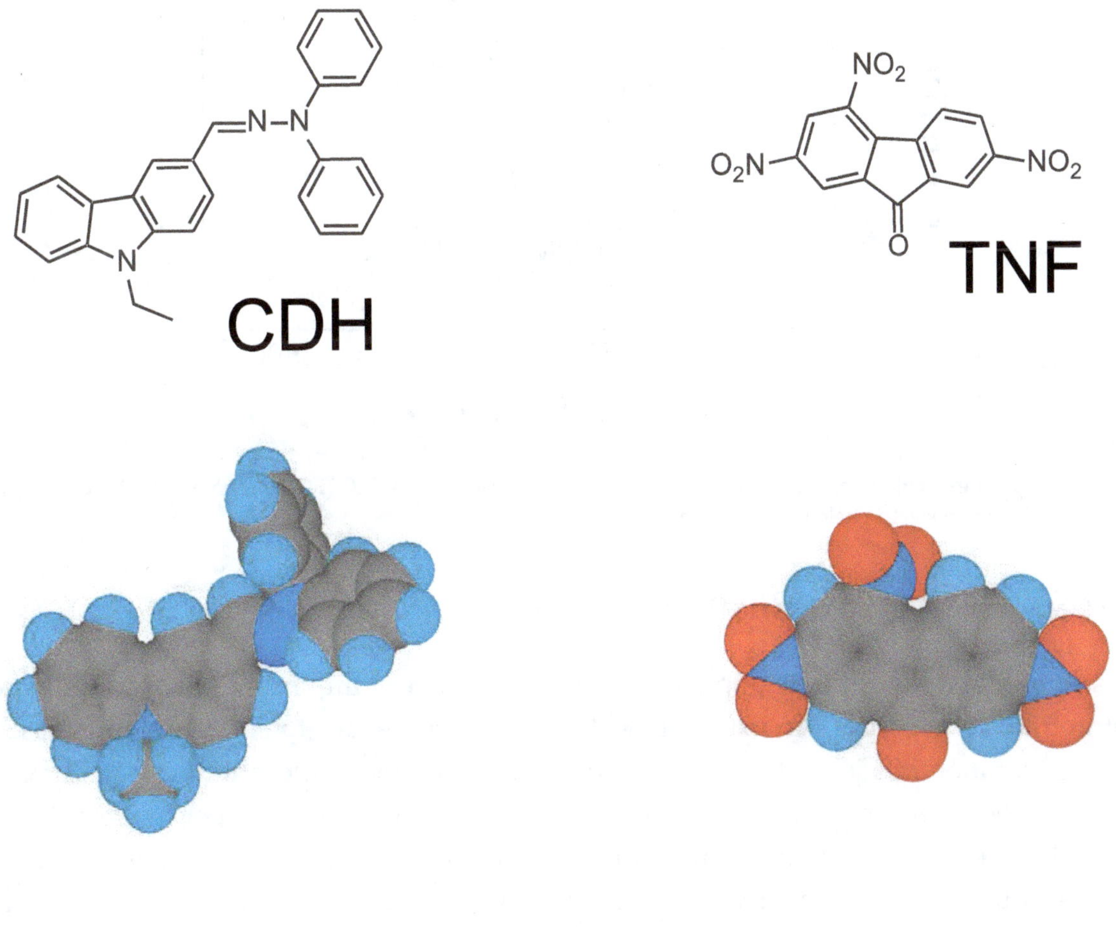


**Figure 10.** Structures of the photoconductive compound CDH and the sensitizer TNF.

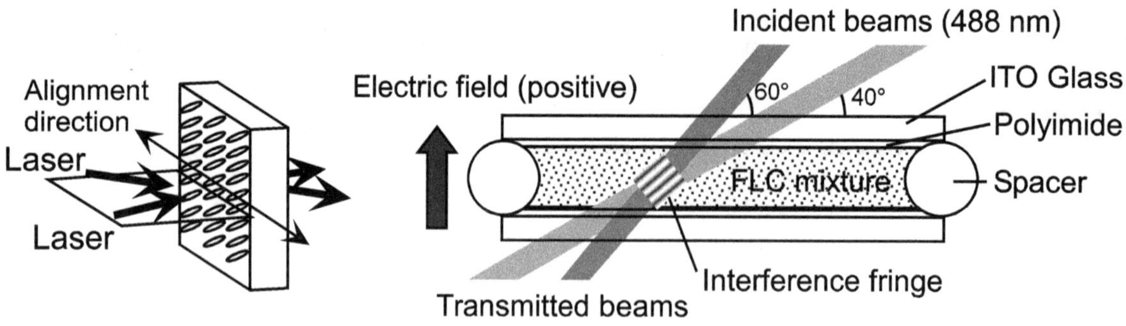


**Figure 11.** Laser beam incidence condition and the structure of the LC cell.

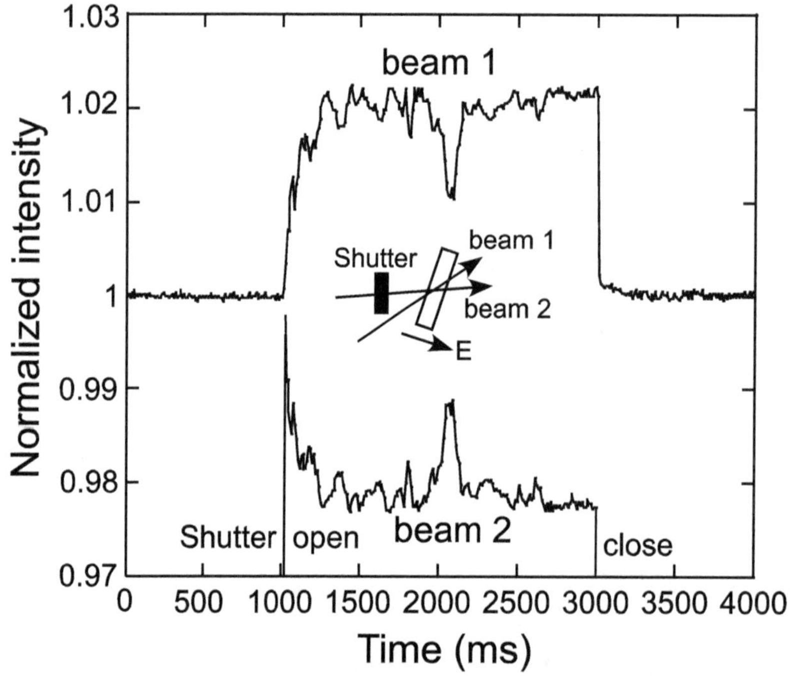


**Figure 12.** Typical example of asymmetric energy exchange observed in an FLC (SCE8) mixed with 2 wt.% CDH and 0.1 wt.% TNF with an electric field of +0.3 V/μm applied to the sample.

### 4.2. Effect of the applied electric field magnitude

The strength of the externally applied electric field is a very important factor for polymeric photorefractive materials. An external electric field is necessary to sufficiently increase the charge separation efficiency to induce a photorefractive effect; the photorefractivity of the polymer is obtained only with an electric field larger than a few volts per micrometer. The typical thickness of the polymer film for investigation of photorefractive effect is 100 μm. The strength of the voltage necessary to activate the photorefractive effect in polymer materials reaches to a few kilovolts. In contrast, the photorefractive effect in FLCs can be activated by a very weak external electric field application. The maximum gain coefficient for the FLC (SCE8)

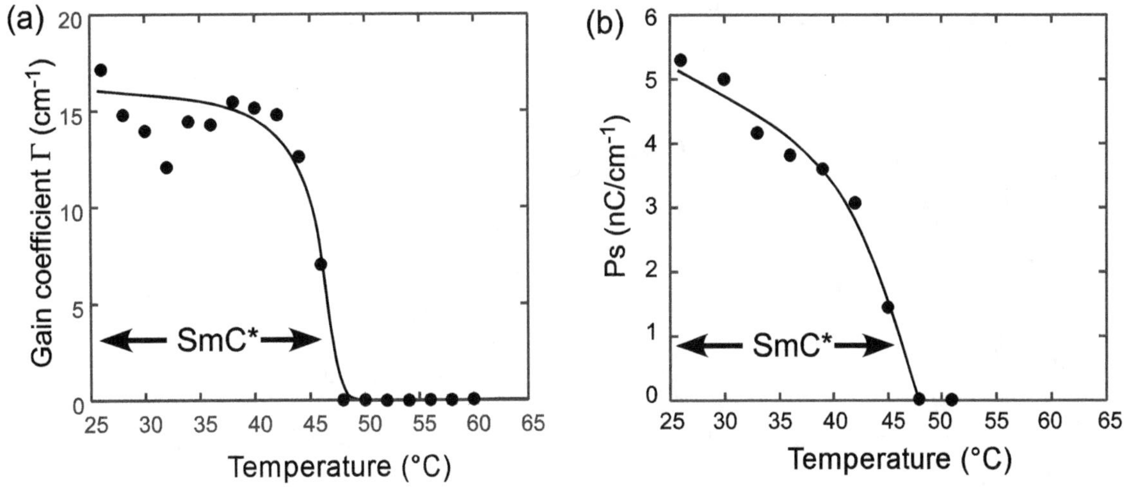


**Figure 13.** Temperature dependence of the (a) gain coefficient and (b) spontaneous polarization for an FLC (SCE8) mixed with 2 wt.% CDH and 0.1 wt.% TNF. For two-beam coupling experiments, an electric field of 0.1 V/μm was applied to the sample.

sample was obtained by only 0.2–0.4 V/μm electric field. The typical thickness of the photorefractive FLC sample is 10 μm; therefore, the voltage necessary to activate the photorefractive effect is only a few volts. Figure 14 shows the electric field dependence of the gain coefficient for a mixture of FLC (SCE8)/CDH/TNF. As the strength of the external electric field increased, the gain coefficient of SCE8 doped with 0.5 wt.% to 1 wt.% CDH increased. On the other hand, the gain coefficient of SCE8 doped with 2 wt.% CDH decreased when the external electric field larger than 0.4 V/μm was applied. The same tendency was also observed for another commercially available FLC; M4851/050 (Clariant, SmC* 65 °C SmA 70 °C N* 74 °C I, spontaneous polarization=14 nC/cm$^2$). The formation of an orientational grating is enhanced when the external electric field is increased from 0 to 0.2 V/μm due to the induced charge separation. However, when the external electric field exceeded 0.2 V/μm, a number of zigzag defects appeared in the SS-state, which caused light scattering and a decrease of the gain coefficient. The gain coefficient of FLC materials reported in the year 2003 (Figure 14) was much smaller than that of polymer materials [16].

### 4.3. Refractive index grating formation time

The formation of a refractive index grating involves charge separation and reorientation. The index grating formation time (response time of the photorefractive effect) is affected by these two processes, and both may act as rate-determining steps. The refractive index grating formation times for the commercially available FLCs examined (SCE8 and M4851/050) were determined on the basis of the simplest single-carrier model of photorefractivity [3, 5], wherein the gain transient is exponential. The rising signal of the diffracted beam was fitted using a single exponential function:

$$\gamma(t)-1=(\gamma-1)\left[1-\exp(-t/\tau)\right]^2, \tag{4}$$

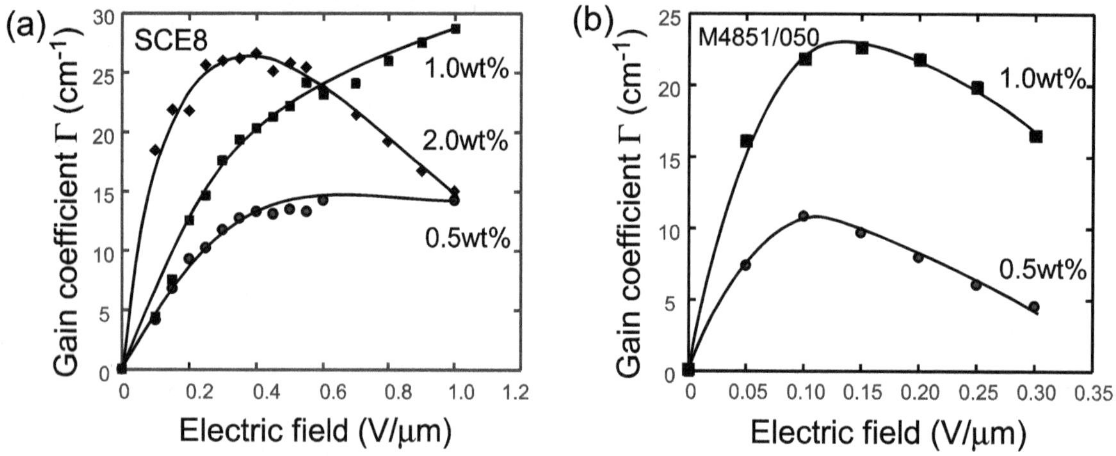


**Figure 14.** Electric field dependence of the gain coefficient for SCE8 and M4851/050 mixed with several concentrations of CDH and 0.1 wt.% TNF in a 10 μm gap cell measured at 30 °C.

where $\gamma(t)$ represents the transmitted beam intensity at time $t$, divided by the initial intensity $[\gamma(t) = I(t)/I0]$, and $\tau$ is the formation time. The grating formation time in SCE8/CDH/TNF is plotted as a function of the external electric field strength in Figure 15. The grating formation time shortened with an increase in the electric field strength because of the increased efficiency of charge generation. The formation time was shorter at higher temperatures, which corresponded to a decrease in the viscosity of the FLC with the increase in temperature. The formation time for SCE8 was 20 ms at 30 °C. The response time of FLC materials is thus faster than those of polymer materials, in which the typical response time is reported to be around 100 ms [5-8].

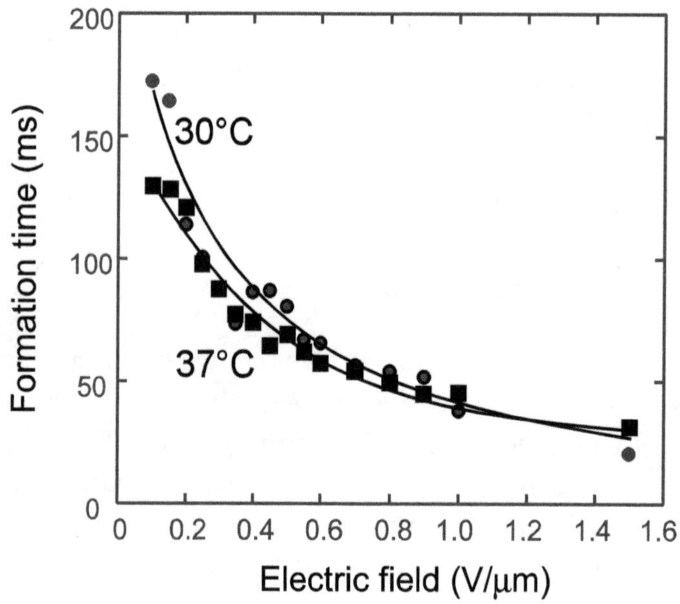


**Figure 15.** Electric field dependence of the index grating formation time. FLC (SCE8) mixed with 2 wt.% CDH and 0.1 wt.% TNF in a two-beam coupling experiment. ●: measured at 30 °C (T/TSmC* - SmA = 0.95); ▪: measured at 36 °C (T/TSmC* - SmA = 0.97).

# 5. Photorefractive effect in photorefractive FLC blends containing photoconductive chiral compounds

## 5.1. Photoconductive chiral dopants

FLCs are more crystalline than liquid, in comparison with nematic LCs; therefore several sophisticated techniques are required to prepare fine FLC films. Preparation of a uniformly aligned, defect-free SS-FLC using a single FLC compound is very difficult. In most cases, mixtures of LC compounds are typically used to obtain fine SS-FLC films. The composition of FLC mixture contains a base LC, which is a mixture of SmC phase forming LC compounds, and a chiral dopant. The chiral dopant introduces a helical structure into the LC phase. Utilization of an FLC as a photorefractive material requires the addition of photoconductive compounds to the FLC. However, the introduction of such non-LC compounds to the FLC often hinders the formation of a uniformly aligned SS-state. Thus, appropriate design of the photoconductive compounds is crucial. The photorefractive effect of FLC blends containing photoconductive chiral dopants has been investigated [17-19]. Terthiophene was selected as a photoconductive chromophore because it is a well-known semiconductor compound and has a rod-like structure (Figure 16), which increases the solubility into the rod-like structured LC material. The structures of the LC compounds, the electron acceptor TNF, and the photoconductive chiral compounds are shown in Figure 17. A ternary mixture of LC compounds was selected as a base LC. The mixing ratio of 8PP8, 8PP10, and 8PP6 was 1:1:2 since the 1:1:2 mixture exhibits the SmC phase over the widest temperature range. The textures of the FLC blends in 10 μm gap cells were observed using polarizing optical microscopy. The alignment of the FLC molecules is dominated not only by the properties of the FLCs but also by the affinity of FLC molecules with the alignment layer (polyimide). A homogeneous, anisotropic film can be obtained through interactions between LC molecules and the alignment layer. The FLC cell is fabricated by the precise assembly of indium tin oxide glasses coated with polyimide alignment layer into a cell of 10 μm gap determined by the diameter of the spacer bead. The appropriate preparation conditions for the fabrication of the LC cell differ from FLC to FLC. The thickness of the polyimide coating (Hitachi Chemicals LX-1400) was 20 to 30 nm and the surface of the polyimide was rubbed with a polyester velvet roll under specific conditions. Typical examples of textures observed in the 3T-2MB and 3T-2OC samples under a polarizing microscope are shown in Figures 18 and 19. On increasing the concentration of the photoconductive chiral dopant, defects appeared in the texture. The uniformly aligned state with few defects was obtained for samples with 3T-2MB concentrations lower than 8 wt.% (Figure 18). However, the 3T-2OC sample retained the uniformly aligned state with few defects for 3T-2OC concentrations less than 6 wt.% (Figure 19). The spontaneous polarization of the 3T-2MB samples was less than 1 nC/cm$^2$. On the other hand, the spontaneous polarization of the 3T-2OC samples was approximately 5 nC/cm$^2$. The smaller spontaneous polarization (and thus smaller intermolecular interactions) of the 3T-2MB sample may be advantageous for the formation of the uniformly aligned SS-state. A texture with a pattern of strips was observed in the 3T-2MB samples (Figure 18). It indicates that a complete SS-state was not formed in the 3T-2MB sample in the 10 μm gap cell and the helical structure existed. Zig-zag defect, which is the typical defect in the SS-state (book shelf structure), was also observed in the texture (Figure 18). The evidence shows that the FLC mixture exhibited the SS-state in the area close to the glass surface and formed a helical structure around the center of the thickness of the 10 μm gap cell.

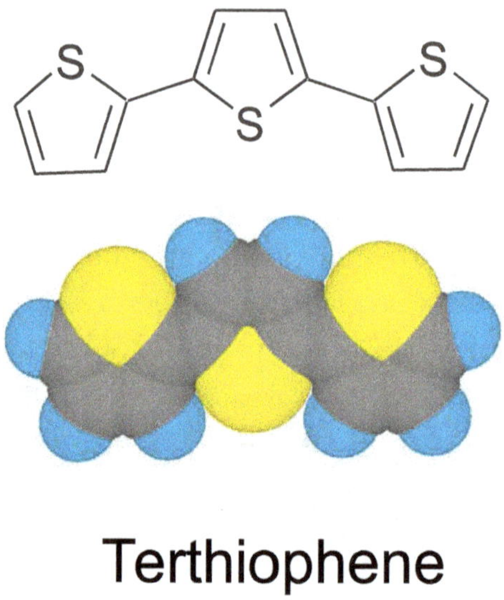


**Figure 16.** Structure of terthiophene.

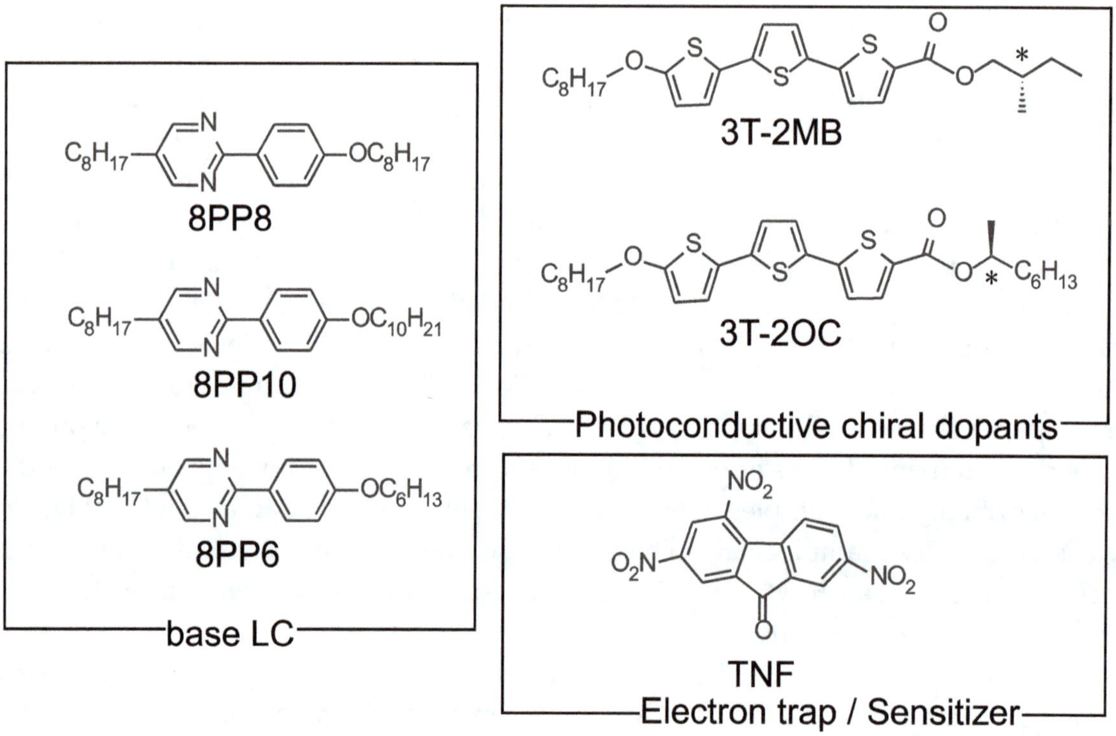


**Figure 17.** Structures of the smectic LCs (8PP8, 8PP10, and 8PP6), photoconductive chiral dopants (3T-2MB and 3T-2OC), and the sensitizer TNF.

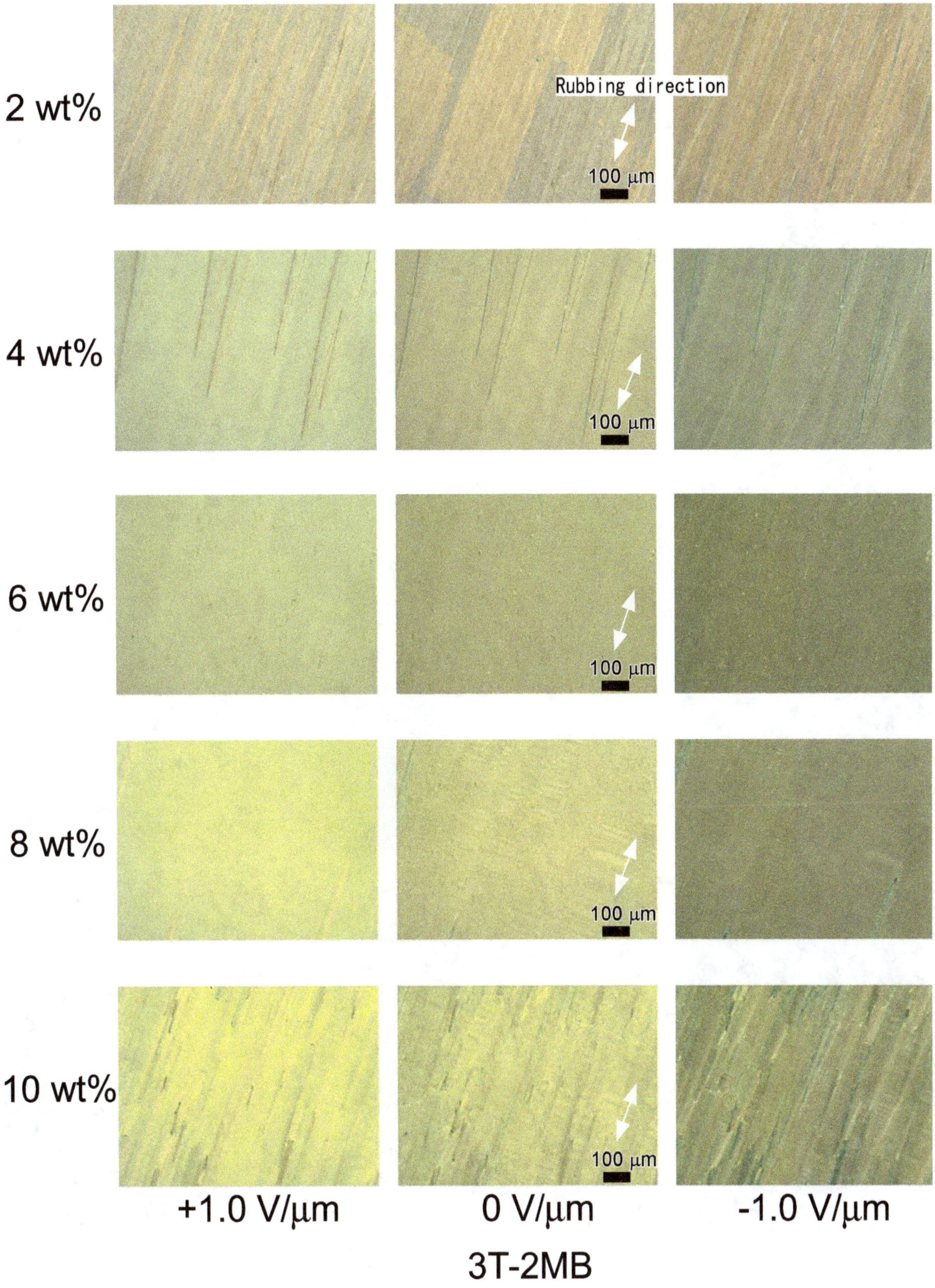


**Figure 18.** Textures of FLC mixtures containing 3T-2MB in a 10 μm gap LC cell observed under a polarizing microscope. The strengths of the external electric field were +1.0, 0, and 1.0 V/μm. The direction of the applied electric field is shown in Fig. 5. The 3T-2MB concentrations were in the range of 2 wt.% to 10 wt.% with the addition of 0.1 wt.% TNF.

**Figure 19.** Textures of FLC mixtures containing 3T-2OC in a 10 μm gap LC cell observed under a polarizing microscope. The strengths of the external electric field were +1.0, 0, and 1.0 V/μm. The direction of the applied electric field is shown in Fig. 5. The 3T-2OC concentrations were in the range of 2 wt.% to 10 wt.% with the addition of 0.1 wt.% TNF.

## 5.2. Asymmetric energy exchange in photorefractive FLC blends

The photorefractive effect of the FLC blends was measured by two-beam coupling experiments. A typical example of the asymmetric energy exchange observed for a mixture of the base LC, 3T-2MB, and TNF at 25 °C with an applied electric field of 1 V/μm is shown in Figure 20. When the laser beams interfered in the sample, the increase in transmitted intensity of one of the beams and the decrease in the transmitted intensity of the other beam were observed. When the polarity of the applied electric field was reversed, these transmittance characteristics were also reversed. Asymmetric energy exchange was not observed when the external electric field was not applied. This indicates that the beam coupling was not caused by a thermal grating or gratings formed through photochemical mechanisms. It was found that approximately 40% of the energy of the $L_2$ laser beam migrated to the $L_1$ beam.

The gain coefficients of the samples were measured as a function of the applied electric field strength (Figure 21(a)). The gain coefficient of 1200 $cm^{-1}$ was obtained in the 10 wt.% 3T-2MB sample with application of 1.5 V/μm. This gain coefficient is much higher than the values for FLCs reported previously [16]. The higher transparency of the LC blend was considered to contribute to the large gain coefficient. The weak external electric field strength required for the photorefractive effect in FLCs is advantageous for the photorefractive applications. The response time decreased with an increase of the electric field strength due to the increased charge separation efficiency. The shortest formation time of 0.9 ms was obtained with application of an external electric field of 1.9 V/μm (Figure 21(b)). The large gain and fast response are advantageous for the realization of optical devices such as real-time image amplifiers and accurate measurement devices.

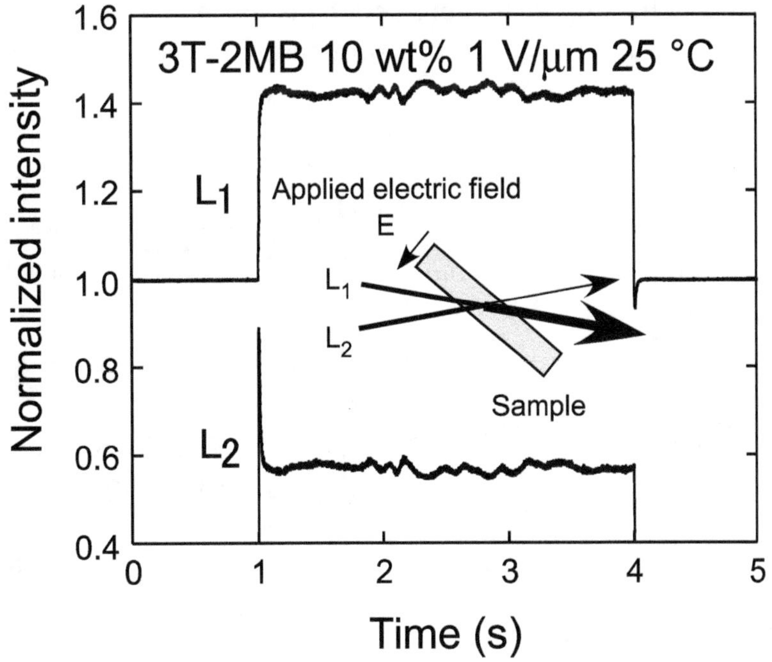


**Figure 20.** Typical results for two-beam coupling experiments with a ternary mixture base LC, 3T-2MB, and TNF measured at 25 °C. The pump beam was incident at 1 s and closed at 4 s.

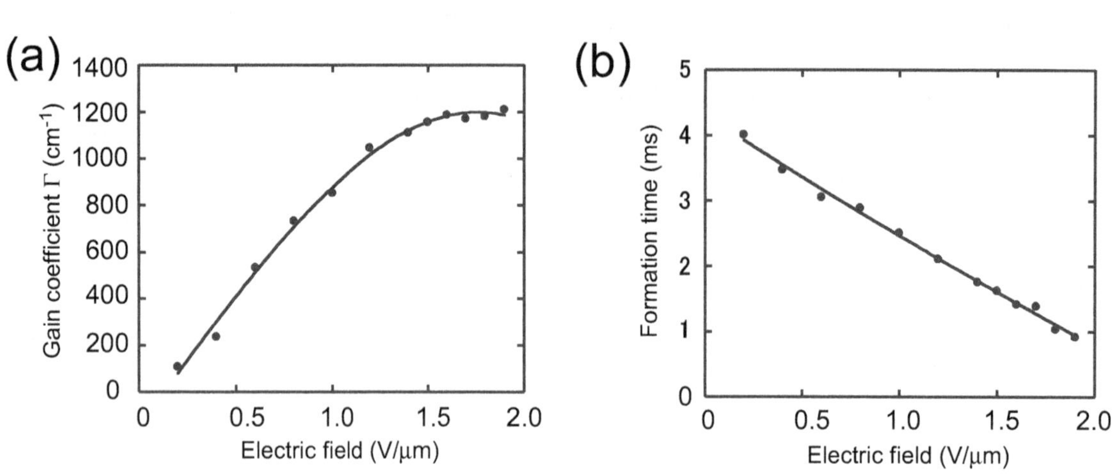


**Figure 21.** Electric field dependence of the (a) gain coefficients and (b) refractive index grating formation times (response time) for mixtures of the base LC, 3T-2MB (10 wt.%), and TNF (0.1 wt.%) measured at 25 °C.

## 5.3. Temperature dependence of the asymmetric energy exchange for photorefractive FLC blends

The temperature dependence of the gain coefficient for a photorefractive FLC blend with 3T-2MB is shown in Figure 22. Asymmetric energy exchange was observed at temperatures below the SmC*-SmA phase transition temperature. Figure 23 shows temperature dependence of the helical pitch of the 3T-2MB samples observed under polarizing microscopy. The helical pitch diverged when the temperature approached the phase transition temperature. It has been reported that the asymmetric energy exchange for an FLC sample was observed only in the temperature range where the sample exhibits spontaneous polarization [8]. Thus, asymmetric energy exchange was observed only in the temperature range where the sample exhibits ferroelectric properties (i.e., the SmC* phase).

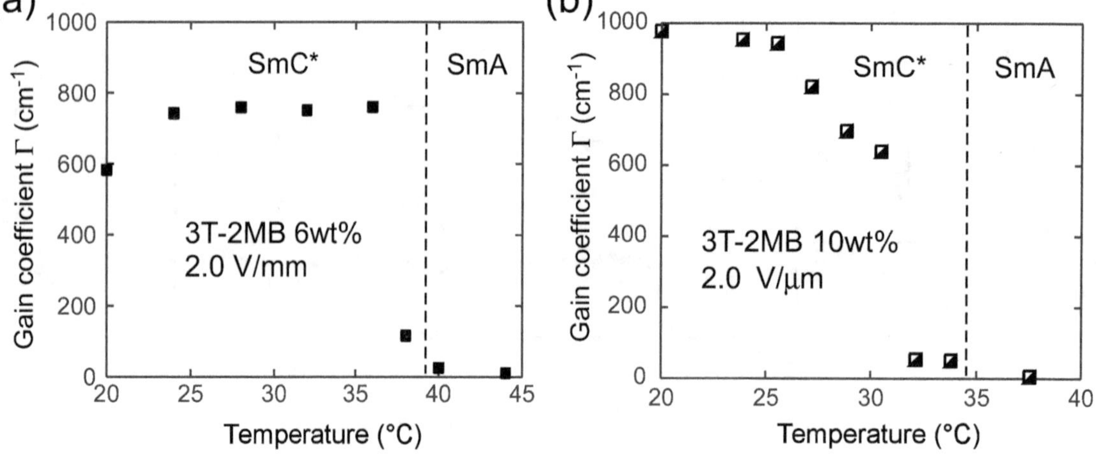


**Figure 22.** Temperature dependence of gain coefficients for mixtures of the base LC with 3T-2MB concentrations of (a) 6 wt.% and (b) 10 wt.%.

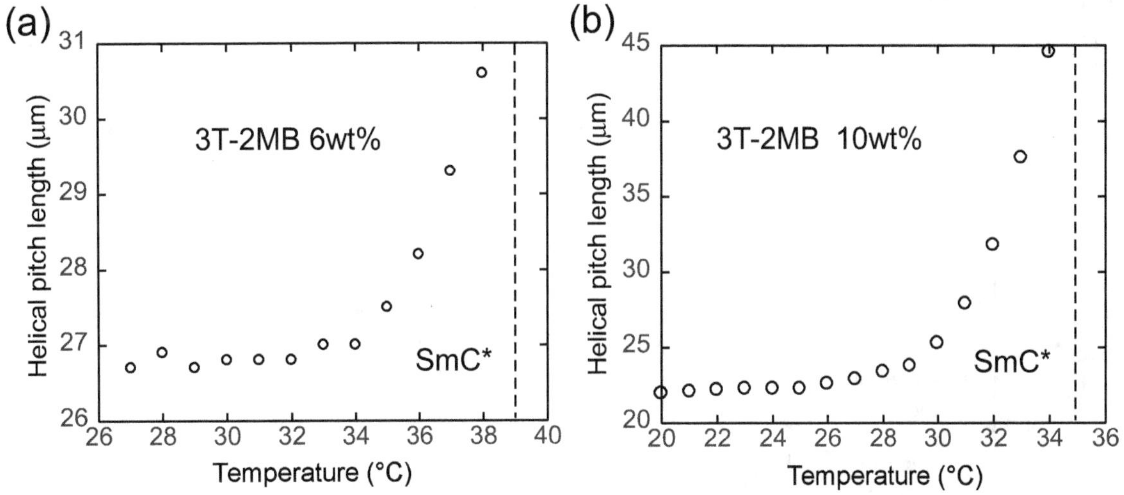


**Figure 23.** Temperature dependence of helical pitch lengths for mixtures of the base LC with 3T-2MB concentrations of (a) 6 wt.% and (b) 10 wt.%.

## 5.4. Effect of the photoconductive chiral dopant concentration

The gain coefficients of samples with various 3T-2MB and 3T-2OC concentrations are plotted as a function of the external electric field magnitude in Figure 24. The gain coefficient increased with the strength of the external electric field up to 1.6 V/μm. The decrease in the gain coefficient at high external electric field is due to the realignment of the FLC molecules being restricted because the strength of the external electric field exceeds that of the internal electric field. The gain coefficient increased with the concentration of the photoconductive chiral dopants. The increase in the concentration of charge carriers in the FLC medium and an increase in the magnitude of spontaneous polarization contributes to a larger gain coefficient. The magnitude of the gain coefficient was independent of the concentration of TNF. It shows that charges are drifted through electric conduction based on a hopping mechanism, where electron holes hop between the photoconductive chiral dopants. The molecular weight of the photoconductive chiral dopant is similar to that of the LC molecules; therefore, in a 10 wt.% doped sample, approximately 10 photoconductive molecules are dispersed in 90 LC molecules. A cube, wherein each side includes 5 LC molecules, contains 125 LC molecules. Thus, the average distance between the photoconductive chiral dopant molecules in the LC is no more than 3 LC molecules. In this case, charge transport based on a hopping mechanism may be feasible. The gain coefficient of the 10 wt.% 3T-2MB sample was 1200 $cm^{-1}$ with an applied electric field of only 1.6 V/μm (Figure 24(a)), which is twice as high as that of the 8 wt.% 3T-2OC sample at a similar external electric field (Figure 24(b)).

The response time of the samples is plotted as a function of the external electric field magnitude in Figure 25. The response time decreased with the increase of the electric field strength due to increased charge separation efficiency. The shortest formation time of 0.93 ms was obtained for the 10 wt.% 3T-2MB sample with an applied electric field of 2.0 V/μm. It was found that the response was faster than that of the 3T-2OC samples. The gain coefficient is lower and the response speed slower for the 3T-2OC sample even though the magnitude of the spontaneous polarization in the 3T-2OC samples (5 nC/cm$^2$) is higher than that in the 3T-2MB sample (less than 1 nC/cm$^2$). The transparency of the FLC film is more important for the photorefractive effect than the magnitude of spontaneous polarization.

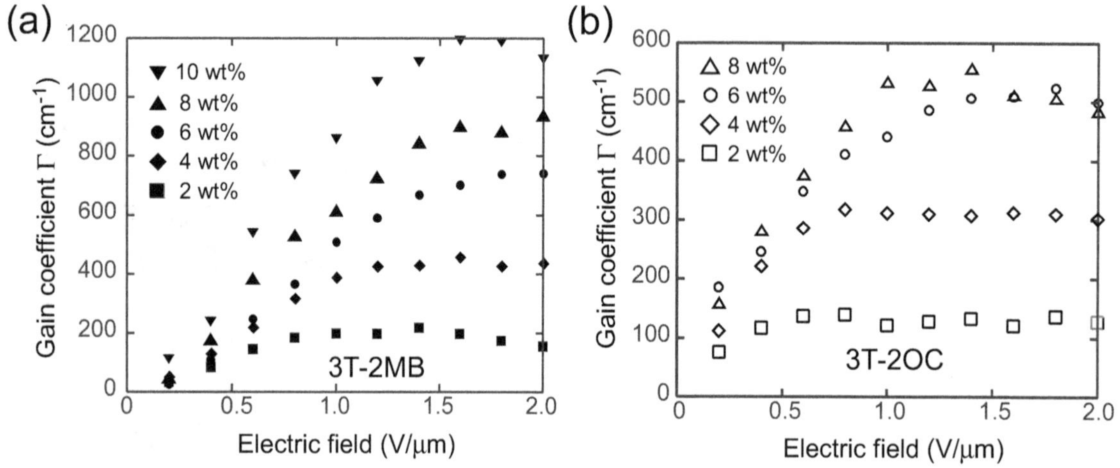


**Figure 24.** Electric field dependence of the gain coefficients for (a) mixtures of the base LC, 3T-2MB (2-10 wt.%), and TNF (0.1 wt.%), and (b) mixtures of the base LC, 3T-2OC (2-8 wt.%), and TNF (0.1 wt.%) measured at 25 °C.

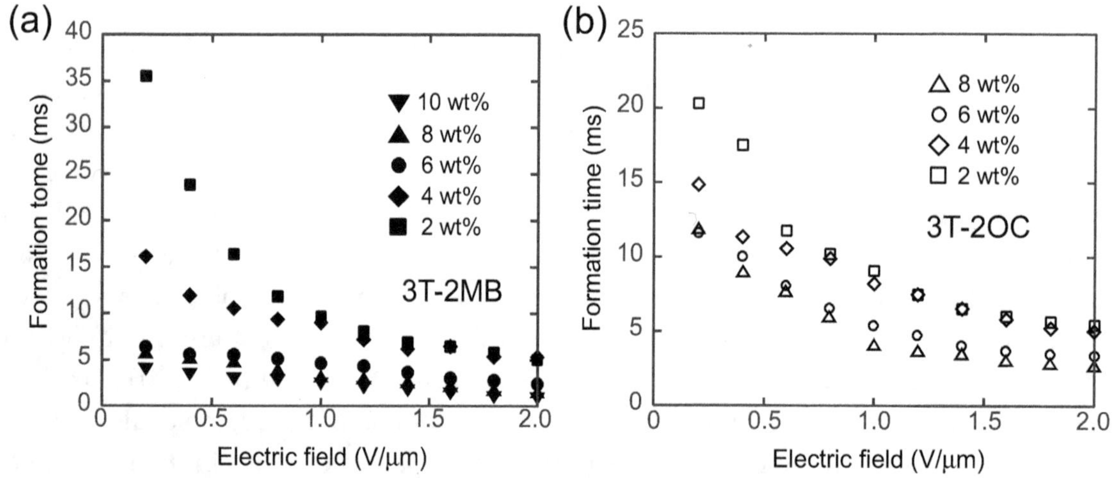


**Figure 25.** Refractive index grating formation times (response time) for (a) mixtures of the base LC, 3T-2MB (2-10 wt. %), and TNF (0.1 wt.%), and (b) mixtures of the base LC, 3T-2OC (2-8 wt.%), and TNF (0.1 wt.%) measured at 25 °C.

## 5.5. Dynamic amplification of optical images in photorefractive FLC blends

The most straightforward application of the photorefractive effect is the amplification of optical signals, which is one of the important elements of optical technologies. Light amplification based on stimulated emission and non-linear optical effects are well known. Distinct from these phenomena, the photorefractive effect enables selective amplification. The photorefractive effect is based on the formation of holograms in a material; therefore, it distinguishes a specific light signal from other light signals based on the difference in wavelength, polarization, and phase. Optical image amplification was demonstrated (Figure 26), where a computer-generated image was displayed on a spatial light modulator (SLM) and irradiated with a 473 nm diode-pumped solid-state laser beam [18]. A signal laser beam carrying a 2D image was transmitted through the FLC sample. The image was monitored with a charge-coupled device (CCD) camera. A pump beam (a beam divided from the signal beam before SLM) was interfered with the signal beam. The amplification of the signal beam transmitted through the FLC was observed. The intensities of the optical signal beam with and without the pump beam are shown in Figure 27. The intensity of the signal beam was amplified six-fold to the value without the pump beam.

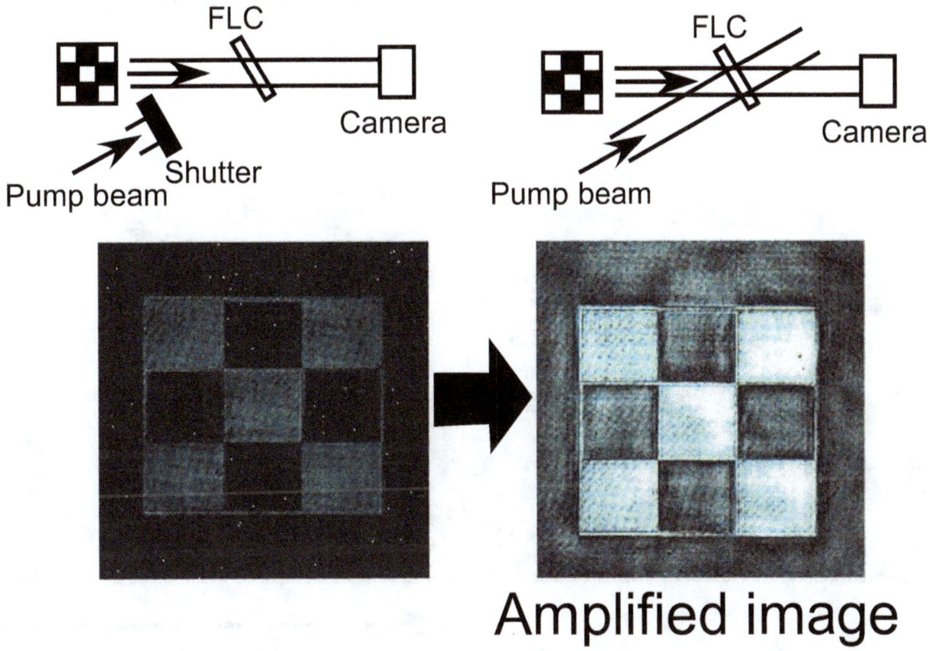


**Figure 26.** Optical image amplification experiment. A computer-generated image was displayed on the SLM. The SLM modulated the object beam (473 nm), which was irradiated on the FLC sample and interfered with the pump beam. The image transmitted through the FLC sample (10 wt.% 3T-2MB) was monitored with a CCD camera.

Dynamic amplification of moving optical signal was demonstrated using a photorefractive FLC blend [18]. A rotating image was displayed on the SLM with the frame rate at 30 fps. A 473 nm beam was irradiated on the SLM and the reflected beam was incident on the FLC sample. A pump beam was then interfered with the beam from the SLM in the FLC sample. The laser beam containing the moving animation image was amplified by the incident pump beam (Figure 28). This result shows that the response of the photorefractive FLC was sufficiently fast to amplify the moving optical image. If a typical photorefractive polymer with a response time of ca. 100 ms was used in place of the FLC sample, then the moving image would not be amplified. In that case, although a still image could be amplified, the intensity of the video-rate moving image would not be amplified.

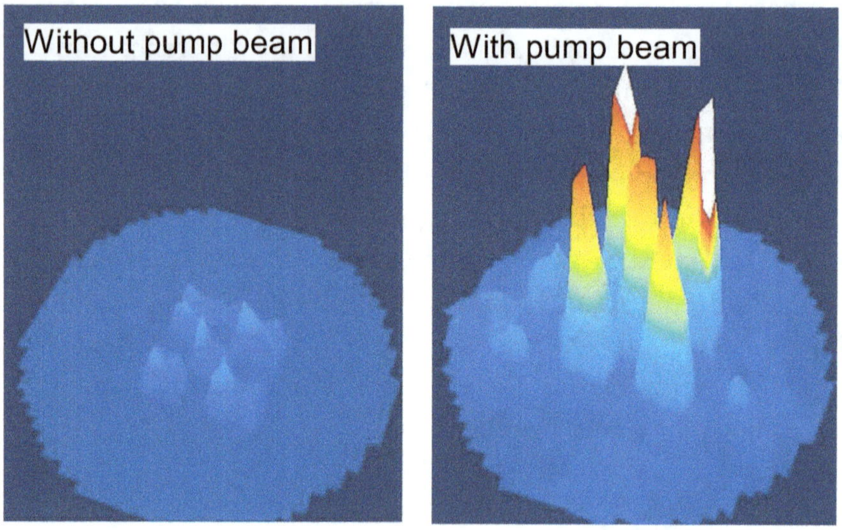


**Figure 27.** Signal beam intensities with and without the pump beam.

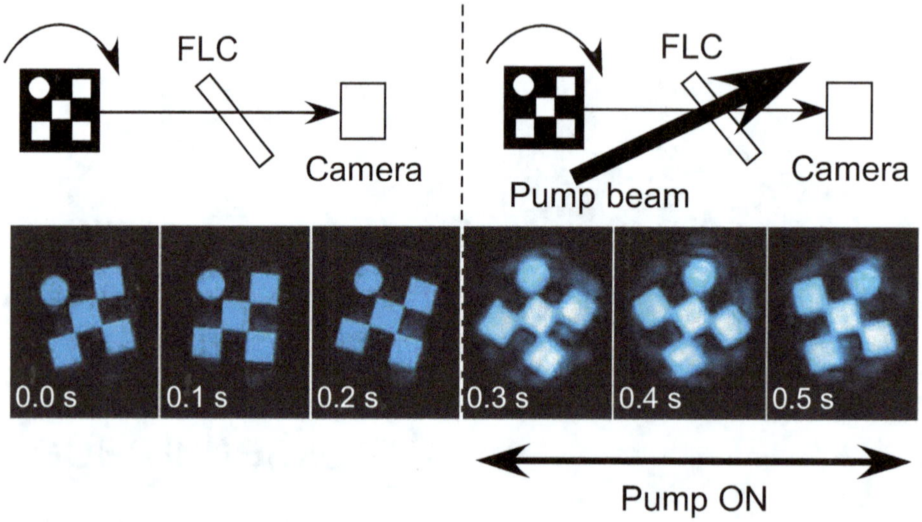


**Figure 28.** Optical image amplification experiment. A computer-generated animation was displayed on the SLM. The SLM modulated the object beam (473 nm), which was irradiated on the FLC sample, and interfered with the reference beam. The image transmitted through the FLC sample (10 wt.% 3T-2MB) was monitored with a CCD camera.

## 5.6. Dynamic holograms formed in FLC blends

Dynamic hologram formation was demonstrated in an FLC blend [19]. A moving image was displayed on an SLM and a laser beam was irradiated onto the SLM. The reflected beam was incident on the FLC sample and a reference beam was then interfered with the beam within the FLC sample. The interference condition was set to a Raman–Nath diffraction regime. The multiple scattering was observed in this condition. A red beam from a He–Ne laser (633 nm) was incident on the FLC sample and the diffraction was observed. A moving image was observed in the diffracted beam (Figure 29). No image retention was observed, which indicates that the hologram image (refractive index grating) formed in the FLC was rewritten with sufficient speed to project a smooth reproduction of the holographic movie. This result shows that a hologram image was formed at the interference area in the FLC material and that contributes to the optical image amplification.

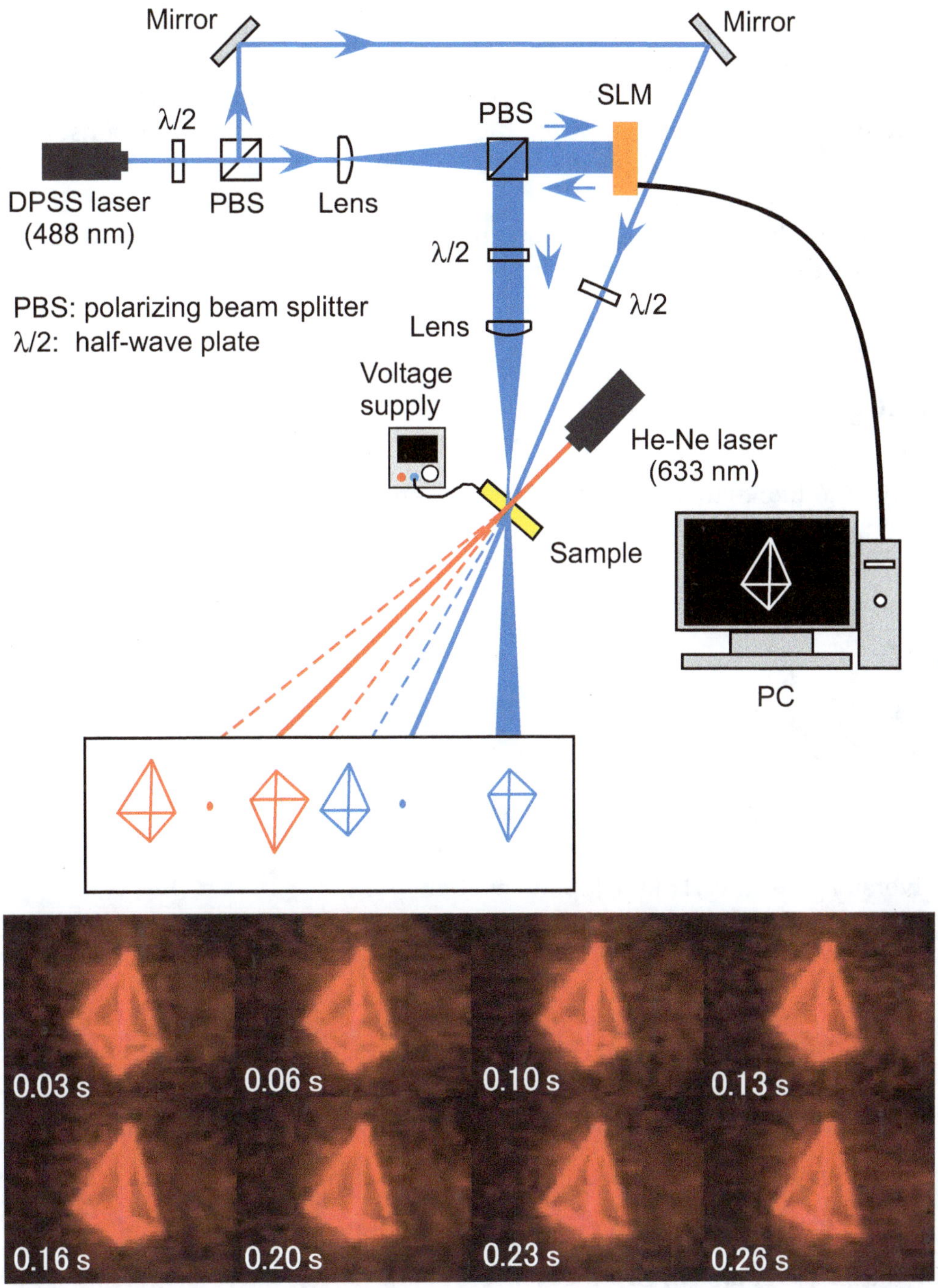


**Figure 29.** Dynamic hologram formation experiment on an FLC sample. A computer-generated animation was displayed on the SLM. The SLM modulated the object beam (488 nm), which was irradiated on the FLC sample and allowed to interfere with the reference beam. The readout beam (633 nm) was irradiated on the FLC and diffraction was observed.

## 6. Conclusions

Real-time dynamic amplification of optical image signals was demonstrated in photorefractive FLC blends. The response time was of sub-millisecond order and was dominated by the formation of an internal electric field. The photorefractive effect of FLCs was significantly dominated by the properties of the FLCs. Aside from spontaneous polarization, viscosity, and the phase transition temperature, the homogeneity of the SS-state was found to be a major factor. The gain coefficient and the response time were also significantly dominated by the homogeneity of the SS-state. Therefore, a highly homogeneous SS-state is necessary to create a photorefractive device. A gain coefficient higher than 1200 $cm^{-1}$ and a response time shorter than 1 ms were obtained with application of only 1.5 V/μm in a photorefractive FLC blend. This response time is sufficiently short for real-time dynamic holograms. FLC mixtures containing photoconductive chiral dopants exhibited high gain coefficients and fast responses, which confirms their usefulness in photorefractive device applications.

## Acknowledgements

The authors would like to thank the Japan Science and Technology Agency S-innovation and the Canon Foundation for support.

## Author details

Takeo Sasaki

Address all correspondence to: Sasaki@rs.kagu.tus.ac.jp

Tokyo University of Science,Tokyo, Japan

## References

[1] Skarp, K.; Handschy, M.A. *Mol Cryst Liq Cryst.* 1988, *165*, 439-569.

[2] Oswald, P.; Pieranski, P. *Smectic and Columnar Liquid Crystals*; Taylor & Francis: New York, 2006.

[3] Solymar, L.; Webb, J.D.; Grunnet-Jepsen, A. *The Physics and Applications of Photorefractive Materials*; Oxford: New York, 1996.

[4] Goonesekera, A.; Wright, D.; Merner, W.E. *Appl Phys Lett.* 2000, *76*, 3358-3361.

[5] Moerner, W.E.; Silence, S.M. *Chem Rev.* 1994, *94*, 127-155.

[6] Kippelen, B.; Peyghambarian, N. *Advances in Polymer Science, Polymers for Photonics Applications II*; Springer: 2002, 87-156.

[7] Ostroverkhova, O.; Moerner, W.E. *Chem Rev.* 2004, *104*, 3267-3314.

[8] Sasaki, T. *Polym J.* 2005, *37*, 797-812.

[9] Tay, S.; Blanche, P.A.; Voorakaranam, R.; et al. *Nature*, 2008, *451*, 694-698.

[10] Blanche, P.A.; Bablumian, A.; Voorakaranam, R.et al. *Nature*, 2010, *468*, 80-83.

[11] Khoo, I.C. *Liquid Crystal Photorefractive Optics.* In: Yu, F.T.S (ed) *Photorefractive Optics*; Academic Press: 2000.

[12] Wiederrecht, G.P., Yoon, B.A., Wasielewski, M.R. *Adv Mater.* 2000, *12*, 1533-1536.

[13] Sasaki, T.; Kino, Y.; Shibata, M.; et al. *Appl Phys Lett.* 2001, *78*, 4112-4114.

[14] Talarico, M.; Goelemme, A. *Nature Mater.* 2006, *5*, 185-188.

[15] Meyer, R.B. *Mol Cryst Liq Cryst.* 1977, *40*, 33-48.

[16] Sasaki, T.; Naka, Y. *Opt Rev.* 2014, *21*, 99-109.

[17] Sasaki, T.; Miyazaki,D.; Akaike, K.; Ikegami, M.; Naka, Y. *J Mater Chem.* 2011, *21*, 8678-8686.

[18] Sasaki, T.; Kajikawa, S.; Naka, T. *Faraday Discuss.* 2014, *174*, 203-218.

[19] Sasaki, T.; Ikegami, M.; Abe, T.; et al. Appl Phys Lett. 2013, 102, 063306 (1-3).

3

# $RE_2O_3$ Nanoparticles Embedded in $SiO_2$ Glass Matrix—A Colossal Dielectric and Magnetodielectric Response

S. Mukherjee, T. H. Kao, H. C. Wu, K. Devi Chandrasekhar and H. D. Yang

**Abstract**

Significant experimental effort has been inspected to consider and implement favorable high-*k* gate dielectrics with magnetodielectric (MD) effect of series of rare earth oxide ($RE_2O_3$, RE ~ rare earth ions) nanoparticles (NPs) embedded in sol–gel derived $SiO_2$ glass matrix. Properly calcined $RE_2O_3$ NP-glass composite systems (in which RE ~ Sm, Gd and Er) show an intriguing colossal enhancement of dielectric constant along with MD effect near room temperature. The enhancement of dielectric constant is closely related to oxygen vacancy induced dielectric relaxation (or, more correctly, particle size effect from different calcined temperature), reconstructed from extended X-ray absorption fine structure. The MD response is strongly depended on the superparamagnetic property of the rare earth ions. From application point of view, the enhancement of dielectric constant associated with MD response can be achieved by tuning the NPs size through varying annealing temperature and/or increasing the doping concentration of magnetic rare earth oxide, which will be the key guidelines to accomplish the compatibility, performance and reliability requirements for future complementary metal-oxide-semiconductor (CMOS) technology.

**Keywords:** Rare earth oxide nanoparticle, high-*k* materials, magnetodielectric effect, diffuse phase transition

## 1. Introduction

Tiny electrical components are now unanimously required to be high in functionality and reliability and low priced in response to progress in the high density mounting technology.

Continued device scaling for future technology nodes requires reduction in equivalent oxide thickness (EOT) of gate dielectrics to maintain electrostatic control of the charges induced in the channel. The use of amorphous $SiO_2$ as a gate dielectric offers several key advantages in complementary metal-oxide semiconductor (CMOS) processing, including thermal and chemical stability as well as superior electrical isolation properties (high band gap of nearly 9 eV, and a Si–$SiO_2$ potential barrier for electrons of about 3 eV). The continuous miniaturization of Si electronics has imposed severe constraints on the performance of the $SiO_2$ gate oxide, with its thickness now approaching the quantum tunneling limit [1,2]. To continue the downward scaling, dielectrics with a higher dielectric constant (high-*k*) are being suggested as a solution to achieve the same transistor performance while maintaining a relatively thick physical thickness. Following this roadway, many materials systems (viz. lead–free non-ferroelectric) are currently under consideration as alternatives to conventional silicon oxide films as the gate dielectric material for sub-0.1 μm CMOS technology. Such an approach allows one to employ the best available materials for each phase, whose properties are known a priority due to the scarcity of high-*k* materials, to suit the desired application. Recent reports of giant dielectric constant have directed considerable attention to several new material systems, such as perovskite–related materials $ACu_3Ti_4O_{12}$ (A = Ca, $Bi_{2/3}$, $Y_{2/3}$, $La_{2/3}$) [3,4], $La_{2/3}Li_xTi_{1-x}Al_xO_3$ [5], $Nd_2O_3$ doped $(1\text{-}x)Bi_{0.5}Na_{0.5}TiO_{3\text{-}x}Bi_{0.5}K_{0.5}TiO_3$ [6], Fe-containing complex perovskites $A(Fe_{1/2}B_{1/2})O_3$ (A = Ba, Sr, Ca; B = Nb, Ta, Sb) [7,8], non-perovskite material $Li_{0.05}Ti_{0.02}Ni_{0.93}O$ [9], percolative $BaTiO_3$-Ni composites [10], electron-doped manganites $Ca_{1-x}La_xMnO_3$ and hole-doped insulators $La_2Cu_{1-x}Li_xO_4$ and $La_{2-x}Sr_xNiO_4$ [11–13]. The sensitivity of these complex oxides to strain, stoichiometry, phase heterogeneities, oxidation state, disorder, etc. can lead to drastic modifications in their magnetic and electric properties at the nanoscale. Besides that, as the key guidelines for replacing alternative dielectrics with high-*k* materials are required to (i) remain thermodynamically and chemically stable between the metal-oxide and Si substrate; (ii) kinetic stability against Si and the metal gate, in particular during high temperature processing and annealing; (iii) insulating properties: band offsets with Si over 1 eV to assure low leakage currents; (iv) a passivated, low-defect-density interface with Si to ensure large carrier mobility in the Si channel and good breakdown properties; and (v) interface quality between the high-*k* dielectrics and Si substrate: a low defect density in the high-*k* dielectric itself to prevent flat band and threshold voltage shifts and instabilities. Many dielectrics appear favorable in some of these areas, but very few materials are promising with respect to all of these guidelines. The ranking of $HfO_2$-based system as a desired high-*k* gate dielectric material to replace amorphous $SiO_2$ drops considerably, as $HfO_2$ suffers crystallization at a relatively low process temperature (< 500°C), resulting high leakage current along the grain boundaries [14]. Therefore, the exploitation of new type of amorphous phase pure high-*k* gate dielectrics candidates as a replacement of $SiO_2$ still faces several daunting challenges.

Besides the aforementioned consideration, the superior electrical characteristics of the Si–$SiO_2$ interface in ideal gate dielectric stack compatible with planarization technology has not achieved with any other alternative semiconductor–dielectric combination. Despite several key advantages of $SiO_2$, the continual scaling of CMOS technologies has pushed the Si–$SiO_2$ system in formidable challenge. One promising alternative approach to overcome the scaling limit has been proposed to substitute by silica-based single-valence nanoparticles (NPs) as gate insulator (interface between silicon and NP-oxides embedded silica), where flexibility, compartibility and functionality may be achieved through different NPs sizes/concentrations. Concentrating on the desired NP-oxides, potentially stable rare earth oxides ($RE_2O_3$, RE ~ rare earth, a series of elements from La to Lu with stable $RE^{3+}$) were chosen, which are attractive materials based on good thermodynamic energy considerations with silicon, highly resistive and a high conduction band offset over 2 eV. We have presented extensive results, providing useful insight into the physics of nano-composite high-*k* gate dielectrics. Sol–gel derived non-magnetic $SiO_2$ glass matrix with magnetic/nonmagnetic rare earth NP-oxides provides a convenient way to tailor desired magnetic, dielectric (in presence of applied magnetic field), and other properties by altering the type, size and concentration of the dopant ions.

## 2. Sample preparation

The preparation of $RE_2O_3$:$SiO_2$ nano-glass composite system (RE ~ La, Nd, Sm, Eu, Gd, Dy, Ho, Er, Tm, Yb and Lu) consists of three consecutive processes: (a) preparation of wet gel in which rare earth ions were doped by sol–gel way, (b) drying of the gel and (c) densification of the dry gel to a dense glass in which $RE_2O_3$ NPs embedded by calcining at selective temperatures [15]. The process was based on the hydrolysis of precursors, such as tetraethylorthosilicate {$Si(OC_2H_5)_4$} (TEOS) and subsequent condensation of hydrolyzed TEOS in a medium containing a hydroalcoholic solution of rare earth salt [16] (Figure 1(a)) having different mol% concentrations following essentially the method developed by Sakka and Kamiya [17]. Water was required for the hydrolysis. The molar ratio of water and TEOS was kept at 20 while that of TEOS and catalyst HCl at 100. Dry ethanol was used for diluting the alkoxide. The following composition of the $Si(OC_2H_5)_4$ solutions used in the study (Table 1):

| $Si(OC_2H_5)4$ (gm) | $H_2O$ (gm) | $C_2H_5OH$ (gm) | Mole ratio of $H_2O$ to $Si(OC_2H_5)_4$ | Volume | $SiO_2$ content (gm/100 cc) |
|---|---|---|---|---|---|
| 169.5 | 292.8 | 37.5 | 20 | 500 | 9.80 |

**Table 1.** Compositions of the $Si(OC_2H_5)_4$ (TEOS), $H_2O$, and $C_2H_5OH$ solutions used in the sol–gel process.

There are two distinct chemical reactions involved in the sol–gel process, describing Eqn. (1) for hydrolysis of the alcohol groups, Eqns. (2) and (3) for polycondensation of hydroxyl groups.

**a.** Hydrolysis:

$$\equiv \text{Si-OR} + H_2O \rightarrow \equiv \text{Si-OH} + \text{ROH} \tag{1}$$

**b.** Condensation (water/alcohol condensation):

**i.** Water condensation:

$$\equiv \text{Si-OH} + \equiv \text{Si-OH} \rightarrow \equiv \text{Si-O-Si} \equiv + H_2O \quad (2)$$

Alcohol condensation:

$$\equiv \text{Si-OH} + \equiv \text{Si-OR} \rightarrow \equiv \text{Si-O-Si} \equiv + \text{ROH} \quad (3)$$

The clear solutions without any precipitaion are prepared with the mixing of half amount of ethanol in alkoxide and the solution consisting of the specified amount of water with another half of the ethanol containing HCl and dopant. The mixure solutions continued stirring for 2–3 hours at room temperature. The clear solution was kept in pyrex beaker at the atmospheric condition for 7/8 days to form stiff monolithic transparent gel. Further, the gels were allowed to dry for 4–5 weeks at room temperature. The dried (liquid removed by thermal evaporation) monolith is termed as xerogel. The oven–dried gel (temperature range 100–200°C) still contains large concentration of chemisorbed hydroxyls. Heat treatment in the temperature range 500–800°C desorbs the hydroxyls, forming a stabilized gel. At 1000°C, it transformed to a dense glass. Heat treatments of samples were performed according to preselected calcination temperature schedule [16] (Figure 1(b)).

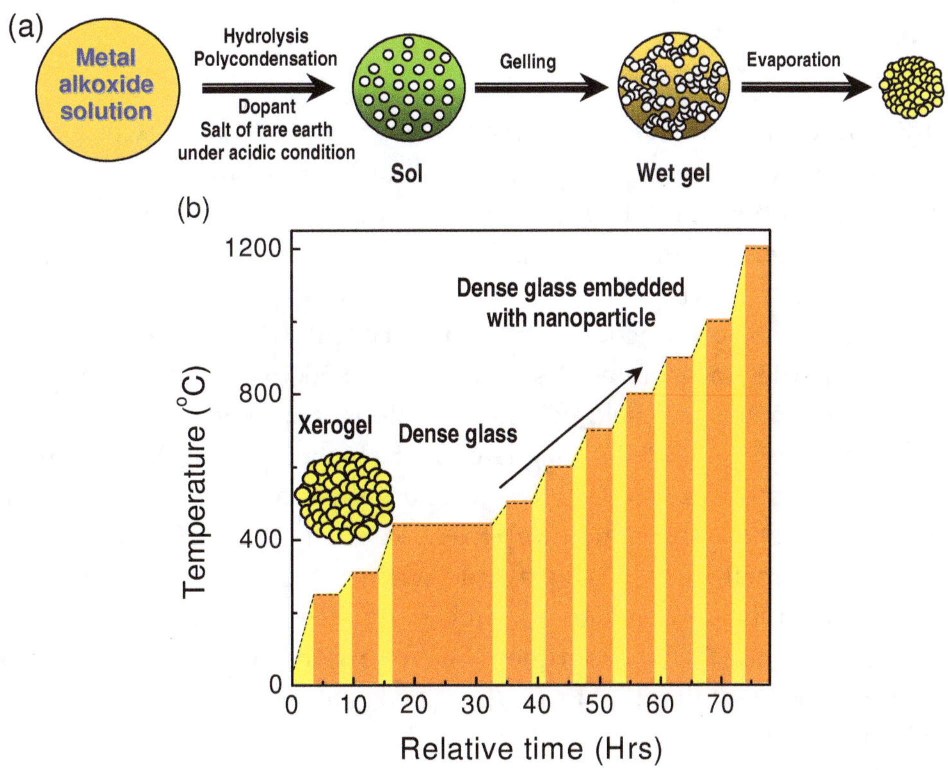


**Figure 1.** (Color online) (a) Sol–gel process. (b) Gel–glass embedded with rare earth nanoparticle calcination process.

It is relevant to mention here the important findings of Raman spectroscopic studies including measurements of pore size, density and specific surface area on the densification of undoped

$SiO_2$ gel as a function of heat treatment up to 900°C [18]. With increasing temperature from 700 to 800°C, the average pore size increases abruptly from 1.0 nm to 2.3 nm, whereas, the specific surface area decreases from 550 $m^2/g$ to 160 $m^2/g$ and the pore volume/gm decreases from 0.19 cc/g to 0.12 cc/g. The surface energy for a siloxane surface is higher than for a hydroxyl surface. The Si-OH groups condense to Si-O-Si bonds with increasing temperature, thus increasing the surface energy as well as enhancing pore collapse. In these rare earth elements doped gel-glass specimens, a large number of small pores collapse at 700°C and an agglomeration of individual $RE^{3+}$ are set free after small pores collapse to form NPs. The remaining pores join to form larger pores. At still higher temperature (i.e., 800°C), collapse of larger pores also takes place with similar observation in undoped sample, which is indicated by rapid fall of pore volume/g from 0.12 cc/g to 0.026 cc/g in going from 800°C to 900°C [18]. Thus it is possible at the highest temperature in the present case (i.e., 1200°C), the rare earth oxide NPs grow to a maximum size because of complete annihilations of pores, leading to a disappearance of Si - OH groups to condensation of Si-O-Si bonds. To this end, we have systematically synthesized nano-glass composites systems with different doping concentrations of rare earth elements to prepare the sols, since the gelation process (rate of the hydrolysis and condensation reactions) strongly affect with varying the kind and/or the amount of the starting solvent and the final outcome of the preparation.

## 3. Experimental details

Powder X-ray diffraction (XRD) of the sample was performed by using Cu- $K_\alpha$ radiation. To analyze the detailed structure of samples, an ultrahigh-resolution transmission electron microscopy (TEM) (Model: JEM-3010, JEOL) was employed of NPs embedded $SiO_2$ glass matrix calcined at different temperatures. Impedance spectroscopic measurements with/without magnetic field are carried out in the temperature range 150–350 K using LCR meter (Model E4980A, Agilent) in conjunction with laboratory built cryostat arrangement integrated to the physical properties measurement system (Model: 6000, Quantum Design). The magnetization (zero-field-cooled and field-cooled conditions) and magnetic hysteresis measurements are performed using a SQUID magnetometer (Model: MPMS-XL, Quantum Design) with temperatures varying from 2 to 350 K with ±1.0 K thermal stability and equipped with a superconducting magnet producing fields up to ±60 kOe. The sample in the powder form was packed in polytetrafluoroethylene (PTFE) capsule, where sample mass was chosen in the range of 8 to 12 mg for obtaining a good signal-to-noise ratio. Room temperature extended X-ray absorption fine structure (EXAFS) experiments were carried out in fluorescence mode (very low concentration of $RE^{3+}$) at the RE ~ Er/Gd/La $L_{III}$-edge at the 17C beamline in the National Synchrotron Radiation Research Center (NSRRC), Hsinchu, Taiwan. The EXAFS analysis was based on a multiparameter single-scattering fit (by the FEFFIT code [19] and using standard IFEFFIT data analysis package) in the *R*-space from the first two coordination shells; the fitting in the *k*-space ($k^2$-weighted $\chi(k)$ with a *k*-cut range from 2.5-8 Å$^{-1}$) led to the same result. A reference sample of bulk $RE_2O_3$ (unsupported $SiO_2$ glass matrix) was used as a model standard for determining co-ordination numbers and inter-atomic distances. In obtaining the EXAFS function $\chi(k)$, the background absorption features which were superimposed on the EXAFS oscillations were removed from the spectrum using spline fit to both the pre-edge and the post-edge regions.

## 4. Results and discussion

**A. Structural studies through transmission electron microscopy (TEM) and X-ray diffraction (XRD):** Typical data are shown for $Er_2O_3$:$SiO_2$ NPs-glass composite system having 0.5 mol% dopant $Er_2O_3$ concentration calcined at different temperatures, namely, 700, 800, 900, and 1200°C (henceforth referred as Er05-7, Er05-8, Er05-9 and Er05-12, respectively) [20]. Here, we discuss the TEM image of the powder specimen of typical Er05-8 (Figure 2(a)), showing nearly spherical NPs of $Er_2O_3$ embedded in the glass matrix in the range of 3–6 nm. The particle size distributions histogram (upper inset of Figure 2(a)) of the sample is calculated by counting onto the the micrograph at least 100 particles from Figure 2. The high-resolution TEM (HRTEM) image (lower inset of Figure 2(a)) of the respective sample also shows lattice fringes with interplanar spacing (3.04 Å). The selected area electron diffraction (SAED) pattern (Figure 2(b)) also shows spotted ring patterns, developing the regions of localized crystallinity.

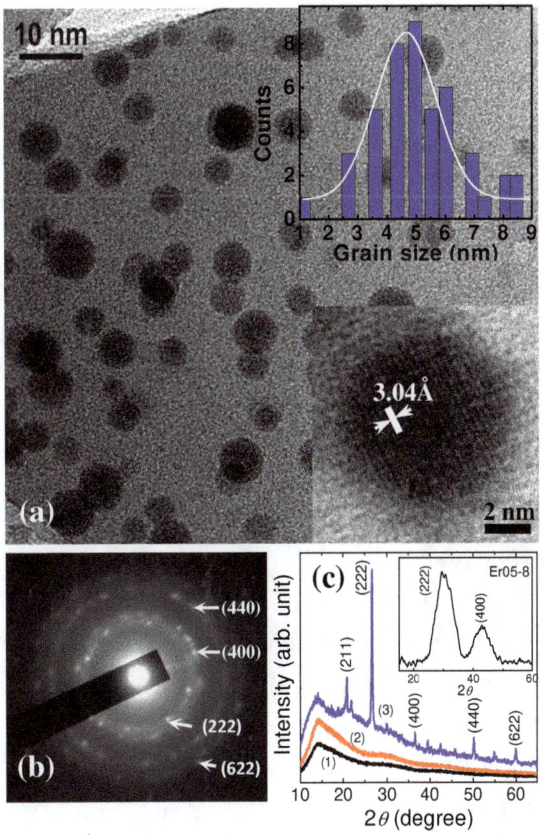


**Figure 2.** (Color online) (a) TEM image of Er05-8, upper inset: the particle size distribution histogram and lower inset: the HRTEM image, (b) electron diffraction and (c) XRD patterns of all the Er05-7, Er05-8 and Er05-12 samples.

The XRD patterns of the $Er_2O_3$ oxide NPs doped $SiO_2$ matrix calcined at 1200°C (Er05-12) show crystalline nature with quite large $Er_2O_3$ NPs. It exhibits clearly in Figure 2(c) the most intense characteristic line of single phase $Er_2O_3$ (JCPDF Card No. 43-1007) at $2\theta \sim 29.30°$ (222) corresponding the unit cell parameter $a = 10.54$ Å (space group: *Ia3*(206)). The sizes of Er-oxide NPs in Er05-12 are also estimated (> 40 nm) by applying the well-known Scherrer's equation from the integral breadths of the XRD lines. XRD patterns cannot be well resolved from feeble broad peaks due to their amorphous-like character of Er05-7 and Er05-8 samples. However, it is significantly mentioned here that the sizes of $Er_2O_3$ NPs grow larger for samples calcined at higher temperatures. In the $SiO_2$ glass matrix, low concentrations of $Er^{3+}$ ions are bound to non-bridging oxygen atoms with quite low solubility [21]. Consequently, at a higher $Er_2O_3$ concentration with $SiO_2$ (1:2), the formation of an $Er_2Si_2O_7$ mixture with $Er_2O_3$ may occur at 1200°C [22]. However, at a higher calcined temperature (~1200°C and above), there is a possibility of the formation of $Er_2Si_2O_7$ mixture with crystalline $Er_2O_3$ even with very low concentration (0.5 mol %) of dopant ions. Herein, the sample obtained at 1200°C are not emphasized because of the observed dielectric behavior (discussed in the next section) is almost comparable with pure bulk $Er_2O_3$ (unsupported with $SiO_2$).

**B. Dielectric and magnetodielectric (MD) effect**

**$Er_2O_3$ NPs-glass composite system: colossal enhancement of dielectric constant and large MD effect**

Among various rare earth oxides, $Er_2O_3$ has been chosen first in the present work as it possesses most appealing properties viz. high resistivity ($10^{12}$-$10^{15}$ $cm^{-3}$), large band gap ($E_g$= 5–7 eV), static dielectric constant ($k \sim 14$) [23,24], good thermodynamic stability with silicon and, moreover, not yet been explored from the viewpoint of observing the high MD effect. Although, present TEM and XRD studies also stimulate us for further exhaustive investigation on typical $Er_2O_3$:$SiO_2$ nano-glass composite system with different $Er_2O_3$ NPs size to throw more light exploring the origin and application feasibility on these rich dielectric materials. Typical data are shown for a $Er_2O_3$:$SiO_2$ nano-glass composite system having 0.5 mol% dopant ($Er_2O_3$) concentration calcined at different temperatures, namely, 700, 800, 900, and 1200°C (henceforth referred as Er05-7, Er05-8, Er05-9 and Er05-12 respectively).

### 4.1. Temperature dependence dielectric response

Figure 3(a) illustrates the relative dielectric constant ($\varepsilon'$) vs. temperature curves of typical Er05-8 sample without applied magnetic field measured at several selective frequencies (1–100 kHz). The shape of the curves with a notable broadening around the well-defined maxima $\varepsilon'_m$ (maximum value of $\varepsilon'$) is indicative of a diffuse phase transition (DPT) presence [25-27] with high $\varepsilon'$ (~ 600 at 1 kHz), quite different from and much higher than pure bulk $Er_2O_3$ ($\varepsilon' \sim 11.5$) [28] and $SiO_2$ ($\varepsilon' \sim 3.9$). Following the concept of DPT, the dielectric constant accords with a modified Curie–Weiss type equation viz. $\varepsilon'^{-1} - \varepsilon'^{-1}_m = C_i(T - T_m)^\gamma$, where $\gamma$ is the diffuseness exponent indicative of degree of disorder, $C_i$ is a temperature independent coefficient (in general, dependent of frequency) and $\varepsilon'_m$ is the maximum value of $\varepsilon'$ at $T_m$. For $\gamma = 1$, normal Curie–Weiss behavior and $\gamma \sim 2$, it implies typical DPT for the ideal ferroelectric relaxor [29].

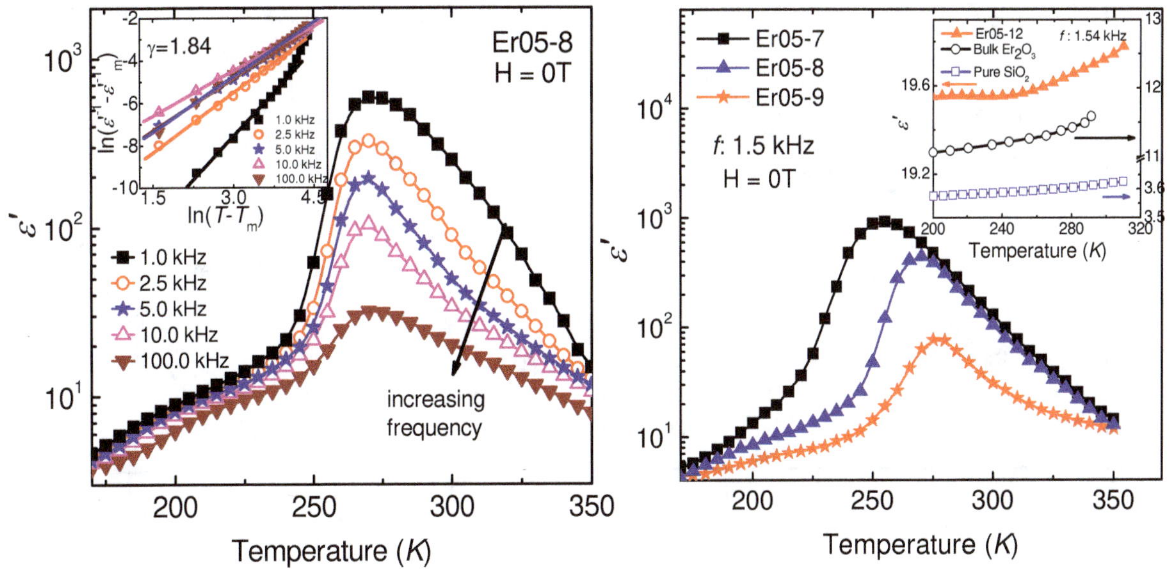


**Figure 3.** (Color online) (a) The ($\varepsilon'$-$T$) curves of Er05-8 at different selective frequencies, inset: representative plots of $\ln(\varepsilon'^{-1}-\varepsilon'^{-1}_m)$ vs. $\ln(T-T_m)$ of Er05-8 at temperatures > $T_m$, (b) ($\varepsilon'$-$T$) curves of $Er_2O_3$ nanoparticles dopes $SiO_2$ glass calcined at different temperatures, inset shows ($\varepsilon'$-$T$) curves of Er05-12 with respect to Bulk $Er_2O_3$ and $SiO_2$ showing non-DPT feature.

The plots of $\ln(\varepsilon'^{-1}-\varepsilon'^{-1}_m)$ vs. $\ln(T-T_m)$ at several frequencies are illustrated in inset of Figure 3(a). We obtained average $\gamma = 1.84$ from different slopes of linear fitting, which is close to the relaxor value like other oxide relaxor systems [30]. The enhancement of dielectric value is observed even at temperatures below $T_m$ (~ 250–260 K), which ruled out the possibility of any space charge or interfacial polarization. The high dielectric value associated with DPT behavior (Figure 3(b)) diminishes with increasing calcined temperature (i.e., the $Er_2O_3$ NPs size). It is relevant to mention here that the temperature and frequency dependent dielectric phenomena of Er05-12 crystalline sample (calcined at 1200°C) without DPT behavior behaves like pure bulk $Er_2O_3$ [28] or $SiO_2$ glass (inset of Figure 3(b)). The critical calcination temperature above which DPT behavior completely diminishes for this typical concentration of $Er_2O_3$ (~ 0.5 mol %) is found to be around 1000°C. The DPT behavior is thus confined to the low temperature calcined (< 1000°C) system only where the NPs are in the 2–10 nm range.

### 4.2. Dielectric relaxation analysis

To shed more light on the relaxation dynamics of rare earth oxide NPs-glass composite systems, temperature dependent dielectric loss tangent (tanδ) are carried out at various frequencies. The appearance of three maxima with strong frequency dispersion located at peak $A$ ~180 K, peak $B$ ~ 260 K, and peak $C$ > 320 K are observed in tanδ vs. temperature curve. The former two peaks (peaks $A$ and $B$) are shifted to high temperature with increasing frequency, whereas, the peak $C$ with high-dielectric leakage (~ 7) is shifted to the lower temperature. In Figure 4(b), the temperature dependence logarithmic plot of the relaxation time (τ), determined as the inverse of the maximum peak frequency exhibit straight line, is shown in an Arrhenius representation $\tau = \tau_o \exp(E_{relax}/kT)$, with an energy barrier $E_{relax}$. Near the DPT temperature ($T_m$), thermally

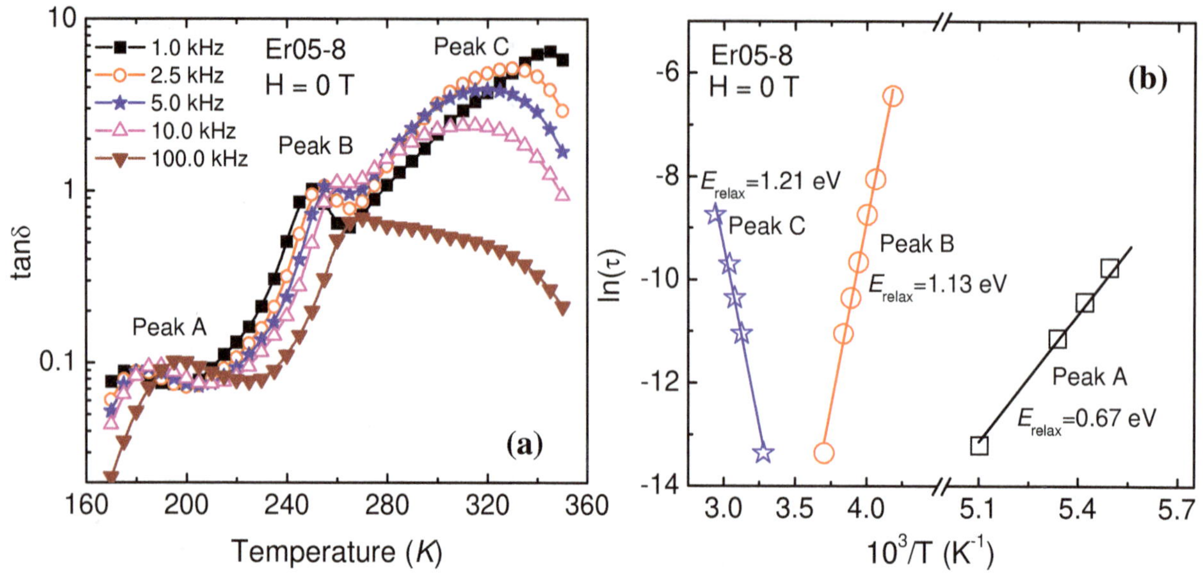


**Figure 4.** (Color online) (a) Dielectric loss tanδ of Er05-8 at different frequency, (b) representative Arrhenius plot of the relaxation time of Er05-8. The calculated activation energy values (in electron volt (eV)) are illustrated in each case.

activated response is described with an energy barrier $E_{relax}$ of about 1.13 eV. However, the temperature dependent relaxation response above 300 K becomes reversed with the activation energy 1.21 eV. These experimental facts suggest the presence of thermally activated oxygen vacancies associated with the dielectric relaxation process as presented earlier with activation energy ~ 0.7-1.2 eV [30,31]. Our present results support the recent experimental finding of perovskite type $ABO_3$ [31] material, closely related to the thermally activated reorientation of dipole moment via the oxygen ion jumping through the oxygen vacancy, which can be controlled by sintering process. The dielectric response and DPT behavior diminishe by long-time annealing of the sample at higher temperature, which might be associated with reduced concentration of oxygen vacancies. Here, electrode effect in dielectric measurement is excluded by using different thickness of the samples with different electrode materials, indicating the intrinsic nature of this system.

### 4.3. Polarization studies

Figure 5 shows the frequency and temperature dependence hysteresis loop (*P-E* curves) of typical Er05-8. The values of remanent polarization ($P_r$ ~ 0.032 $\mu$C/cm$^2$) and coercive field ($E_c$ ~ 0.78 kV/cm) of relatively narrow *P-E* loop near $T_m$ (~270 K) without saturation are attributed to noncanonical ferroelectric-like (FEL) correlation in the sample, similar to those commonly observed in $ABO_3$ perovskites [32]. However, the present NP-glass composite system has very high magnetic dilution of the NPs $Er_2O_3$ concentration (0.5 mol% $Er_2O_3$ : 99.5 mol% $SiO_2$) and hence small amount of dipole moment per unit volume are not high enough to induce significant changes in the polarization. The spurious hysteresis loop reveals some contribution of lossy dielectric (space charge such as oxygen vacancies) or nicknamed as "banana loops," the terminology recently coined by Scott [33]. At lower frequency, the hysteresis loop becomes slightly fatter. However, to check the possible FEL correlation in the sample, temperature

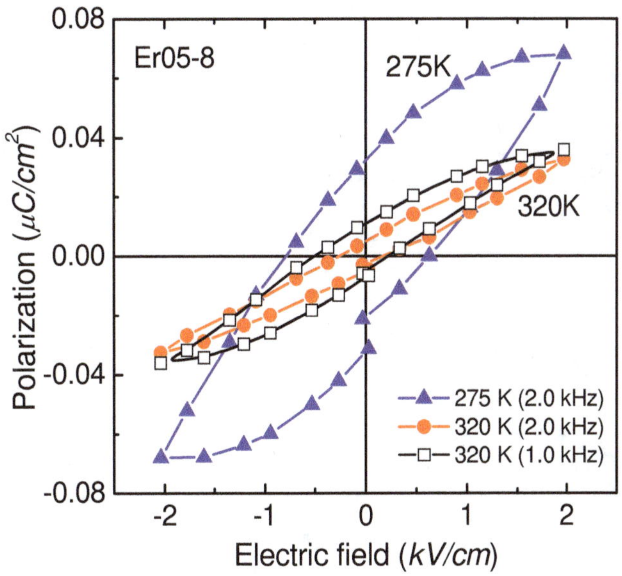


**Figure 5.** (Color online) Dielectric hysteresis loop of Er05-8, measured near DPT (275 K) and above room temperature (320 K) using 2.0 and 1.0 kHz polarization frequency.

dependent *P-E* characteristics are carried out at highest polarization frequency (2.0 kHz), obtained in our instrument (Precision LC meter, Radiant Technologies). It is noted that the measured hysteresis loop at high frequency is closely related with the intrinsic ferroelectric switching processes of the system [34]. Although, the values of remanent polarizaion and coercive field of *P-E* curves becoming more pronounced with decreasing temperature from 320 to 275 K suggesting ferroelectric-like ordering in NPs-glass composite system, further investigations are certainly needed to delineate it.

### 4.4. Equivalent circuit analysis

Materials exhibiting colossal enhancement of dielectric value are usually adopted to explain by Maxwell–Wagner (MW) mechanism. The present NPs-glass composite system is basically NPs grain of rare earth oxide (uniformly distributed) embedded in more insulating $SiO_2$ matrix. The enhancement of dielectric constant along with DPT behavior might be a signature of the effect of internal barrier layer capacitance depending on the ration of grain size and the grain-boundary thickness. The complex impedance curves in Figure 6 have also been analyzed using an equivalent circuit, consisting of the two inclined semicircular arc (deviation from the ideal Debye response). Thus, the two depressed semi-arc in the Nyquist plot (complex impedance $Z''$-$Z'$plane) of the impedance data could be modeled on two parallel resistor–capacitor (RC) networks connected in series, one corresponds to the conducting part in high frequency region assigned to the intrinsic effect of grain (typical $Er_2O_3$ NPs) and the other arc in low frequency side corresponds to the more resistive part ($SiO_2$ matrix) of the sample. Interestingly, the entire measured frequency region ($20 - 2 \times 10^6$ Hz) at the temperature below $T_m$ (<270 K) is governed by the grain response (intrinsic effect). The temperature dependence of grain ($Er_2O_3$) resistance ($R_g$) values are obtained from equivalent circuit model with the help of commercial software (Z-VIEW, version 2.9c). The contribution of grain resistance (intrinsic response of NPs-glass systems) in the presence of magnetic field effect are discussed in the next section.

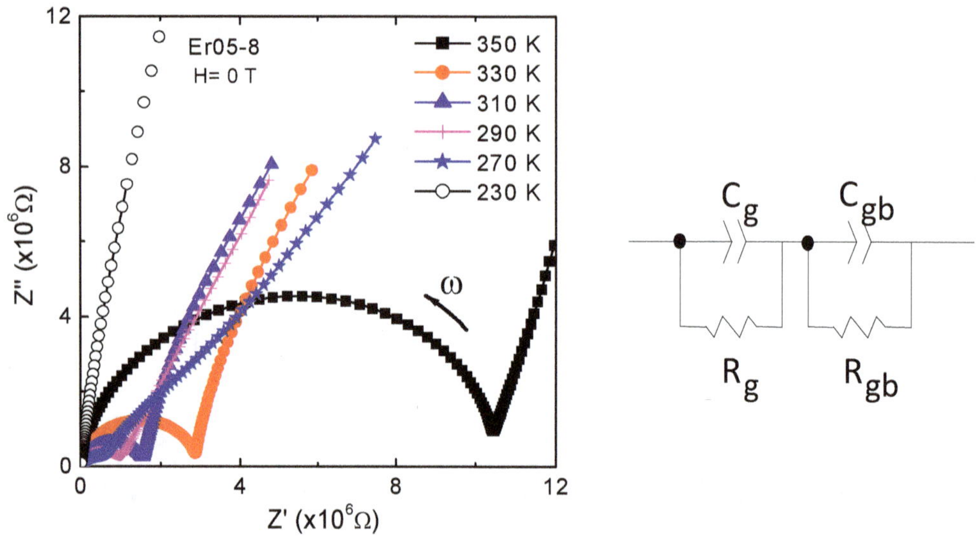


**Figure 6.** (Color online) (a) Complex plane plots, Z″-Z′, of Er05-8 at several temperatures and (b) schematic model of equivalent electrical circuits indicating of two parallel resistor–capacitor (RC) combinations [($R_gC_g$): $Er_2O_3$ nano-grain, ($R_{gb}C_{gb}$): $SiO_2$ matrix] connected in series.

## 4.5. Magnetodielectric effect

The observation of colossal MD effect is the most interesting finding of Er05-8 system as shown in Figure 7(a) at a specific frequency of 2.5 kHz. The large enhancement of dielectric constant (~2.75 times) is observed around the transition regime 260–300 K under 9 T magnetic field. The inverse of dielectric constant with temperature under magnetic field (upper inset of Figure 7(a)) are also fitted by Curie–Weiss law with Curie constant (C) (3968.82, 6211.29 and 6918.04 K for 0, 5 and 9 T, respectively) and Curie–Weiss temperature ($T_o$) (260.06, 270.12 and 271.64 for 0, 5 and 9 T, respectively). It is obvious that both dielectric temperatures ($T_m$ and $T_o$) are shifted to higher temperatures with increasing magnetic field, indicating the occurrence of magnetic spin-ordering at higher temperature under magnetic field and hence exhibit a reduced spin-lattice coupling strength under magnetic field. Temperature and frequency dependent dielectric constant of Er05-7 and Er05-8 are measured at a typical higher magnetic field (~9 T), shown in Figures 7(b) and 7(c). One may speculate about particle size dependent field effect playing a role of the larger ε′ response in the lower temperature annealed samples [35]. Within this scenario, the system in which we observed MD effect as well as Curie–Weiss behavior is the single phase of amorphous $Er_2O_3$ NPs of 2–10 nm size embedded in $SiO_2$ glass calcined at 700–900°C. The estimated field dependent MD response (MDR) near $T_m$ (~275 K) is defined by $\{\Delta\varepsilon'(H)/\varepsilon'(0)\% = [[\varepsilon'(H)-\varepsilon'(0)]/\varepsilon'(0)]\times 100\}$ as a function of the square of the magnetization as shown in the lower inset of Figure 7(a). Strikingly, the fractional change of the magnetic field induced change in the dielectric constant can be well approximated by the scaling function, $\Delta\varepsilon'/\varepsilon' \approx \alpha M^2$, where magnetoelectric interaction constant, $\alpha$ is estimated at 0.782.

Similar behavior of the change in dielectric constant on the square of the magnetization is also observed in several materials including intrinsic multiferroics, such as $BiMnO_3$ [36], suggesting MD response in the present system (magnetic NPs of the guest oxide and $SiO_2$ host glass) are closely related to the magnetism, typical size and concentration of the $Er_2O_3$ NPs.

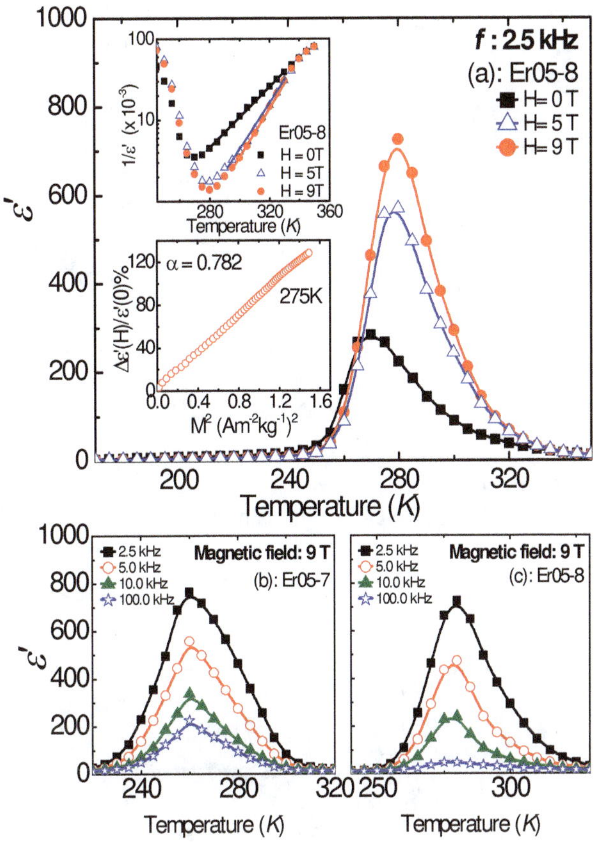


**Figure 7.** (Color online) (a) The magnetic field dependence of ($\varepsilon'$-$T$) curves of Er05-8 at a fixed frequency 2.5 kHz. Upper inset of (a): inverse of $\varepsilon'$ with temperature under magnetic field exhibiting the Curie–Weiss behavior and the lower inset of (a): the fractional change of the magnetic field induced change in the dielectric constant ($\Delta\varepsilon'/\varepsilon'$) of Er05-8 showing linear variation with the square of magnetization $M^2$, measured in the vicinity of $T_m$ (~ 275K). [(b), (c)] ($\varepsilon'$-$T$) curves of Er05-7 and Er05-8 samples measured with several selective frequencies under 9 T applied magnetic field.

### 4.6. Micro-structural correlated resistivity analysis

Figure 8 shows the contribution of amorphous NP $Er_2O_3$ grain resistance $R_g$ (calculated from equivalent circuit element analysis in Figure 6) of Er05-8 under external magnetic field as a function of measuring temperature. The temperature dependence of *ac* conductivity ($\sigma_{ac}$) at various frequencies is demonstrated in Figure 9. In the inset of Figure 9, the *ac* conductivity as a function of temperature under external magnetic field is illustrated. Concomitantly, the grain resistance $R_g(T)$ in Figure 8 exhibits a metal to insulator like transition coinciding with the dielectric maxima temperature $T_m$ of $\varepsilon'(T)$ (Figure 3(a)) as well as $\sigma_{ac}(T)$ (Figure 9). Interestingly, the $R_g$ decreases under magnetic field, similarly observed in colossal magnetoresistive materials [25]. These experimental facts truely corroborate that the nature of charge carriers responsible for *dc* conduction in the grain interior and the dielectric relaxation maxima belongs to the same category.

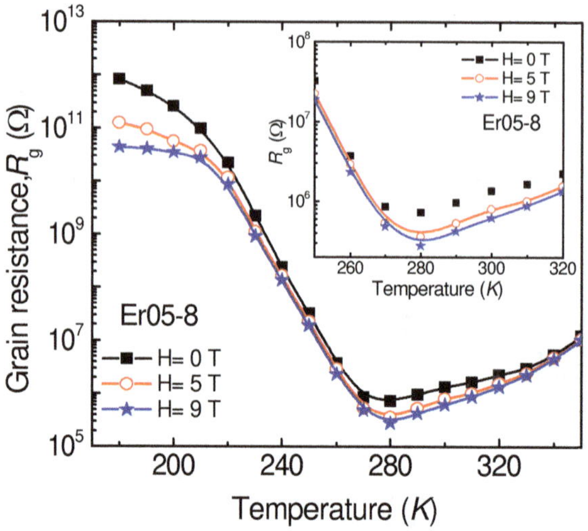


**Figure 8.** Color online) Temperature dependence of grain resistance ($R_g$) calculated from impedance complex plane plots with external magnetic field. Inset: The region close to $T_m$ is highlighted.

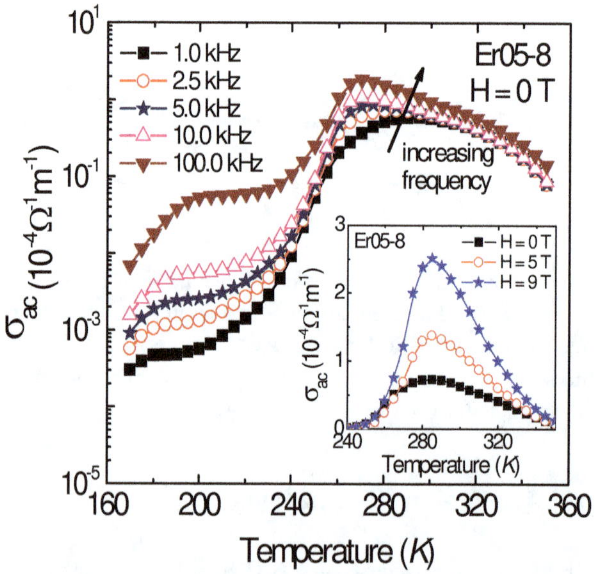


**Figure 9.** (Color online) Temperature dependence of *ac* conductivity ($\sigma_{ac}$) of Er05-8 for various frequencies. Inset: temperature dependence of *ac* conductivity ($\sigma_{ac}$) at 2.5 kHz with external magnetic fields.

The magnetoresistive property of magnetic NPs is attributed by spin-polarized tunneling [37]. Although, the observed strong positive magnetoelectric interaction constant ($\alpha$~ 0.782) has a similar appearance to intrinsic multiferroics, the MD effect can also be achieved through a combination of magnetoresistance and the Maxwell–Wagner effect, as predicted by Catalan

[38]. Since the current results suggest that MD behavior is probably a manifestation of magnetoresistance changes, depending on the NP size and separation. Enhancement of MD response (i.e., positive MD effect) through the decreases of NPs $Er_2O_3$ resistance under external magnetic field, (i.e., negative magnetoresistance) might imply the possible tunability of the resistive MD effect.

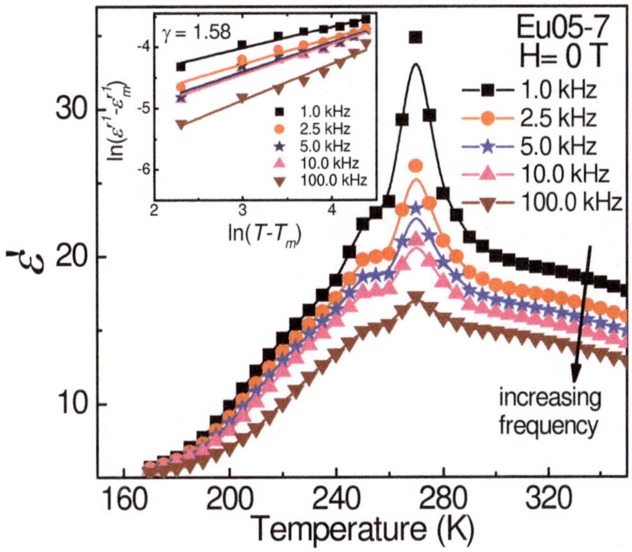


**Figure 10.** (Color online) (a) The ($\varepsilon'$-$T$) curves of Eu05-8 at different frequency, inset: representative plots of $\ln(\varepsilon'^{-1} - \varepsilon'^{-1}_m)$ vs $\ln(T - T_m)$ at temperatures higher than $T_m$ for the Eu05-8 at different frequency values.

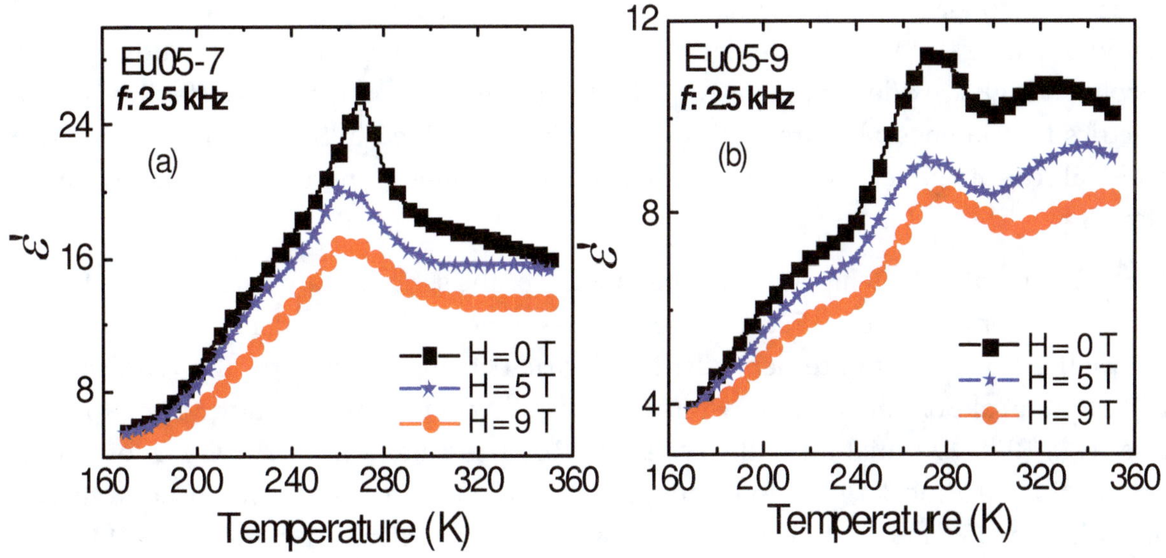


**Figure 11.** (Color online) (a), (b) The ($\varepsilon'$-$T$) curves of Eu05-7 and Eu05-9 measured under different applied magnetic fields at a fixed frequency 2.5 kHz.

## 5. $Eu_2O_3$ NPs-glass composite system: Smaller dielectric response and negative MD effect

Typical data is shown for $Eu_2O_3$:$SiO_2$ NPs-glass composite system having 0.5 mol% dopant $Eu_2O_3$ concentration calcined at different temperatures, namely, 700, 800 and 900°C (henceforth referred as Eu05-7, Eu05-8 and Eu05-9 respectively). Figure 10 represents the temperature dependence of $\varepsilon'$of Eu05-7 in the absence of the magnetic field. The nature of the variation of the ($\varepsilon'$-$T$) curves with maxima at $T_m \sim 270$ K, represents as "diffuse phase transition" (DPT) around $T_m$ with smaller $\varepsilon'$ than Er05-7 system, quite different and larger than that of bulk $Eu_2O_3$ [28] or pure $SiO_2$ (~ 3.9). Such system also follows a modified Curie–Weiss relationship [29], obtaining diffuseness exponent $\gamma$ =1.58 (inset of Figure 10). The $\varepsilon'$-$T$ plot (at 2.5 kHz), around the transition regime 240-320 K of Eu05-7 and Eu05-9 shows a decrease of $\varepsilon'$ under a magnetic field of 9 T (Figure 11). Here, we also observed particle size dependent effect, associated with DPT behavior as well as MD effect.

## 6. Effect of different rare earth oxides on the dielectric properties

In previous sections, we have presented interesting particle size dependent colossal dielectric response along with MD effect in $Er_2O_3$ and $Eu_2O_3$ nano-glass composite systems. From the obtained results, there are colossal enhancement of dielectric constant and large MD effect in $Er_2O_3$ case [20], while those in $Eu_2O_3$ case, smaller responses were observed [39]. Obviously, the different electronic and magnetic properties for $Er_2O_3$ and $Eu_2O_3$ play a crucial role. However, these results suggest great promise in further systematic investigation to distinguish the mechanisms that contribute to colossal dielectric responses along with MD effect in other $RE_2O_3$:$SiO_2$ nano-glass composite systems ($RE_2O_3$, RE ~ rare earth, a series of elements from La to Lu with stable $RE^{3+}$) with different $RE_2O_3$ NPs size.The purpose of such study is to find amorphous high-*k* oxide candidates and MD effect with superior phase stability for gate dielectrics from a lineup of rare earth metal oxides embedded in $SiO_2$ matrix, and to find a sequential coupling between different constituents among these nano-glass composite materials.

Figure 12 illustrates the temperature dependent $\varepsilon'$ of the series of rare earth oxide NPs-glass composite systems calcined at 700°C in the absence of the magnetic field. Here, we concentrate the variations of temperature dependent dielectric behavior mainly on the 700°C calcined sample for the sake of clarity. Besides that, the possibility of formation of other rare earth oxide phase (e.g., $RE_2Si_2O_7$) mixture with crystalline $RE_2O_3$ is ruled out at 700°C for low dopant concentration (0.5 mol%) [22]. Interestingly, $RE_2O_3$:$SiO_2$ nano-glass composite systems in which RE ~ Sm, Gd and Er show colossal enhancement of dielectric constant ($\varepsilon' \sim 10^3$) around room temperature. The nature of the variation of the ($\varepsilon'$-$T$) curves represents well-defined maxima and notable dielectric broadening around $\varepsilon'_m$ (maximum value of $\varepsilon'$) with high $\varepsilon'$ and different from pure bulk $RE_2O_3$.

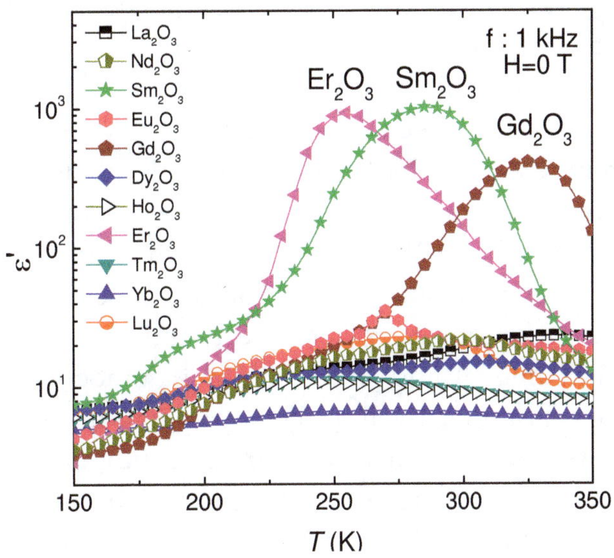


**Figure 12.** (Color online) The $\varepsilon'$-$T$ curves of $RE_2O_3$ (gel-glass calcined at 700°C) at 1 kHz without applied magnetic field.

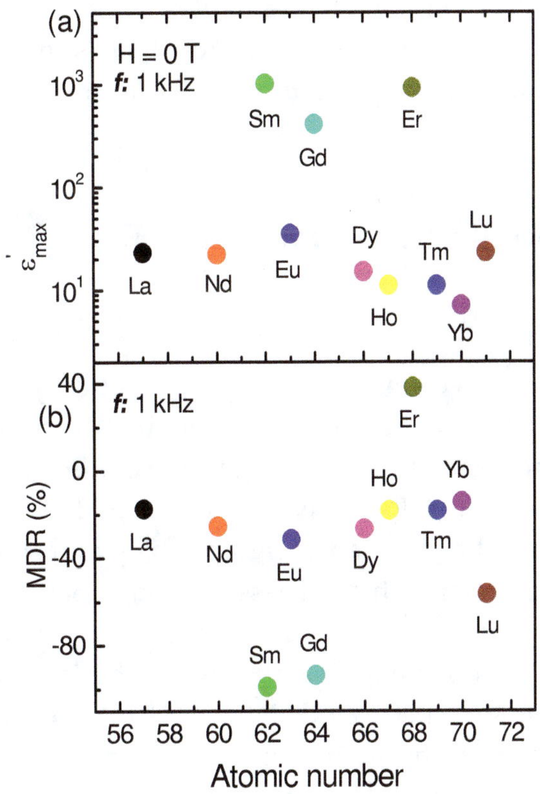


**Figure 13.** (Color online) (a) Maximum value of dielectric constant, and (b) MDR under 5 T applied field of $RE_2O_3$:$SiO_2$ nano-glass composite systems calcined at 700°C with rare earth atomic number.

The amorphous self-organized rare earth oxide nano-glass composite systems may be the promising high-*k* gate dielectrics due to its reproducible high dielectric constant (Figure 13(a)), single-stage process in air at moderate temperature and good compatibility with modern microelectronics processing technique. The present systems also show the MD effect around the transition temperature. The MDR at 1 kHz is plotted as a function of atomic number of the rare earth elements near $T_m$ as shown in the Figure 13(b). The $RE_2O_3$:$SiO_2$nano-glass composite systems in which RE ~ Sm, Gd, Er and Lu show colossal response of dielectric constant under applied magnetic field. Sol–gel process provides a convenient way for tailoring phase pure, self-organized NPs of nearly uniform sizes (particle size distribution histogram from TEM image) and for facilitating homogeneous dispersion of these metal-oxide NPs in the silica matrix. It is believed that the sol–gel derived NP-glass composite systems prepared and calcined in identical condition (Figure 1(b)), the particle size distributions of all rare earth oxide at specific calcined temperature is nearly equal (say 700°C, we have checked for Er, Eu, Gd and La systems). These rare earth oxide NPs are rigidly fixed within the insulating silica matrix at all temperatures. So this dielectric behavior does not arise from the physical motion of the NPs. This feature takes place inside the rare earth NPs grain and they are very much conditioned by magnetic property of NPs grain, the potential barriers in the grain boundaries, the degree of deformation of the lattice and the crystallites, as well as the grain size and their constituent host. Why RE ~ Sm, Gd and Er shows much larger effects than other rare earths? It needs further investigation using magnetic and non-magnetic rare earth oxide NPs together with different doping concentrations to explore the mechanism and application feasibility on these rich dielectric materials.

## 7. Effect of molar concentration of magnetic NPs on the observed dielectric and MD properties

As discussed in the above section, we have systematically investigated the colossal responses of dielectric behavior along with MD effect in rare earth oxide ($RE_2O_3$, RE ~ rare earth, a series of elements from La to Lu with stable $RE^{3+}$) NPs embedded $SiO_2$ glass composite systems with different $RE_2O_3$ NPs size. Properly annealed NPs-glass composites, where RE ~ Sm, Gd and Er, show an intriguing colossal response of dielectric behavior and MD effect near room temperature. These reproducible experimental facts suggest simultaneously a question why only these three magnetic rare earth elements have larger effects than others. Herein, we have systematically synthesized together via sol–gel route the magnetic $Gd_2O_3$ and non-magnetic $La_2O_3$ NPs with different doping concentrations and size embedded in $SiO_2$ matrix. The doping concentration and the corresponding sample index are highlighted in Table II [40]. Here, we report that the high-*k* and MD of these NP-glass composite systems are very much conditional by magnetic property of $Gd_2O_3$ NPs size, concentration, and the degree of deformation of the host matrix. To improve the dielectric tunability in presence of external magnetic field, the crucial magnetic properties of dielectrics are necessary for the application of the devices.

| Sample name | Non-magnetic $La_2O_3$ (mol%) | Magnetic $Gd_2O_3$ (mol%) |
|---|---|---|
| LGS1 | 0.150 | 0.000 |
| LGS2 | 0.120 | 0.030 |
| LGS3 | 0.090 | 0.060 |
| LGS4 | 0.075 | 0.075 |
| LGS5 | 0.060 | 0.090 |
| LGS6 | 0.000 | 0.150 |
| LGS7 | 0.000 | 0.500 |

**Table 2.** Different doping concentrations of non-magnetic $La_2O_3$ and magnetic $Gd_2O_3$ (mol%) in $SiO_2$ NP-glass composite systems.

Figure 14 illustrated the $\varepsilon'$-$T$ curves of $La_2O_3/Gd_2O_3$ NPs embedded $SiO_2$ glass composite systems (henceforth referred to as LGS) with different doping concentrations in the absence of the magnetic field. All the curves have well-defined diffuse phase transition-like maxima at $T_m$ ~320 K characteristic with oxygen vacancy-induced dielectric relaxation, thoroughly discussed in our previously reported sections [16,20,39] of other rare earth systems. Interestingly, at higher magnetic $Gd_2O_3$ doping concentrations, the LGS NP-glass composite systems show colossal enhancement of dielectric behavior near room temperature. Even, a very high magnetic dilution of $Gd_2O_3$ NPs (LGS2 sample ~ doping level 0.03 mol%) system, the dielectric value is higher than that of pure bulk crystalline counterpart [28].

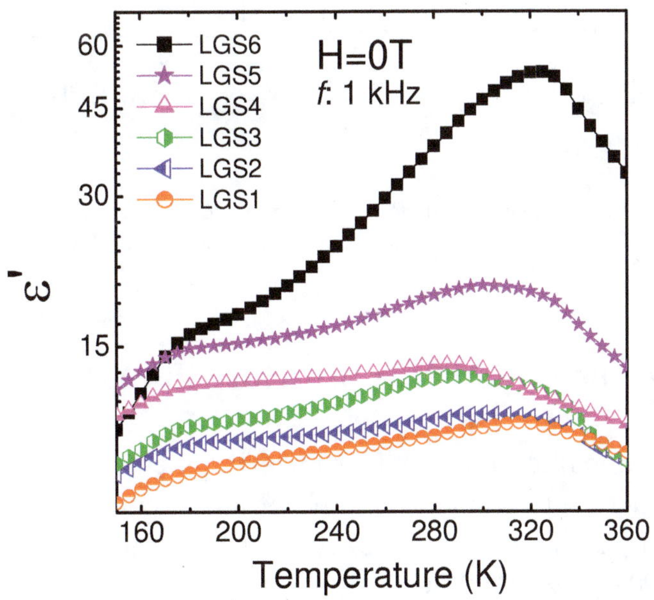


**Figure 14.** (Color online) The $\varepsilon'$-$T$ curves of LGS NP-glass composite systems with different doping concentrations calcined at 700°C.

Figure 15(a) illustrates the $\varepsilon'_{max}$ vs. $Gd_2O_3$ doping concentrations of NP-glass composite systems calcined at 700°C in the absence of the magnetic field. Here, $\varepsilon'_{max}$ increases with higher doping concentration of magnetic $Gd_2O_3$ NPs. The present systems also show strong MDR around the transition temperature ($T_m$). The MDR at several selective frequencies near $T_m$ are plotted as a function of doping concentration of $Gd_2O_3$, shown in the Figure 15(b). MDR enhances with faster rate at relatively lower doping concentration (≤0.1 mol%) of magnetic $Gd_2O_3$, whereas at moderate concentration range (≥0.2 mol%), the colossal response are more pronounced. A profound analogy may be expected between the colossal MDR in magnetic NP-glass composite systems and the inhomogeneous magnetoelectric interaction, inducing through magnetic spin modulation (flexomagnetoelectric polarization) [41]. Depending on the characteristic size (particle radius) and magnetization in amorphous-like nanosized systems, the flexomagnetoelectric effect induces linear magnetoelectric tunability [42].

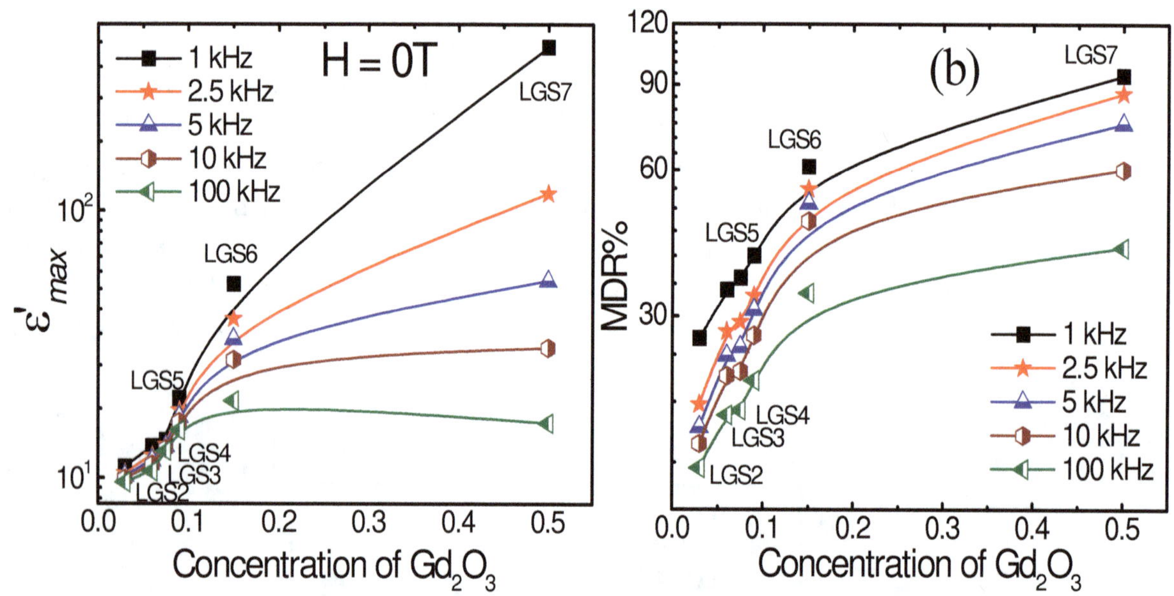


**Figure 15.** (Color online) (a) Maximum value of dielectric constant, and (b). MDR under 5 T applied field of LGS NP-glass composite systems with different $Gd_2O_3$ doping concentrations calcined at 700°C.

## 8. Extended X-ray Absorption Fine Structure (EXAFS) experiments

Extended X-ray absorption fine structure (EXAFS) experiments were performed at the Gd $L_{III}$ edge to obtain further detailed information on the local surroundings of the chemical elements of the Gd-oxide doped (0.5 mol%) nano-glass composite system calcined at different temperatures, namely, 700, 800, 900, and 1200°C (henceforth referred as Gd05-7, Gd05-8, Gd05-9, and Gd05-12, respectively). The broad dielectric anomaly in $Gd_2O_3$:$SiO_2$ can be plausibly attributed to oxygen vacancy defects, which are implied by the EXAFS measurements. Of particular interest are the coordination number of oxygen around the Gd atom and type of neighbors, the interatomic distance between the Gd atom and the surrounding oxygen, and the Debye–Waller factors which contain the mean-square relative displacements due to static disorder or thermal vibration. Figure 16 depicts the room temperature experimental EXAFS spectra of samples Gd05-7, Gd05-8 and Gd05-9. The EXAFS signals weighted by $k^2$ [i.e., $k^2\chi(k)$] are shown in Figure 16(a), and the moduli of their Fourier transforms $|F(R)|$ are presented in Figure 16(b). The EXAFS spectra in Figure 16(a) look very much the same, on the first glance, and thus one would not expect a major difference in the Fourier transform data. The Fourier transform modulus (Figure 16(b)) exhibits two main coordination peaks: the first one, located at $R \sim 1.8$ Å corresponds to the interatomic distances of $Gd^{3+}$–O (first coordination shell), while the second one (not growing properly) is located at $R \sim 2.8$ Å, indicating mostly the amorphous environment of Gd [21], as expected for a Gd–Gd coordination (second coordination shell) in a crystal. However, the TEM results confirm the presence of NPs $Gd_2O_3$ with localized crystallinity in the glass specimen calcined at 700°C. The fitting results of $Gd^{3+}$–O interatomic distances obtained from Figure 16 are summarized as a function of calcination temperatures in Table III. The values obtained from the EXAFS spectra are not significantly different in all the glass specimens with very low doping concentration. However, on close inspection it is evident that the average $Gd^{3+}$–O interatomic distance of Gd05-7 is shorter than that of Gd05-8, Gd05-9, and bulk $Gd_2O_3$ (unsupported $SiO_2$ glass matrix), suggesting that oxygen ions surrounding the oxygen vacancies around Gd ions should concurrently involve a relaxation toward their adjacent vacancies and, as a result, shorter bond length can be attributed [43]. This argument can, presumably, also be applied to the $Er_2O_3$:$SiO_2$ NP-glass system. It has been found that the porous glass are formed at ~ 400°C and annihilation of pores start at ~ 700°C, completed at ~ 800°C [44]. Therefore, silica gel-glass doped with low concentration of rare earth ions are subjected to systematic heating, the collapse of the silica pores is initiated near 700°C (Si-OH groups condense to Si-O-Si bonds). In this calcined glass, agglomeration of individual rare earth oxide ions loosely attached to the pores to be detached to form clusters. The dimensions of the clusters reside in the nanometer range. At 700°C calcined sample, in rare earth oxide NPs, oxygen vacancies could be easily created by loss of oxygen at low oxygen partial pressure, according to $O_0 \leftrightarrow V_0 + 1/2O_2$. Based on these considerations, the formation of oxygen vacancies inside the NPs influence the bond-length change, and the number of oxygen atoms surrounding the Gd atom should also decrease. This argument is in fair agreement with the result of a relative decrease in the number of nearest neighbors (Table III) in Gd05-7 in comparison with other high temperature calcined samples. At higher temperature, namely, 800°C, collapse of larger pores also takes place, agglomerate more $RE_2O_3$ to form bigger size NPs [18]. This implies that annealing at higher temperatures increase the particle size with lower $V_0$ concentration. These results are consistent with dielectric studies, resulting in the decrease in dielectric constant along with DPT behavior by annealing the sample at higher temperatures. Likewise, the Debye-Waller factor increases with decreasing NPs size, indicating that the disorder in oxygen environment increases with lower calcination temperatures. In this connection, it is relevant to refer to the important findings of $RE^{3+}$ doped EXAFS studies by Rocca *et al.* [45] on the densification of silica xerogels as a function of heat treatment up to 900°C. Their measurements have shown that the densification of xerogel without evidence of clustering of $RE^{3+}$ induces a decreasing of the co-ordination number and a shortening of the main RE–O distance and absence of RE–RE correlation. Nevertheless, our TEM studies in densified glass specimens (700°C and above) have revealed clear detectable NPs with separation. So, it is reasonable to assume that the shortening of $Gd^{3+}$–O and the decreasing of the coordination number with lower calcined nano-glass specimens depends on NPs size (or, more correctly, the oxygen vacancy) for which further experiments of all the rare earth with different doping concentrations are still necessary. Therefore, our EXAFS findings provide additional insight into the origin of the unusually colossal room-temperature dielectric response in rare earth oxide NPs embedded oxide glasses.

| Sample | Bond type | EXAFS | | |
|---|---|---|---|---|
| | | $N$ | $R+\Delta$ (Å) | $\sigma^2$ (Å$^2$) |
| Gd05-7 | First coordination shell ($Gd^{3+}$–O) | 5.3±0.2 | 2.216±0.012 | 0.016±0.002 |
| Gd05-8 | | 5.6±0.2 | 2.234±0.011 | 0.015±0.003 |
| Gd05-9 | | 5.8±0.3 | 2.248±0.013 | 0.014±0.002 |
| $Gd_2O_3$ | | 6.0±0.2 | 2.265±0.012 | 0.007±0.003 |

**Table 3.** Results of the quantitative analysis of the first coordination shell derived from EXAFS filtered data of $Gd_2O_3$:$SiO_2$ nano-glass composite systems at different calcinations temperatures. $N$, $R$ and $\sigma^2$ are the average coordination number, interatomic distance, and relative mean square displacement (Debye-Waller factor), respectively.

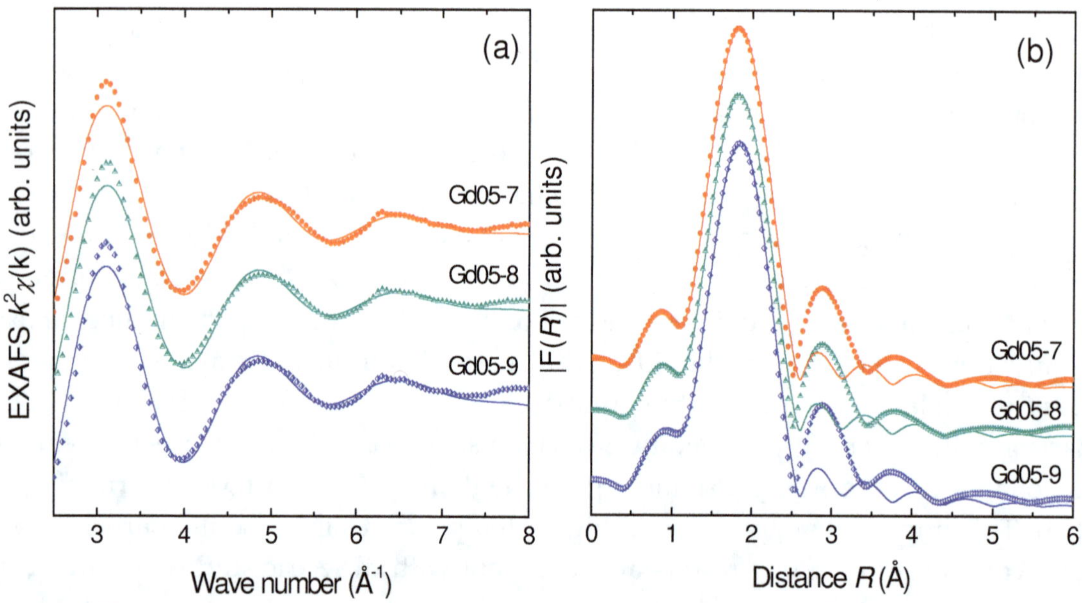


**Figure 16.** (Color online) Gd $L_{III}$-edge EXAFS spectra of $Gd^{3+}$-doped $SiO_2$ glass samples calcined at different temperatures. Spectra are vertically shifted for clarity. (a) $k^2$-weighted EXAFS signals. (b) Fourier transforms moduli radial distribution functions. Both experimental data (symbols) and the best-fit theoretical curves (dashed) are also reported. The transformation range is $k$ = 2.5–8 Å$^{-1}$ for all the spectra and the range for the first coordination shell fit is $R$ = 1–3 Å.

Figure 17 depicts the room temperature experimental EXAFS spectra of LGS systems ((La, Gd)$_2$O$_3$:$SiO_2$ NP-glass composite systems) with different doping concentrations (Table II) and particles size (different calcined temperatures). The first coordination peak located at ~1.8 Å (Figures 17(a), (b)) with the interatomic distance of $La^{3+}$-O/$Gd^{3+}$-O looks very much the similar without any perceptible shift at different doping concentrations. However, the interatomic distances of the first coordination peak (~1.8 Å) of LGS4 with different calcination temperatures (Figures 17(c), (d)) are shifted significantly even with very low doping concentration of $La_2O_3$/ $Gd_2O_3$. It reveals significantly that the average $La^{3+}$–O/$Gd^{3+}$–O interatomic distances of LGS4 samples at lower calcination temperature is shorter, suggesting higher oxygen vacancies around La/Gd ions, supported with our previously reported article [16]. Therefore, the dielectric value decreases by annealing the sample at higher temperatures (or, more correctly, with higher NPs size) with identical molar concentration of dopant element. However, identical particle size (magnetic and/or non-magnetic NPs) with concentration dependence does not affect the oxygen vacancies. In other words, oxygen vacancies depend only on the particle size but not its magnetic phase.

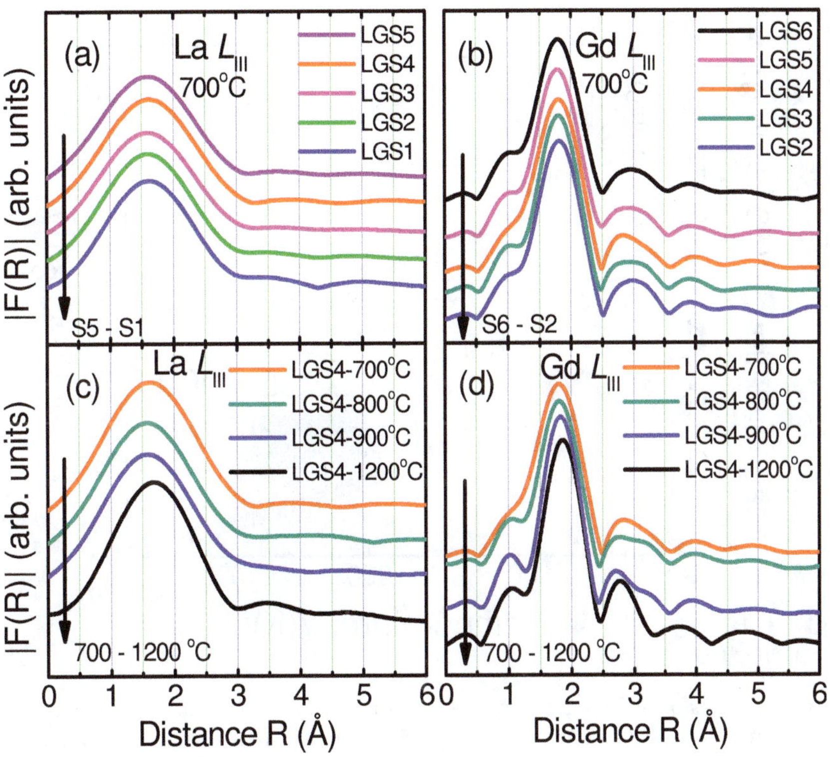


**Figure 17.** (Color online) Room temperature Fourier transforms moduli radial distribution functions EXAFS spectra of LGS systems at (a) La $L_{III}$-edge with different $La_2O_3$ concentrations, (b) Gd $L_{III}$-edge with different $Gd_2O_3$ concentrations calcined at 700°C, (c) La $L_{III}$-edge of LGS4 sample at different calcined temperatures, and (d) Gd $L_{III}$-edge of LGS4 sample at different calcined temperatures. Spectra are vertically shifted for clarity.

## 9. Magnetic measurements

The *dc* magnetization (after diamagnetic correction) of LGS5 sample under zero-field-cooled (ZFC) and field-cooled (FC) condition as a function of temperature (2–350 K temperature range) in the presence of an applied magnetic field of 200 Oe are shown in Figure 18. It is noteworthy that the observed temperature dependent magnetic feature of LGS5 sample is attributed from magnetic transition of $Gd_2O_3$ NPs only. On close inspection, the divergent behavior between ZFC and FC data occurs at low temperatures (irreversibility temperature, $T_{irr}$) with a rounded maximum for ZFC curve is obtained at 58 K (identified as the blocking temperature, $T_B$). A spread in the blocking temperature may be rightly assumed with NPs size distribution. Such behavior is akin to superparamagnetic phase of $Gd_2O_3$ NPs in the nano-glass composite system similarly observed in oxide glasses containing $Gd_2O_3$ [46]. This typical characteristic temperature would be unlikely arising from a magnetic transition in oxygen contaminant [47]. The FC magnetization increases continuously with the lowering of temperature below the irreversibility temperature, consistent with ferromagnetic-like ordering of $Gd_2O_3$ NPs. The shape of the inverse susceptibility data (right axis of Figure 18) reveal a quite different behavior in comparison with usually found Curie-type behavior in bulk $Gd_2O_3$ [48].

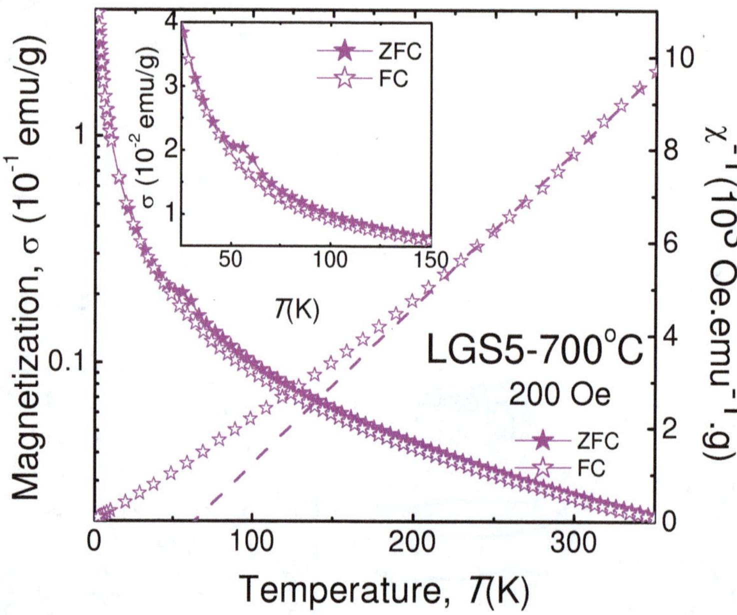


**Figure 18.** (Color online) ZFC and FC magnetization versus temperature curves of LGS5 sample calcined at 700°C, right axis shows the temperature dependent inverse susceptibility curve. In the inset, the region close to the superparamagnetic transition is highlighted.

The data obtained for the temperature dependence magnetization of LGS4 samples calcined at 700, 800, 900 and 1200°C are graphically depicted in Figure 19. The superparamagnetic blocking temperature cannot be traced in low accuracy (or resolution) of measurement with very low percentage-doping (~ 0.075 mol%) of magnetic $Gd_2O_3$. However, from the observed continuous increase in ZFC and FC curves at low temperature indicating the ferromagnetic nature of the LGS4 sample. Magnetic properties with size dependency are also observed for LGS4 samples calcined at different temperatures, related with the uncompensated surface spins present on the $Gd_2O_3$ NPs. It is likely that the $Gd_2O_3$ NPs with smaller size (i.e., higher surface-to-volume ratio) contain larger proportion of uncompensated surface spins and

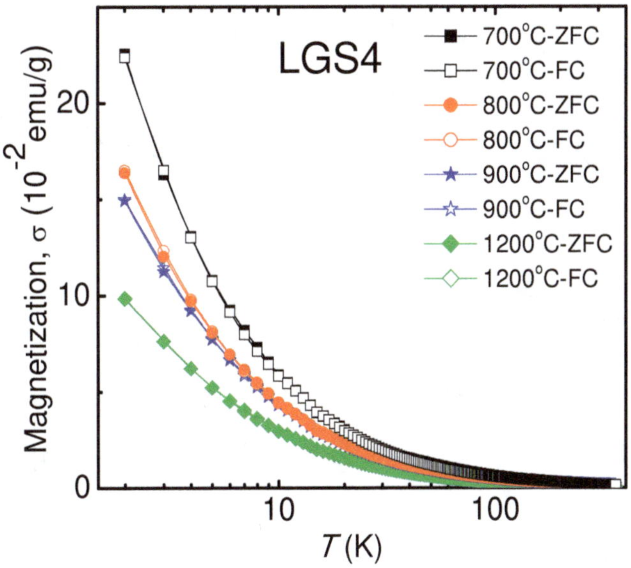


**Figure 19.** (Color online) ZFC and FC magnetization versus temperature curves of LGS4 sample at different calcination temperatures. Applied *dc* magnetic field 200 Oe.

consequently reveal higher ferromagnetic values than larger NPs (higher calcined temperature). Temperature dependent inverse susceptibility data for LGS4 samples calcined at different temperatures with respect to bulk $Gd_2O_3$ can be fitted by the Curie–Weiss law (Figure 20(a)) having different slopes of straight lines. The intersection points of fitted lines with *x*-axis exhibit the Curie–Weiss temperatures, found to be 32.7, 16.6, 12.1 and 4.8 K for 700, 800, 900 and 1200°C respectively. These positive values for LGS4 sample calcined at different temperatures indicate weak ferromagnetic behavior, whereas, bulk $Gd_2O_3$ shows antiferromagnetism with negative Curie–Weiss temperature at -16.3 K. The current tendency of lowering Curie-Weiss temperatures with increasing calcined temperature for LGS4 samples infers that lager sized $Gd_2O_3$ NPs possess toward bulk crystalline counterpart.

Isothermal magnetization-field sweeps were performed to further investigate the nature of the superparamagnetic state and the ferromagnetism below transition temperature. Figure 21 displays the magnetic field dependence of the magnetization (*M-H*) curves for LGS4 sample calcined at 700°C at 300, 200, 100 and 5 K in the *dc* magnetic field range ±60 kOe. Defining the magnetic characteristics, magnetic hysteresis curve obtained at 300 K has zero area, whereas, there is a dramatic change both in magnitude (enhancement of magnetic moment/unit mass) and shape (deviate from linearity) with measurable finite areas at 5 K (lower inset of Figure 21). This constitutes strong evidence that at 5 K the $Gd_2O_3$ NPs are going to the magnetically ordered state or in ferromagnetic nature. Moreover, magnetization vs *H*/*T* curves plotted at different temperatures are linear and collapse to a single curve (upper inset of Figure 21) confirming the existence of superparamagnetic phase of $Gd_2O_3$ NPs embedded in $SiO_2$ glass matrix. It is abundantly clear that $Gd_2O_3$ NPs grown with high magnetic dilution in glass matrix are best described as an assembly of non-interacting superparamagnetic NPs. The hysteresis curves have no magnetic saturation in the magnetic field range of ±60 kOe, considering large anisotropic fields appears in the $Gd_2O_3$ NPs systems [49].

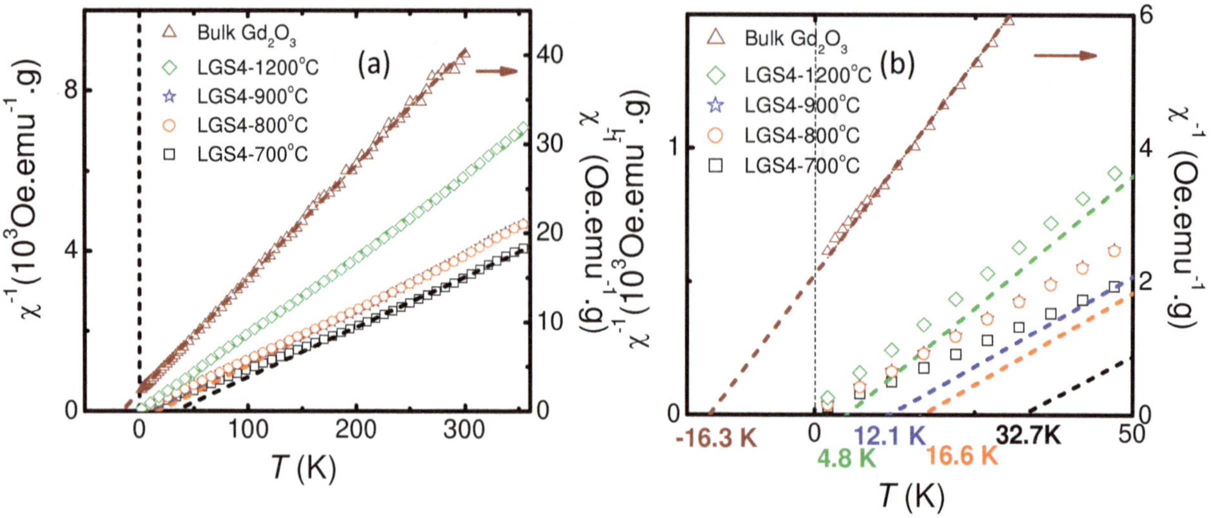


**Figure 20.** (Color online) (a) Inverse susceptibility versus temperature curves of LGS4 samples at different calcination temperatures with respect to bulk $Gd_2O_3$, (b) the region close to the extrapolated lines intersect with temperature axis is highlighted.

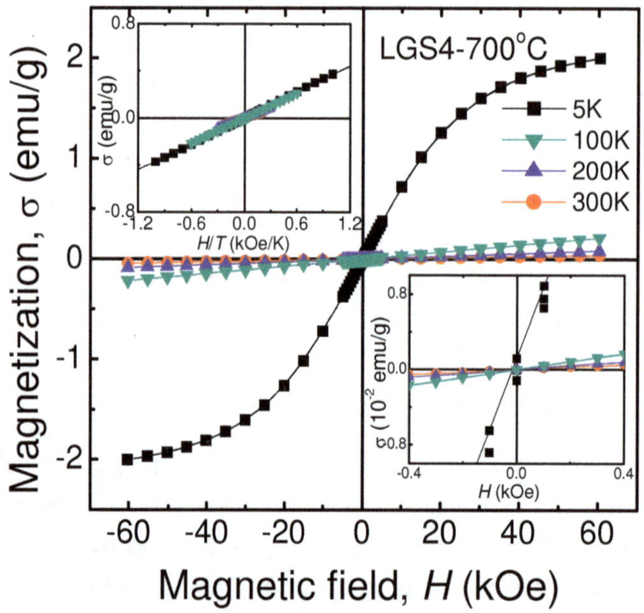


**Figure 21.** (Color online) Hysteresis loop of LGS4 sample calcined at 700°C, lower inset: the region close to the coercive field value is highlighted, upper inset: magnetization vs. H/T of LGS4 sample.

## 10. Conclusions

We have synthesized self-organized $RE_2O_3$ NPs with almost equal size and separation embedded in $SiO_2$ glass matrix by the sol–gel method.

Principal findings may be summarized below:

a. Presence of superparamagnetic phase occurs in magnetic rare earth oxide NPs doped glass samples.

b. Properly annealed sol–gel glass (in which RE ~ Sm, Gd and Er) (Fig. 2) shows an interesting colossal response of dielectric constant along with DPT and MD behavior around room temperature.

c. The experimental facts strongly suggest that the dielectric anomaly with DPT behavior is related to oxygen vacancy-induced dielectric relaxation in the material without ferroelectric phase transition.

d. The MDR observed in this glassy composite is considered to be associated with the direct consequence of magnetoresistance changes depending on the calcination temperatures (magnetic NPs size).

e. However, keeping the NPs size constant, the increase in dielectric constant and MDR strongly depends on the magnetic property (superparamagnetism) of the rare earth ions.

## Acknowledgements

This research was partially supported by the Ministry of Science and Technology, Taiwan under Grant No. NSC 103-2112-M-110-010-MY3.

## Author details

S. Mukherjee[1,2], T. H. Kao[1], H. C. Wu[1], K. Devi Chandrasekhar[1] and H. D. Yang[1*]

*Address all correspondence to: yang@mail.nsysu.edu.tw

1 Department of Physics and Center for Nanoscience and Nanotechnology, National Sun Yat-Sen University, Kaohsiung, Taiwan

2 UGC-DAE Consortium for Scientific Research, Mumbai Centre, Bhabha Atomic Research Centre, Trombay, Mumbai, India

## References

[1] Chau R, Datta S, Doczy M, Doyle B, Kavalieros J, Metz M. High-k/Metal gate stack and its MOSFET characteristics. IEEE Electron Device Lett 2004;25:408–10.

[2] Lo SH, Buchanan DA, Taur Y, Wang W. Quantum-mechanical modeling of electron tunneling current from the inversion layer of ultra-thin-oxide MOSFETs. IEEE Electron Device Lett 1997;18:209–11.

[3] Homes CC, Vogt T, Shapiro SM, Wakimoto S, Ramirez AP. Optical response of high-dielectric-constant perovskite-related oxide. Science 2001;293:673–6.

[4] Liu J, Duan CG, Yin WG, Mei WN, Smith RW, Hardy JR. Large dielectric constant and Maxwell-Wagner relaxation in $Bi_{2/3}Cu_3Ti_4O_{12}$. Phys Rev B 2004;70:144106(7).

[5] Martín SG, Orrantia AM, Aguirre MH, Franco MÁA. Giant barrier layer capacitance effects in the lithium ion conducting material $La_{0.67}Li_{0.25}Ti_{0.75}Al_{0.25}O_3$. Appl Phys Lett 2005;86:043110(3).

[6] Yang Z, Hou Y, Liu B, Wei L. Structure and electrical properties of $Nd_2O_3$-doped $0.82Bi_{0.5}Na_{0.5}TiO_3$–$0.18Bi_{0.5}K_{0.5}TiO_3$ ceramics. Ceram Int 2009;35:1423–7.

[7] Wang Z, Chen XM, Ni L, Liu XQ. Dielectric abnormities of complex perovskite $Ba(Fe_{1/2}Nb_{1/2})O_3$ ceramics over broad temperature and frequency range. Appl Phys Lett 2007;90:022904(3).

[8] Raevski IP, Prosandeev SA, Bogatin AS, Malitskaya MA, Jastrabik L. High dielectric permittivity in $AFe_{1/2}B_{1/2}O_3$nonferroelectric perovskite ceramics (A=Ba, Sr, Ca; B=Nb, Ta, Sb). J Appl Phys 2003;93:4130–6.

[9] Wu JB, Nan CW, Lin YH, Deng Y. Giant dielectric permittivity observed in Li and Ti doped NiO. Phys Rev Lett 2002;89:217601(4).

[10] Pecharromán C, Esteban-Betegón F, Bartolomé JF, López-Esteban S, Moya JS. New percolative $BaTiO_3$–Ni composites with a high and frequency-independent dielectric constant ($e_r \approx 80000$). Adv Mater 2001;13:1541–4.

[11] Cohn JL, Peterca M, Neumeier JJ. Low-temperature permittivity of insulating perovskite manganites. Phys Rev B 2004;70:214433(6).

[12] Park T, Nussinov Z, Hazzard KRA, Sidorov VA, Balatsky AV, Sarrao JL, Cheong SW, Hundley MF, Lee JS, Jia QX, Thompson JD. Novel dielectric anomaly in the hole-doped $La_2Cu_{1-x}Li_xO_4$ and $La_{2-x}Sr_xNiO_4$ insulators: signature of an electronic glassy state. Phys Rev Lett 2005;94:017002(4).

[13] Rivas J, Rivas-Murias B, Fondado A, Mira J, Señarís-Rodríguez MA. Dielectric response of the charge-ordered two-dimensional nickelate$La_{1.5}Sr_{0.5}NiO_4$. Appl Phys Lett 2004;85:6224–6.

[14] Zhu WJ, Tamagawa T, Gibson M, Furukawa T, Ma TP. Effect of Al inclusion in $HfO_2$ on the physical and electrical properties of the dielectrics. IEEE Electron Device Lett 2002;23:649–651.

[15] Mukherjee S, Pal AK, Bhattacharya S, Raittila J. Magnetism of $Mn_2O_3$ nanocrystals

dispersed in a silica matrix: Size effects and phase transformations. Phys Rev B 2006;74:104413(10).

[16] Mukherjee S, Lin YH, Kao TH, Chou CC, Yang HD. Searching for high-k $RE_2O_3$ nanoparticles embedded in $SiO_2$ glass matrix. J Appl Phys 2012;111:064103(6).

[17] Sakka S, Kamiya K. The sol–gel transition in the hydrolysis of metal alkoxides in relation to the formation of glass fibers and films. J Non-Cryst Solids 1982;48:31–46.

[18] Krol DM, van Lierop JG. The densification of monolithic gels. J Non-Cryst Solids 1984;63:131–44.

[19] Ankudinov A, Ravel B, Rehr JJ, Newville M. FEFFIT Manual within the FEFF Project (Seattle, WA: University of Washington, USA); 1992–1999.

[20] Mukherjee S, Chen CH, Chou CC, Tseng KF, Chaudhuri BK, Yang HD. Colossal dielectric and magnetodielectric effect in $Er_2O_3$ nanoparticles embedded in a $SiO_2$ glass matrix. Phys Rev B 2010;82:104107(7).

[21] Afify ND, Dalba G, Kuzmin A. Local structure around $Er^{3+}$ in $SiO_2$-$HfO_2$glassy waveguides using EXAFS. Phys Rev B 2007;76:024114(8).

[22] Maqsood A. Phase transformations in $Er_2Si_2O_7$ ceramics. J Mater Sci Lett 1997;16:837–840.

[23] Losurdo M, Giangregorio MM, Capezzuto P, Bruno G, Toro RG, Malandrino G, Fragalà IL, Armelao L, Barreca D, Tondello E, Suvorova AA, Yang D, Irene EA. Multifunctional nanocrystalline thin films of $Er_2O_3$: interplay between nucleation kinetics and film characteristics. Adv Funct Mater 2007;17:3607–12.

[24] Chen S, Zhu YY, Xu R, Wu YQ, Yang XJ, Fan YL, Lu F, Jiang ZM, Zou J. Superior electrical properties of crystalline $Er_2O_3$ films epitaxially grown on Si substrates. Appl Phys Lett 2006;88:222902(3).

[25] Rivas J, Mira J, Rivas-Murias B, Fondado A, Dec J, Kleemann W, Señarís-Rodríguez MA. Magnetic-field-dependent dielectric constant in $La_{2/3}Ca_{1/3}MnO_3$. Appl Phys Lett 2006;88:242906(3).

[26] Jana A, Kundu TK, Pradhan SK, Chakravorty D. Dielectric behavior of Fe-ion-doped $BaTiO_3$ nanoparticles. J Appl Phys 2005;97:044311(6).

[27] Rivera I, Kumar A, Ortega N, Katiyar RS, Lushnikov S. Divide line between relaxor, diffused ferroelectric, ferroelectric and dielectric. Solid State Commun 2009;149:172–6.

[28] Lal HB. Low temperature dielectric studies of some rare-earth oxides. J Phys C: Solid State Phys 1980;13:3969–76.

[29] Yu Z, Ang C, Guo R, Bhalla AS. Ferroelectric-relaxor behavior of $Ba(Ti_{0.7}Zr_{0.3})O_3$ ceramics. J Appl Phys 2002;92:2655–7.

[30] Raymond O, Font R, Suárez-Almodovar N, Portelles J, Siqueiros JM. Frequency-temperature response of ferroelectromagnetic $Pb(Fe_{1/2}Nb_1/_2)O_3$ ceramics obtained by different precursors. Part I. Structural and thermo-electrical characterization. J Appl Phys 2005;97:084107(8).

[31] Ang C, Yu Z, Cross LE. Oxygen-vacancy-related low-frequency dielectric relaxation and electrical conduction in $Bi:SrTiO_3$. Phys Rev B 2000;62:228–36.

[32] Samara GA. The relaxational properties of compositionally disordered $ABO_3$ perovskites. J Phys: Condens Matter 2003;15:R367–R411.

[33] Scott JF. Ferroelectrics go bananas. J Phys: Condens Matter 2008;20:021001(2).

[34] Pintilie L, Alexe M. Ferroelectric-like hysteresis loop in nonferroelectric systems. Appl Phys Lett 2005;87:112903(3).

[35] Yáñez-Vilar S, Mira J, Sánchez-Andújar M, Castro-García S, Fondado A, Rivas J, Senarís-Rodríguez MA. Particle size reduction: a way to enhanced dielectric properties of magnetocapacitive $La_{2/3}Ca_{1/3}MnO_3$. Appl Phys Lett 2010;96:162904.

[36] Kimura T, Kawamoto S, Yamada I, Azuma M, Takano M, Tokura Y. Magnetocapacitance effect in multiferroic $BiMnO_3$. Phys Rev B 2003;67:180401(R)(4).

[37] Tang J, Feng L, Wiemann JA. Negative magnetoresistance of $\gamma$-$Fe_2O_3$ observed in $\gamma$-$Fe_2O_3$/Ag granular nanocomposite. Appl Phys Lett 1999;74:2522–4.

[38] Catalan G. Magnetocapacitance without magnetoelectric coupling. Appl Phys Lett 2006;88:102902(3).

[39] Mukherjee S, Chen CH, Chou CC, Yang HD. Anomalous dielectric behavior in nanoparticle $Eu_2O_3$:$SiO_2$ glass composite system. Euro Phys Lett 2010;92:57010(4).

[40] Kao TH, Mukherjee S, Yang HD. Magnetic nanoparticles induced dielectric enhancement in $(La,Gd)_2O_3$:$SiO_2$ composite systems. J Magn Magn Mater 2013;346:11–5.

[41] Lukashev P, Sabirianov RF. Flexomagnetic effect in frustrated triangular magnetic structures. Phys Rev B 2010;82:094417(6).

[42] Eliseev EA, Glinchuk MD, Khist V, Skorokhod VV, Blinc R, Morozovska AN. Linear magnetoelectric coupling and ferroelectricity induced by the flexomagnetic effect in ferroic. Phys Rev B 2011;84:174112(16).

[43] Foster AS, Lopez Gejo F, Shluger AL, Nieminen RM. Vacancy and interstitial defects in hafnia. Phys Rev B 2002;65:174117(13).

[44] Nogami M, Moriya Y. Glass formation through hydrolysis of $Si(OC_2H_5)_4$ with $NH_4OH$ andHCl solution. J Non-Cryst Solids 1980;37:191–201.

[45] Rocca F, Ferrari M, Kuzmin A, Daldosso N, Duverger C, Monti F. EXAFS studies of the local structure of $Er^{3+}$ ions in silica xerogel co-doped with aluminium. J Non-

Cryst Solids 2001;293–295:112–7.

[46] Kliava J, Malakhovskii A, Edelman I, Potseluyko A, Petrakovskaja E, Melnikova S, Zarubina T, Petrovskii G, Bruckental Y, YeshurunY. Unusual magnetic transitions and nature of magnetic resonance spectra in oxide glasses containing gadolinium. Phys Rev B 2005;71:104406(9).

[47] Kliava J, Malakhovskii A, Edelman I, PotseluykoA, Petrakovskaja E, Melnikova S, Zarubina T, Petrovskii G, Bruckental Y, Yeshurun Y. Reply to "Comment on 'Unusual magnetic transitions and nature of magnetic resonance spectra in oxide glasses containing gadolinium.'" Phys Rev B 2006;74:026404(2).

[48] Bud'ko SL, Canfield PC. Evaluation of a long-time temperature drift in a commercial quantum design MPMS SQUID magnetometer using $Gd_2O_3$ as a standard. J Magn Magn Mater 2006;299:281–7.

[49] Hadjipanayis G, Sellonyer DJ, Brandt B. Rare-earth-rich metallic glasses. I. Magnetic hysteresis. Phys Rev B 1981;23:3349–54.

# 4

# Biasing Effects in Ferroic Materials

Vladimir Koval, Giuseppe Viola and Yongqiang Tan

**Abstract**

In this chapter we present an overview of some important concepts related to the processes and microstructural mechanisms that produce the deformation of hysteresis loops and the loss of their symmetry characteristics in ferroelectric, ferroelastic and ferromagnetic systems. The most discussed themes include: aging and fatigue as primary mechanisms of biased hysteresis loops in ferroelectric/ferroelastic materials, imprint phenomenon as an important biasing process in ferroelectric thin films, the development of an exchange bias field and of specific spin states, such as spin canting and spin-glass-like phases, as the main causes of biased hysteresis loops in different types of magnetic heterostructures. The present discussion leads to the identification of the main differences and possible analogies in the underlying mechanisms of possible biasing effects occurring in the different ferroic systems, which can benefit the theoretical description, modelling, and engineering of multifunctional devices based on ferroic systems experiencing the internal bias phenomena.

**Keywords:** Aging, fatigue, imprint, internal field, exchange bias

## 1. Introduction

Ferroic materials such as ferroelectrics, ferroelastics and ferromagnetics exhibit two or more possible equivalent stable states with uniform orientation of the spontaneous polarization, spontaneous strain and spontaneous magnetization in absence of an external electric field, mechanical stress and magnetic field, respectively. These thermodynamic equilibria correspond to minima in the free energy as a function of the order parameter (P – polarization, S – strain and M – magnetization). Under the action of external fields, these equilibrium states can be switched from one to another, giving rise to symmetrical polarization-electric field (P-E), strain-stress (S-σ) and magnetization-magnetic field (M-H) hysteresis loops. It is often the case where two or more order parameters are coupled within the same phase and a change in one parameter produces a variation in another, as it occurs in multiferroic systems. The free energy profile can be influenced by many factors, such as compositional and structural inhomogeneity, defects, internal fields, thermal and loading history. These can induce preferential states of the order parameters, leading to the appearance of deformed and asymmetric hysteresis loops. The understanding of these biasing effects and the study of their possible advantages/disadvantages for specific applications, with the relative elaboration of strategies to induce/avoid them, is of crucial importance to maximizing the performance of functional devices based on ferroic systems.

The present chapter aims to provide an overview of different biasing effects which can occur in different types of ferroic materials, with particular focus on the phenomenology and on the underlying microscopic mechanisms of the anomalies observed in hysteresis loops. The chapter is organized in three main parts. The first part describes the distortions of P-E and S-E hysteresis loops in ferroelectric/ferroelastic materials. The two main phenomena, which cause biased hysteresis loops, namely aging and fatigue, are comprehensively reviewed by describing the most important features in bulk systems and the inherent microscopic mechanisms. Attention is also given to imprint phenomena in thin films, with description of the most important models developed to explain the shifting of the polarization hysteresis loops. The second part is focused on biasing effects in ferroelectric/ferroelastic materials under mechanical stress, with detailed description of the influence of composition, poling state and temperature on the stress-strain curves of perovskite systems. Additionally, biased stress-strain loops in shape memory alloys are also briefly reviewed, highlighting the different mechanical behavior of the martensite and austenite phases. This section is concluded with a summary on Bauschinger effect observed in ferroelastic materials during cyclic mechanical loading. The third and last part describes the biasing processes occurring in ferromagnetic materials, with the main focus on the asymmetric M-H loops caused by the exchange bias effect, and by the coexistence of different magnetic phases in inhomogeneous magnetic systems.

## 2. Ferroelectric materials

### 2.1. Deformations of hysteresis loop shape of ferroelectrics: pinching and asymmetries

The study of the hysteresis loops, namely current-electric field (I-E), polarization-electric field (P-E) and strain-electric field (S-E), is one of the most important tools to investigate the behavior and to assess the properties of ferroelectric/ferroelastic materials.[1-3] These curves usually show certain symmetry properties with respect to one or both the two axes, but they are susceptible to substantial modifications of their shapes, frequently observed in certain structures and compositions, which have undergone a certain thermal and electrical history. The most relevant examples of shape modifications of hysteresis loops are represented by *pinched loops* (constricted P-E loops in the region $E \approx 0$, remanent polarization approaching to zero), *asymmetric loops* (shift of the P-E loops along the E- and P-axis, suppression of left or right wing of the S-E "butterfly" loop) and *polarization gaps* (relaxation of the remanent polarization at E = 0), which have been found in different ferroelectric/ferroelastic systems in both polycrystalline ceramics and single crystal. It can be generally said that these features can be caused by the development of an *internal bias field* $E_{int}$, whose magnitude and distribution depend on the composition/microstructure and on the thermal/electrical history of the system. In most cases, the presence of an internal field in ferroelectric materials can arise during processing (in this case often referred to as *built-in field*) or generated by two thermodynamic processes generally classified as *aging* and *fatigue*. They both share similar features, but they are triggered under different conditions and they can be influenced by different factors, as it will be discussed in the following sections.

## 2.2. Aging

### 2.2.1. *Definition and introduction*

The term *aging* in ferroelectric/ferroelastic systems usually refers to the process that produces variations of dielectric, piezoelectric and ferroelectric properties over time and in absence of an external field. These variations are usually undesired and the understanding of the controlling mechanism is crucial to maximize the performance of materials in most electronic applications. Although the prevailing microscopic mechanisms of aging are still under debate, there is a general agreement that aging is due to the stabilization of charged species or defect complexes in certain locations and configurations. Referring to the most common perovskites ($ABO_3$ chemical formula), aging is absent or weak in materials with high purity and in compounds in which the A- and B-site ions are partially substituted by donor species (i.e. soft compositions). On the other hand, aging phenomena are much more pronounced when the ions are partly replaced by acceptor dopants with a lower valence (i.e. hard compositions).[4]

The effect of aging often results in the presence of pinched P-E hysteresis loops schematically shown in Fig. 1a, and/or in one of the two possible cases of asymmetric P-E loops schematically shown in Figs. 1b and 1c. The internal bias field $E_{int}$ can be estimated in each case using the position of the current peaks (Fig. 1) as [4]:

$$E_{int} = \frac{E_c^+ + E_c^-}{2} \tag{1}$$

where $E_c^+$ and $E_c^-$ are the fields corresponding to positive and negative current neighboring peaks in the current-electric field (I-E) curve.

### 2.2.2. *Effect of thermal and electrical history on the hysteresis loops*

The presence of pinched and asymmetric hysteresis loops depends on the specific thermal and electrical history of the system. Fig. 2 shows the aging characteristics of CuO-modified $BaTiO_3$ ceramics.[5] It can be seen that unpoled aged ceramics display pinched P-E and S-E loops, while poled ceramics display asymmetric P-E and S-E hysteresis curves.

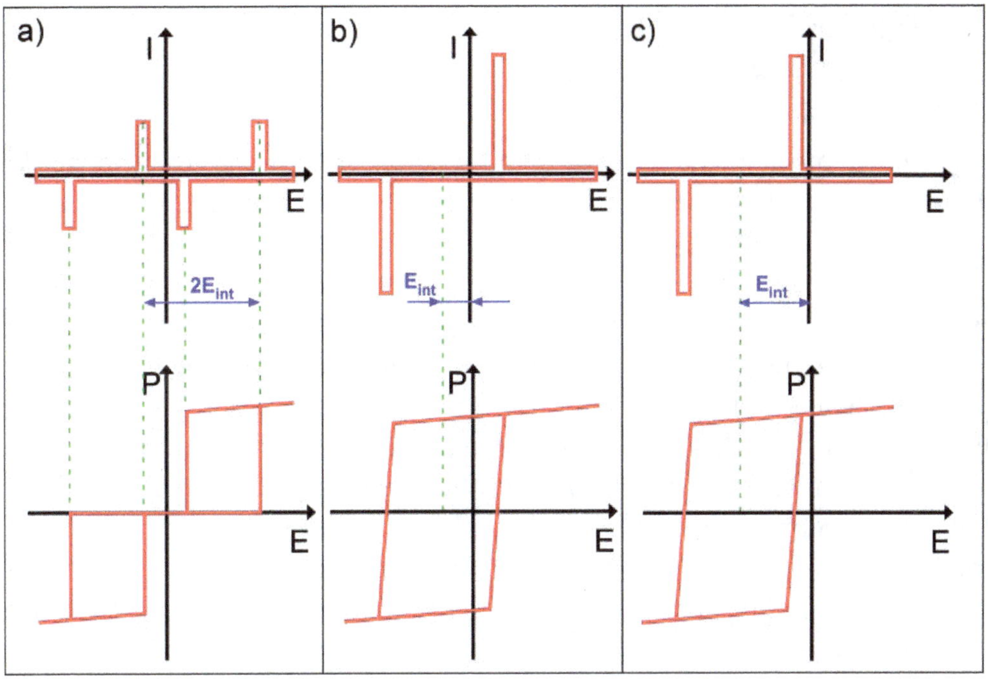


**Figure 1.** Definition of the internal bias field, $E_{int}$, in (a) pinched and (b, c) asymmetric P-E loops. After [4].

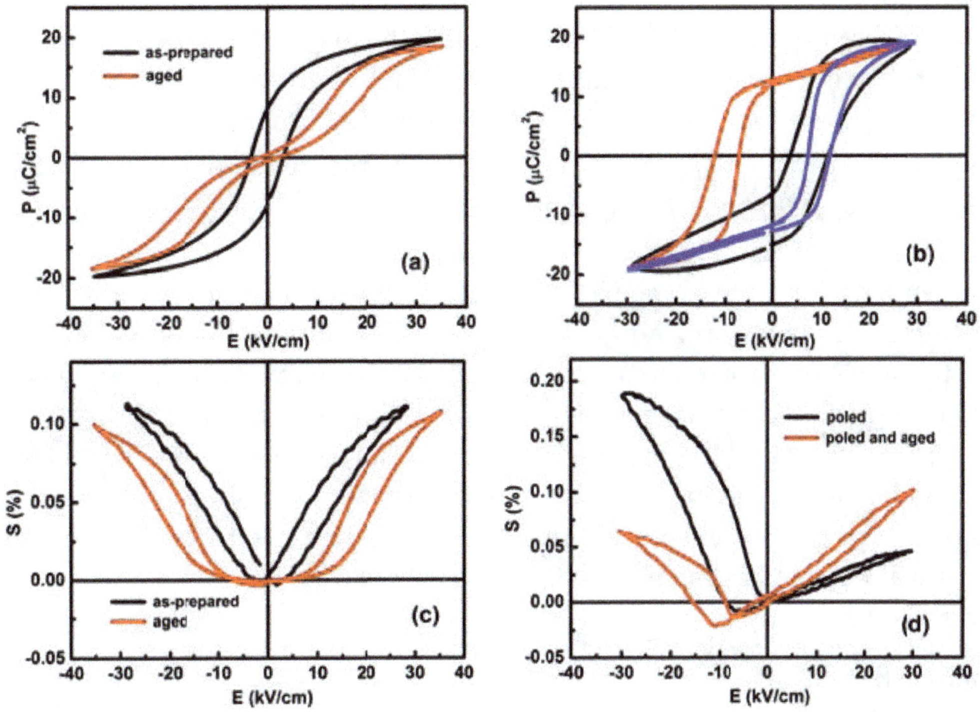


**Figure 2.** Hysteresis loops comparison of CuO-modified $BaTiO_3$ ceramics: (a) P-E loops of the as-prepared ceramics and aged ceramics (aging at room temperature for 24h); (b) the black line corresponds to a poled non-aged sample, red and blue curves are relative to poled and subsequently aged ceramics. The red curve was generated by applying a positive field parallel to the poling, while the blue curve was obtained with a positive electric field antiparallel to the previous poling field; (c) S-E loops of the as-prepared and aged ceramics; (d) S-E loops of poled non-aged, and poled and subsequently aged ceramics (positive field parallel to poling). After [5].

### 2.2.3. *Microscopic mechanisms of aging*

Three main microscopic mechanisms have been proposed for the origin of aging, namely a) *volume effect*, b) *domain effect*, and c) *grain boundary effect*. Up to the present, a definitive agreement on which mechanism is the main cause of the aging process has not been achieved, also due to the fact that often these mechanisms can possibly overlap.

#### a. Volume effect

According to the volume effect model, unintentional and intentional defects, which can be present in the system, are stabilized in preferred lattice sites, creating an internal bias that constrains the switching of the polarization along specific directions. One of the first micro-structural schemes of the volume effect model was proposed by Arlt and Neumann [6], and later extended by Ren with the *symmetry-conforming property of point defects*, which was used to describe aging phenomenon in different crystal structures [7]. According to the symmetry-conforming property principle, point defects (e.g., oxygen vacancies in acceptor-doped perovskites) distribute in the lattice with the same symmetry of the crystal structure, leading to a *defect dipole* $P_D$ parallel to the spontaneous polarization (Fig. 3). During cooling from $T > T_c$ ($T_c$ is a Curie temperature), the crystal structure abruptly changes into a lower symmetry structure stable at $T < T_c$, while the existing cubic structured defects are still arranged in their original symmetry. The reason is that the kinetics of defects migration is slower compared to the movement of ions driven by the phase transition. In the $ABO_3$ perovskite structure, the presence of an acceptor element in the B-site generates oxygen vacancies, which enable to accomplish charge neutrality. The lattice sites, where the defects can be located soon after cooling from $T > T_c$, are represented by the octahedron's corners, which all have the same probability of occupancy $\left(P_1^V = ... = P_6^V\right)$, as shown by the equal white proportions at the octahedron corners in Fig. 3a. However, over time, the defects tend to rearrange assuming the same symmetry of the crystal structure. The relaxation time of this process reduces with increasing temperature (in the range $T < T_c$), because thermal energy promotes short range migration of the defects towards a more stable configuration. After aging, the probability of occupancy for all the octahedron's corners is not the same anymore; the location with highest probability of oxygen-vacancy formation is the octahedron's corner that allows the formation of a defect dipole (acceptor ion-oxygen vacancy) parallel to the spontaneous polarization. In Figs. 3b, 3d and 3f, this location is identified by the octahedron corner with the largest white portion. Strong experimental evidence supporting the volume effect model against other aging mechanisms was obtained from targeted tests carried out on a single crystal-single domain of Mn-doped $BaTiO_3$.[8]

#### b. Domain effect

The domain effect model is based on the concept that mobile defects can diffuse towards domain walls to minimize the local depolarizing fields and may act as pinning agents on domain walls. By performing in situ high-energy X-ray Bragg scattering experiments during the application of an electric field, Tutuncu *et al.* [9] showed that in a pre-poled sample of $0.36BiScO_3$-$0.64PbTiO_3$, the volume fractions of domains oriented along the electric field is different when the electric field is increased along the previous poling direction and when

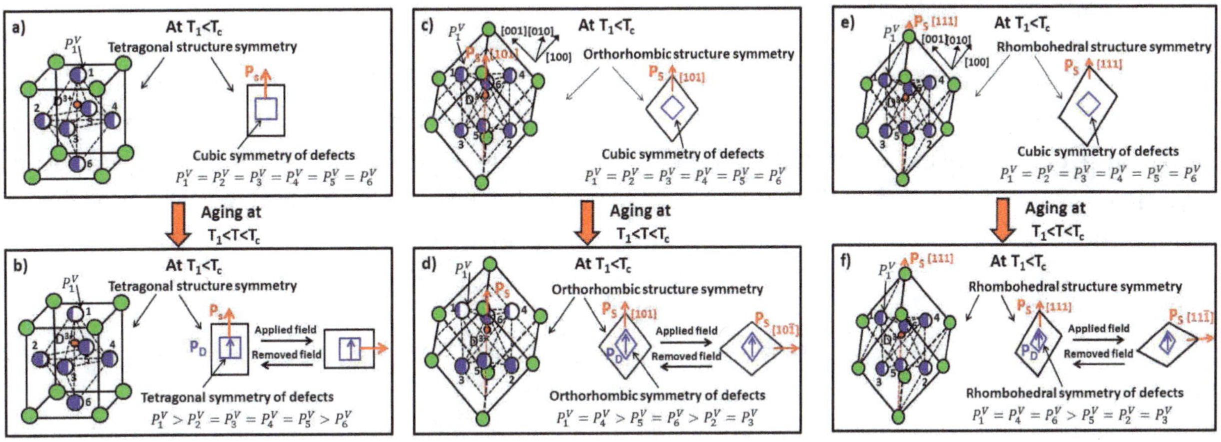

**Figure 3.** Scheme of symmetry conforming property of defects in different crystal structures: tetragonal (a, b); orthorhombic (c, d); rhombohedral (e, f). The scheme refers to the presence of an acceptor $D^{3+}$ in the $B^{4+}$ site of the $ABO_3$ perovskite-type compounds. After [7].

applied in the reverse direction. Indirect experimental proof of the domain effect model is based on the measurement of the dielectric permittivity and loss. Acceptor-doped aged systems usually show a reduction of permittivity and loss compared to the non-aged specimens. In early studies this was attributed to the clamping of domain walls due to the presence of an internal bias, which results in a reduction of the extrinsic contribution to dielectric loss.[10] Theoretical models based on the drift/diffusion of charge carriers, driven by the compensation of the depolarizing field and by the spatial gradient of their concentration, indicate that mobile charged species can migrate to domain walls and hinder domain wall movement.[11]

c. **Grain boundary effect**

The interface regions between dissimilar phases, such as undesired secondary phases, pores and electrodes, are often the location of space charge accumulation, which could represent another source of internal bias field responsible for deformations and asymmetries in hysteresis loops. Systematic investigations of the effect of impurities on the efficiency of the poling process have been presented in early reports [12], where it was proposed that in presence of certain impurity species the poling efficiency decreases due to the effect of space charges.

### *2.2.4. Rejuvenation or de-aging*

The aged state is an out-of-equilibrium state and therefore it can be destabilized by several processes. These include: i) electrical bipolar cycles at sufficient amplitude; ii) heating/quenching from $T > T_c$ to $T \ll T_c$; iii) light illumination in some cases. The recovery process from the aged state results in the re-establishment of unbiased hysteresis loops, and is usually called *rejuvenation* or *de-aging*. The study of the kinetics of de-aging, either electric field-induced or thermally-induced, can contribute to further understand the microscopic mechanisms of aging.

### 2.2.4.1. *De-aging by electric field cycles*

The reduction of the internal bias field during de-aging by electrical cycling depends on: i) time exposure to the field (so frequency and number of cycles), ii) temperature, and iii) electric field amplitude. Early experiments on the kinetics of de-aging in acceptor-doped lead zirconate titanate (PZT) suggested that the de-aging is a thermally activated process.[10] The kinetics of internal field relaxation follows an exponential function of the type:

$$E_{int}(t) = E_{int}^{0} exp\left(-\frac{t}{\tau}\right) \tag{2}$$

where $E_{int}^{0}$ represents the internal bias field before the electrical cycling and $\tau$ is the time needed for the relaxation of the internal bias. These two parameters can be obtained from the slope and the intercept of the $ln(E_{int})-t$ plot. Carl and Hardtl [10] found that the activation energies in undoped PZT lie between 0.25 and 0.5 eV, those in specimens doped with Mn or Fe range between 0.6 and 0.7 eV, while that found in the Al-doped specimens is about 0.8 eV. The authors also noticed that these activation energy values are similar to the activation energy of the electrical conductivity and came to the conclusion that the mechanisms of aging/de-aging should be somehow connected to the property of electrical charges transport. This idea has been further explored by Morozov and Damjanovic [13], who performed a systematic study of charge migration processes in hard, undoped and soft PZT ceramics. They found that in hard compositions, the activation energy of alternating current (AC) conductivity is similar to the activation energy of the de-pinching process found by Carl and Hardtl in the same compositions.[10] It was concluded that the aging/de-aging process in acceptor doped ferroelectrics should be based on charge transport by the local movements of defect dipoles through short range migration of oxygen vacancies. The activation energy of long-range charge migration under constant electric field (DC conductivity) was found to be significantly larger.[13]

### 2.2.4.2. *De-aging by heating-quenching*

The aging effects can be relaxed by heating the system at $T > T_c$ and by subsequently quenching it to a temperature much lower than $T_c$. The heating process produces a thermal disordering of defects at high temperature and the quench allows keeping the defects in a disordered configuration also at lower temperatures. The relaxation of the internal bias in aged samples upon heating-quenching has been observed in several experimental studies; see for instance Ref. 14 among others.

## 2.3. Fatigue

The term *fatigue* in ferroelectric/ferroelastic materials indicates the degradation of the switchable polarization after a certain number of electrical cycles. Additionally, fatigue can determine the appearance of asymmetric loops due to the effect of biasing mechanisms, which induce preferential polarization and strain states. Besides, asymmetric hysteresis loops often observed in DC poled systems have been also classified in the literature as fatigue-like effects.[15] The biasing processes under repeated AC electric field cycles or DC poling fields depend in turn on several other factors, such as the composition and stoichiometry of the material, temperature, geometry of sample and electrodes, electrical loading characteristics, including the electric field amplitude and frequency. Hence a succinct classification of the entire phenomenology becomes challenging.

### *2.3.1. Hysteresis loops after different types of fatigue-like loading*

Fig. 4 shows the typical hysteresis loops observed after different types of fatigue-like electrical loading. Fig. 4a is relative to the effects of fatigue under unipolar AC electrical loading after $10^9$ cycles on the bipolar P-E and S-E curves of $Pb_{0.99}[Zr_{0.45}Ti_{0.47}(Ni_{0.33}Sb_{0.67})_{0.08}]O_3$ (PIC 151) ceramics.[15] Fig. 4b shows the P-E and S-E bipolar loops of $0.94Bi_{1/2}Na_{1/2}TiO_3$–$0.06BaTiO_3$ (0.94BNT-0.06BT) after 1, $10^4$ and $10^7$ electric field bipolar cycles.[16] Fig. 4c displays the bipolar P-E and S-E curves of PZT 4D ceramics previously poled under a DC poling field of ± 2.5kV/mm applied at T = 125 °C for 5 minutes.[17] It can be seen that the sign of the DC poling field significantly affects the shape of the P-E and S-E loops. In particular, samples pre-poled under positive DC field show P-E loops shifted towards left and S-E loops with a suppression of the left wing; opposite effects can be noticed in samples poled under negative DC field. However, it is worth recalling that to correctly establish the sign and value of polarization and strain in samples subjected to previously electrical loading, polarization and strain must be monitored for the entire electrical history. During DC poling of PZT 4D, polarization and strain were not monitored, therefore, the sign and values in Fig. 4c have relative validity; only the polarization and strain amplitudes are meaningful. Fig. 4d shows the P-E and S-E curves of $BaTiO_3$ ceramics sintered at different temperatures after one loading-unloading cycle up to approximately 3 kV/mm field amplitude and 10 Hz frequency. It can be concluded that the deformation of the hysteresis loops is caused by the presence of an internal bias field which forces the systems into preferred polarization and strain states.

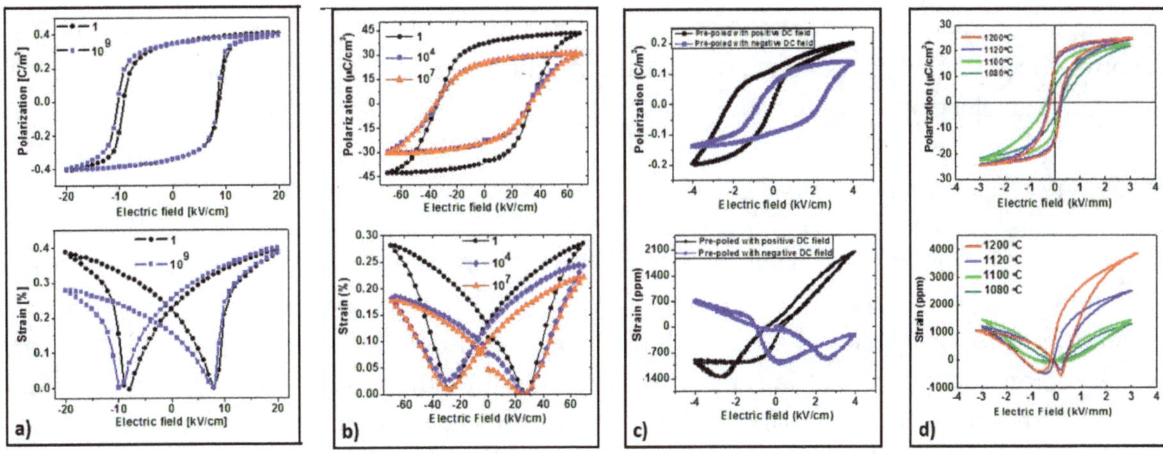


**Figure 4.** Fatigue-like effects on P-E and S-E hysteresis loops: (a) after unipolar electrical fatigue cycles in PIC 151 ceramics, courtesy of N. Balke, after [15]; (b) after unipolar electrical fatigue cycles in 0.94BNT-0.06BT ceramics, courtesy of Z. Luo, after [16]; (c) PZT 4D after DC poling of 2.5kV/mm applied for 5 minutes at 125°C [17]; (d) after one AC poling cycle in $BaTiO_3$ ceramics sintered at different temperatures [17].

### 2.3.2. *Formation of cracks*

Fatigued specimens often show the presence of microcracks, although it is hard to establish whether microcracking is a cause or a consequence of fatigue. It has been conveyed that microcracking can occur in samples with low density, in compositions with large grain size, with large unit cell distortion, and in compositions near the phase boundaries, where electric field-induced transitions can be activated during electrical loading [18]. In addition, macroscopic cracks can develop in the interface region between sample and electrode. In particular, two types of cracks were observed in $Pb_{0.99}[Zr_{0.45}Ti_{0.47}(Ni_{0.33}Sb_{0.67})_{0.08}]O_3$ ceramics: i) *edge cracks* (propagating obliquely from the electrode inside the material) and ii) *delamination cracks* (forming beneath the electrode and propagating parallel to the electrode during bipolar cycling. The latter appear at a later fatigue stage than the former due to a reduced amount of switchable domains during fatigue.[19]

### 2.3.3. *Fatigue effects in soft and hard compositions*

A number of studies have been devoted to the understanding of fatigue effects in soft and hard ferroelectric compositions, but at present the differences in bulk materials are not completely understood mainly due to the fragmentary character of the published literature often presenting contradicting results. For PZT bulk ceramics [20], it was found that soft composition samples (sintered from commercial PZT 5A powder obtained from Morgan Electroceramics Inc.) exhibited a faster fatigue degradation compared to that of the hard PZT ceramics (sintered from commercial APC-841 powder obtained from American Piezoceramics Inc.). For $BaTiO_3$ bulk ceramics, it was shown that a partial substitution of $Ti^{4+}$ ions in the B-site with donor species (e.g., $Nb^{5+}$ ions) leads to a higher fatigue strength compared to that of the acceptor-doped ($Ca^{2+}$ and $Al^{3+}$) and of the undoped specimens.[21] This is in agreement with the concept that fatigue characteristics can be improved by reducing the content of oxygen vacancies. On the contrary, the addition of Fe-acceptor in the B-site of $[(Na_{0.5}K_{0.5})_{0.96}Li_{0.04}][(Nb_{0.86}Ta_{0.1}Sb_{0.04})_{1-x}Fe_x]O_3$ ceramics has produced improved fatigue characteristics.[22] This effect was attributed to the increased mechanical strength of the grain boundaries upon Fe addition, which had overcome the effect of the expected increase of the oxygen vacancies content due to the presence of the electron acceptor species.

### 2.3.4. *Fatigue effects in relaxors*

Fatigue effects in relaxor materials can be different from those of ferroelectrics in terms of phenomenology and microscopic origin.[23] Additionally, non-ergodic and ergodic relaxors have shown different electrical fatigue characteristics. In non-ergodic relaxors, a long-range ferroelectric order is stabilized after the application of a sufficiently large electric field and polarization reversal occurs during further electric field cycles. Ergodic relaxors undergo a weakly polar-to-polar state transition during electrical loading and they return into the weakly polar state during electric field unloading. This behavior has been widely observed in several bismuth-based perovskites.[24-27] The $0.94(Bi_{0.5}Na_{0.5})TiO_3$-$0.06BaTiO_3$ (BNT-BT) system can be classified as a non-ergodic relaxor. An addition of $K_{0.5}Na_{0.5}TiO_3$ to BNT-BT determines a crossover to an ergodic relaxor behaviour, which can be observed also at room temperature.

The non-ergodic relaxor BNT-BT has shown domain fragmentation during fatigue [28], while the ergodic compositions exhibit significantly higher fatigue resistance [27]. The current understanding is that the domain wall pinning effects become less significant in ergodic relaxor phases.[23] In addition, it can be considered that the ergodic relaxors return to a weakly polar state with low remanent polarization and low remanent strain during electric field unloading. This yields smoother variations of polarization and strain during cycling, which could probably be one of the reasons of the less pronounced fatigue effects. However, further studies are needed to better elucidate the mechanisms of the increased fatigue resistance in lead-free ergodic relaxors.

#### *2.3.5. Microscopic mechanisms of fatigue-like effects*

Macroscopic fatigue-like effects such as asymmetric hysteresis loops are often similar in different material systems, possibly due to a common origin represented by the presence of an internal bias field. However, the microscopic origin of the internal bias could be different in different systems and strongly dependent on the type of electrical loading.

##### *2.3.5.1. Development of an internal bias*

It is well known that domain switching occurs through a nucleation-growth process of reversed domains driven by the *local electric field* $E_{loc}(r, t)$, which is given by the sum of the following fields: i) the external applied electric field $E_{app}$; ii) the *depolarization field* $E_{dep}(r, t)$ due to the polarization changes produced by the formation of reversed nuclei (nucleation) at the location r and time t during the application of the external field $E_{app}$; iii) the sum of the *screening fields*, which can be divided into: a) *external screening field* $E_{es}(r, t)$ produced by free charges on electrodes, and b) *bulk screening field* $E_{bs}(r, t)$ caused by the rearrangement of charge carriers in a ferroelectric [29]:

$$E_{loc}(r,t) = E_{app} - \{E_{dep}(r,t) - [E_{es}(r,t) + E_{bs}(r,t)]\} \quad (3)$$

In an ideal case of a perfectly insulating poled ferroelectric placed between two conductive plates of a capacitor, the polarization bound charges in the ferroelectric will cause an accumulation of free electronic charges on the electrodes in the region nearby the ferroelectric to counterbalance the polarization charges, thereby leading to the presence of *a depolarizing field* inside the ferroelectric. By short circuiting the capacitor plates, the compensation charges will flow within the circuit in such a way that the depolarizing field disappears (*the external screening*). In real ferroelectric capacitors, however, the electrodes are not perfect conductors and the ferroelectric would present the so called *dielectric gaps* in the interior and close to the interface with the electrodes, where the spontaneous polarization is significantly suppressed or absent. [30] The presence of dielectric gaps determines a separation between the compensation charges and the polarization bound charges in the ferroelectric. The former become trapped at the dielectric gap-ferroelectric interface forming a space charge layer, which, together with the polarization bound charges in the ferroelectric, generates a *depolarizing field,* which can only be partially compensated by external screening in the short circuit condition. The time constant of the external screening process is determined by the parameters of the external circuit and it is typically in the order of micro/nano seconds. [30] The unscreened part of the depolarizing field in the interior of the ferroelectric, *a residual depolarizing field* $E_{res}$, is antiparallel to the polarization, and triggers the so called *bulk screening process,* which involves a rearrangement of the charge carriers inside the ferroelectric to reduce the residual depolarizing field. The bulk screening process can occur through: i) redistribution of charge carriers, ii) alignment of dipolar defects, or iii) irradiation-induced charge injection. The time constant of the bulk screening process is typically several orders of magnitude higher than that of the external screening and it often exceeds the switching time of the ferroelectric polarization. Therefore, the separated and trapped screening charges at the gap/ferroelectric interface can cause an internal bias field which points in the direction of the polarization.

In uniaxial ferroelectric ceramics where the dipolar defects are constrained, the main physical mechanism responsible for aging and fatigue phenomena is attributed to the compensation of the residual depolarizing field through the redistribution of the screening charges. The idea employed in this scenario assumes that free charges present in the material migrate to minimize the depolarizing field with the consequent development of an internal bias field that hampers domain wall movement and generates biased polarization and strain states (Fig. 5). Balke *et al.* [30] proposed that the depolarizing field $E_{dep}$ surrounding a given grain in samples fatigued under an applied field $E_{app}$ can be estimated as:

$$E_{dep} = -\alpha_{angle} \frac{\Delta P}{\varepsilon_0 \varepsilon_{33}} \tag{4}$$

where $\Delta P$ is the variation of polarization during the increase of the applied field from $E = 0$ to $E = E_{app}$ and $\varepsilon_{33}$ is the permittivity at $E = E_{app}$. The factor $\alpha_{angle}$, which can assume only values between 0.15 and 0.5, takes into account the partial compensation of the depolarization field based on the orientation of the polarization in neighboring grains. This model was able to predict the range of internal bias fields observed in DC poled samples of $Pb_{0.99}[Zr_{0.45}Ti_{0.47}(Ni_{0.33}Sb_{0.67})_{0.08}]O_3$ at room temperature.[11, 30] Additionally, this mechanism was invoked to explain the fatigue-like effects after unipolar cycling in $Pb_{0.99}[Zr_{0.45}Ti_{0.47}(Ni_{0.33}Sb_{0.67})_{0.08}]O_3$ at room temperature [11] and it can be also applied to rationalize the biasing effects observed in DC poled hard piezoceramics, as shown in Fig. 4c.

In the case of bipolar fatigue, the bulk screening process under an alternating electric field leads to an inhomogeneous internal field distribution, which yields the development of the so called *frozen domains.* These get locked and do not switch after a certain number of electric field cycles, giving rise to heterogeneous fatigue effects. The value of $E_{bs}(r, t)$ in the Eq. 3 is determined by the sample history, including sintering/annealing, the subsequent deposition of electrodes, electrical cycles characteristics (frequency and time exposure) and time intervals between cycles.[31] These factors have been taken into account in the *kinetic imprint approach,* which was successfully employed in describing the hysteresis loops after fatigue in different materials. [32]

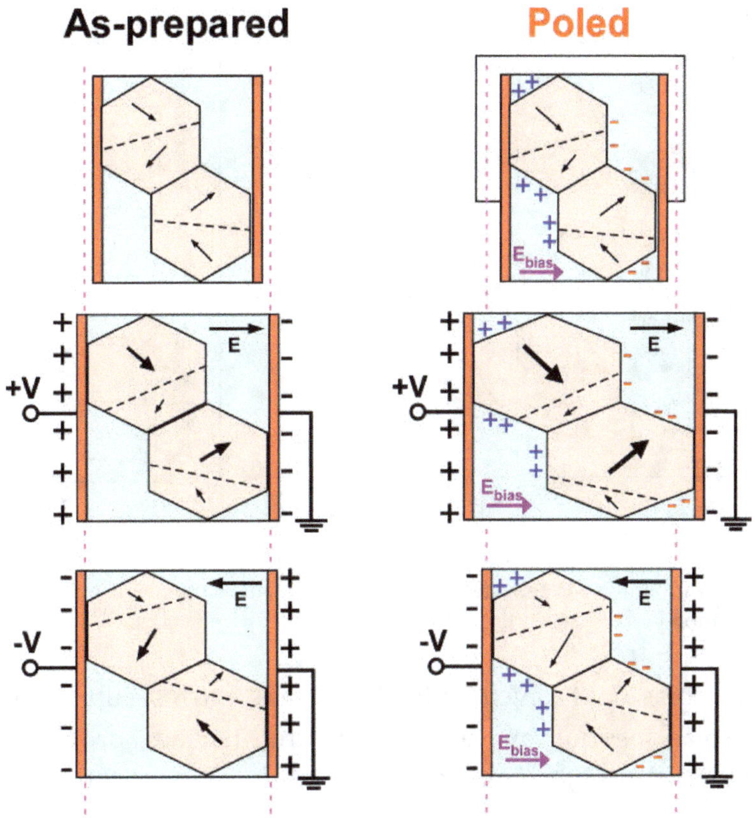


**Figure 5.** Schematic of the formation of an internal bias after poling.

## 2.4. Imprint and biasing effects in ferroelectric thin films

An internal field developed by charged defects is believed to play an important role in the *imprint phenomenon* in ferroelectric thin films. Imprint generally refers to the preference of a certain polarization state, the so called "hard" state, over the other "ease" state in ferroelectric bi-stable states. The preferential polarization may be induced within the ferroelectric by locking-in a particular domain orientation during electrical, thermal or optical (UV light) treatments.[33] The latter two effects are customarily called *thermal imprint* and *optical imprint*.

The polarization imprint of ferroelectric thin film capacitors, which manifests itself macroscopically by a shift of the polarization hysteresis loop along the electric field axis (Fig. 6), is one of the most serious reliability issues for ferroelectric nonvolatile memories applications. An incomplete polarity reversal of the ferroelectric thin film, which also gives rise to an asymmetry in the remanent polarization and an effective variation of the coercive field, can lead to either "read" or "write" failure of the memory cell.[34] The voltage offset of the imprinted loop is often used as a measure of imprint. The sign of the field offset determines the direction of the preferential polarization in the interior of the ferroelectric thin film capacitor.

An asymmetric hysteresis behavior has also been reported for as-grown ferroelectric thin films in an initial state, i.e. without heat treatment, ultraviolet radiation or application of an external electric field.[35] However, unlike the preferred polarization generated from capacitor

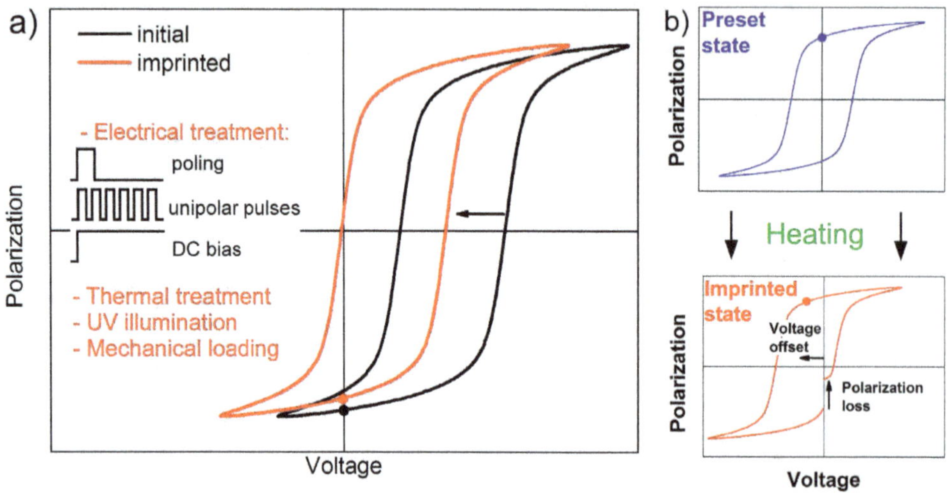


**Figure 6.** (a) Shift of polarization of a ferroelectric in the negative state of remanent polarization due to imprint. (b) Thermal imprint – polarization loss due to shift of the P-E hysteresis loop on heating.

structure, an internal bias field develops in these films intrinsically during the fabrication process. The built-in bias does not vary during switching, instead it provokes polarization back switching and causes the *retention loss*, i.e. a gradual decrease of the remanent polarization with time in poled ferroelectric capacitors. Optical and thermal imprint have been shown to improve the retention performance or aging in PZT film capacitors.[36]

*2.4.1. Physical models of imprint and microscopic mechanisms for asymmetrical switching*

In the last two decades, several phenomenological scenarios and microscopic models have been proposed in the literature to explain the asymmetric switching behavior and the related imprint phenomena in ferroelectric thin film capacitors. Among them, two mechanisms seem to be able to consistently describe the polarization imprint and its dynamics: the *defect-dipole alignment model* [37] and the *interface screening model* [38].

**a. Defect-dipole alignment model**

There have been several indications that defects play an essential role in polarization degradation phenomena in bulk ceramics. To explain the interplay between the voltage shift, defects and polarization, Warren *et al.* [39] proposed the concept of aligning defect dipoles and their involvement in the aging process. By using electron paramagnetic resonance they demonstrated that the defect dipoles in $BaTiO_3$ ceramics align along the direction of the spontaneous polarization. The asymmetric distribution of trapped charge and the alignment of defect dipoles near the electrode-ferroelectric interface have been suggested to be responsible for the imprint phenomena in ferroelectric thin film capacitors.[37] The dominating role of defect dipole alignment for producing internal bias field in PZT films was evidenced by the reduction of the thermally induced voltage shifts upon donor doping at the (Zr, Ti) sites. Similar findings on donor ($La^{3+}$) doping effects have been also reported by Kim *et al.* [40], and the improved imprint behavior of La-doped PZT films was attributed to the reduction of the concentration of defect dipoles induced by the doping. Recently, Folkman *et al.* [41] by monitoring the built-in bias field in epitaxial $BiFeO_3$ thin films have shown that the defect dipole pairs can realign and ultimately disassociate upon electrical cycling. The gradual dissociation of complex oxygen vacancy defects was accounted for a drastic reduction in the built-in bias field and an improvement of imprint characteristics of the films as measured after $10^4$ cycles.

### b. Bulk and interface screening models

In contrast to Warren's and Kim's observations [37, 40], the results obtained by Grossmann *et al.* [38] showed only a slight or negligible impact of Nb and Fe addition on the hysteresis and imprint behavior of PZT capacitors. Instead, the experiments with ultraviolet illumination revealed an enhancement of the polarization imprint of the PZT films. It was proposed that the redistribution of electronic species, such as electrons and/or holes, might be the main cause for the imprint phenomena in ferroelectric thin film capacitors, and the defect dipole alignment, if any, is only a secondary imprint mechanism.

Currently, there are two reasonable approaches explaining the imprinted hysteresis loops of ferroelectric capacitors within the general concept of generation, separation and subsequent trapping of electronic charges: the *bulk screening model* [42] and the *interface screening model* [38]. Both models assume the existence of a thin layer at the interface between the electrode and ferroelectric in which the spontaneous polarization is absent or suppressed. However, they differ in modeling the driving force responsible for the charge separation. In the bulk screening model, the driving force is thought to be the *residual depolarizing field* ($E_{res}$) generated in the interior of the film due to incomplete cancellation of the space and polarization charges by free electronic (screening) charges on the electrodes.

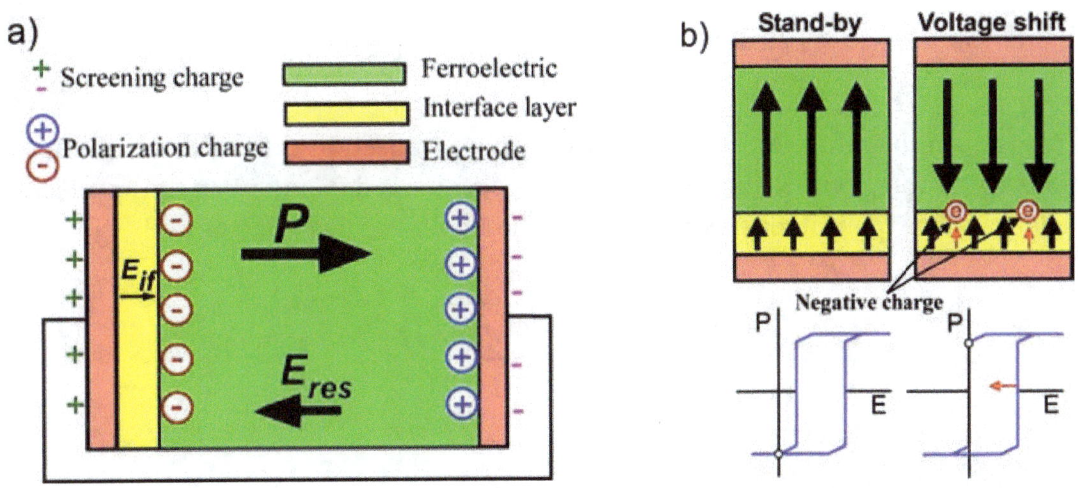


**Figure 7.** (a) A sketch of ferroelectric thin film capacitor under imprint experiment, and (b) a voltage shift of the imprinted thin film capacitor according to the interface screening model. Reproduced with permission from Abe *et al.* [48].

The concept employed in the interface screening model suggests that the imprint effects might have originated from a large electric field $E_{if}$ that develops within the interface layer of the ferroelectric thin film capacitor during the accelerated aging (Fig. 7). This field points in the direction of the polarization and in the opposite direction of the $E_{res}$ field. A gradual shifting of the polarization hysteresis loop is thought to be governed by $E_{if}$ due to the consecutive trapping of electronic charges injected from the electrode into the film. As shown in Figure 7, the injected charges have a polarity that is opposite to the dipole of the ferroelectric, hence, the induced internal field screens the external electric field and a higher applied voltage is required to induce polarization reversal. The experimental observations presented in Ref. 38 indicate that imprint in ferroelectric thin films is governed by the magnitude and orientation of the ferroelectric polarization and the voltage shift follows the direction of the polarization.

Within the interfacial screening model, several charge injection mechanisms have been suggested for the imprint phenomena in ferroelectric capacitors, including the *field enhanced thermionic emission* from the electrodes *(Schottky effect)* and *Pool-Frenkel emission* from traps. The Schottky emission current density can be expressed as follows [43]:

$$J_S = A^*T^2 \exp\left(\frac{-q\varnothing_B}{k_B T}\right) exp\left(\frac{q\Delta\varnothing_S}{k_B T}\right) \tag{5}$$

where $A^*$ is the effective Richardson constant, $q$ is the eletronic charge, $k_B$ is the Boltzmann constant, $T$ is the absolute temperature, and $q\phi_B$ is the Schottky barrier height (i.e., conduction band offset). $\Delta\varnothing_S$ denotes the barrier lowering:

$$\Delta\varnothing_S = \sqrt{\frac{qE}{4\pi\epsilon_r\varepsilon_0}} \tag{6}$$

where $E$ is the applied electric field, $\varepsilon_0$ is the vacuum permittivity and $\varepsilon_r$ is the dynamic dielectric constant of the ferroelectric.

The equation implies that the interfacial-layer charge injection due to thermal emission is exponentially proportional to the barrier lowering, which is dependent on the applied electric field. The standard quantitative expression for the Poole-Frenkel effect is [44]:

$$J \propto E\ exp\left(\frac{-q\left(\varnothing_B - \sqrt{qE/\pi\epsilon_r\epsilon_0}\right)}{k_B T}\right) \tag{7}$$

According to Eq. 7, the displacement of the polarization hysteresis loop along the field axis is expected to be independent of the film thickness, and the voltage offset would increase as the thickness of the sample increases. This effect has been experimentally evidenced in ferroelectric thin film capacitors with variable thickness.[45, 46]

Although the bulk screening model reasonably explains the imprinted hysteresis loops for films illuminated by UV light during poling, it cannot describe an enhancement of the polarization imprint under external bias, as observed by Grossmann *et al.* [38]. The scenario proposed in the interface screening model has consistently and repeatedly been successful in describing, both qualitatively and quantitatively, a wide variety of experimental data on imprint behavior of ferroelectric thin film capacitors, including a logarithmic-type time dependence of imprint [47] and the effects of temperature, illumination and of an externally applied bias [36, 38, 42].

#### 2.4.1.1. *The surface layer(s)*

The idea of the existence of a surface layer with suppressed ferroelectric properties implemented in physical models has demonstrated considerable advantage in explaining asymmetric switching and other anomalies in the behavior of ferroelectric materials. In the models, the interface thin layer is considered as *a non-switchable "dead" layer* [48], *a space charge layer* [49], *a low-dielectric constant layer* [50] or *an insulating region* [51] adjacent to the electrode. Pertsev *et al.* [46] have shown that in ultrathin PZT films the nearby-electrode layer does not behave as an insulator, and proposed that a finite electrical conduction exists in this layer. Also, Tagantsev *et al.* [52] have indicated that the anomalous hysteresis behavior displayed by ferroelectric thin film capacitors might be caused by the effect of a conductive non-ferroelectric layer, the so-called *"passive layer"*, formed at the interface between the film and electrode. This layer is supposed to behave as a space charge layer (charged positively or negatively), operating as a serial capacitor connected to the film capacitor. A high electric field in the surface layer will cause charge transport across it, resulting in charge separation. The built-in potential developed by the trapped and separated charges in the layer due to depletion effects at the ferroelectric-electrode contacts would prevent polarization reversal to occur in a symmetrical manner, even though the driving voltage is perfectly symmetrical.[49]

A direct evidence of an oxygen-deficient surface layer in epitaxial (Ba, Sr)$TiO_3$ [50] thin films and Pb(Zr, Ti)$O_3$ nanostructures [53] has been provided using high-resolution transmission electron microscopy. However, the origin of these layers remains yet to be elucidated. Some researchers connected it with *a partial depletion* [54] or *exhaustion* of charge carriers at the Schottky contacts [52], while others related the existence of the surface layer to extrinsic effects such as *structural disordering* [49, 55], *chemical inhomogeneity* (e.g., oxygen deficiency) [41] or *mechanical distortion* [50]. Abe *et al.* [48] have proposed, based on measurement of switching currents in the heteroepitaxial $BaTiO_3$ thin films, that the passive layer is possibly formed by the relaxation of lattice misfit strain developed during film growth. The existence of the interfacial stresses due to elastic interface deformations in thin films has been evidenced by Spaepen [56]. The oxygen-loss related strain gradient at the interfaces of the PZT capacitor was also reported by Wu *et al.* [57]. The role of strain gradients in the asymmetric hysteresis behavior will be discussed in the following section.

### 2.4.2. *Mechanical stress effects*

It is well established that internal mechanical stresses and their effects on domain-wall movement are the main cause of the significant difference between the physical properties of ferroelectric thin films and bulk ceramics. Experimental results presented in the literature show that the *residual stresses* developed in the films during fabrication are often larger than hundreds of megapascals and create a considerable clamping effect on domains, thereby hindering polarization switching.[58] The misfit strains arising from the mismatch of the lattice parameters and the coefficients of thermal expansion of the underlying substrate and thin film can lead to the lattice distortion or formation of *misfit dislocations* (Fig. 8), which in turn causes the creation of a lattice strain gradient across the film. This strain gradient may induce a linear polarization response through the *flexoelectric effect*.[59] Tuttle *et al.* [60] have reported that the sign of the film stress at the Curie point controls the orientation of the domain structure, and hence the flexoelectric effect, unlike piezoelectricity, can trigger switchable polarization. Under certain misfit strain-temperature conditions, a strong coupling between the polarization and the stress field of the dislocation can appear, resulting in spatial inhomogeneity of the polarization near the dislocation (Fig. 9).[61] These polarization gradients produce the accumulated interfacial charge, which allows for the development of the internal bias field.

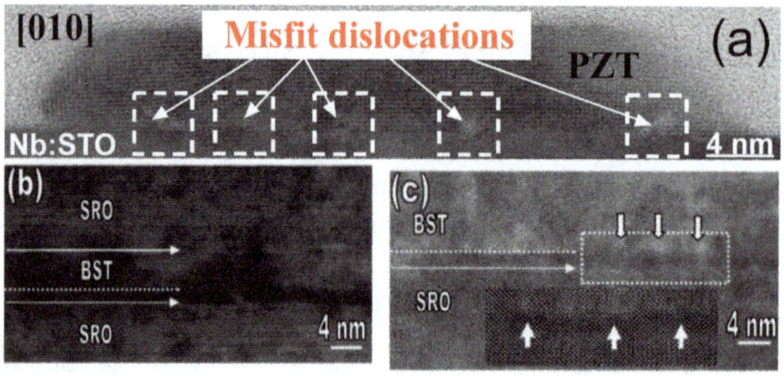


**Figure 8.** High-resolution TEM cross-sectional images of the interfacial layer in the PZT nanostructure (a) and $Ba_{0.7}Sr_{0.3}TiO_3$ thin films (b, c) showing a significant distortion of the ferroelectric lattice around the dislocation cores (labelled by arrow tags). Reproduced with permission from D. Hesse [53] and He *et al.* [50].

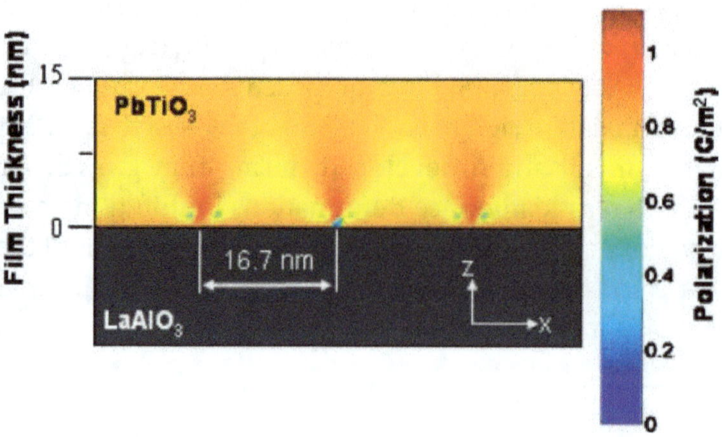


**Figure 9.** Spatial inhomogeneity of the ferroelectric polarization around periodic misfit dislocations due to a strain gradient in a $PbTiO_3$ film on a $LaAlO_3$ substrate. Reproduced with permission from Alpay *et al.* [61]. Copyright 2004, AIP Publishing LLC.

Stress-induced changes in the local asymmetric switching behavior and piezoelectric properties of the sol-gel $Pb(Zr_{0.30}Ti_{0.70})O_3$ and Mn-doped $Pb(Zr_{0.30}Ti_{0.70})O_3$ ferroelectric films have been investigated by Koval *et al.*[62] A modified nanoindentation system with a conductive spherical indenter tip was used for the simultaneous application of driving voltage and mechanical loading. It was shown that the switching charge versus applied voltage (Q-V) hysteresis loops shift gradually along the voltage axis with increasing indentation force (100 – 500 mN). The effect of spherical nanoindentation on the asymmetric switching behaviour is shown in Fig. 10a, which compares the Q-V loops obtained at different indentation loads for the Mn-doped PZT (PMZT) film. In addition, a progressive hysteresis gap of the charge – voltage loops is displayed in the figure inset. The parameter of horizontal loop asymmetry $\delta$ (Fig. 10b) was found to increase almost linearly with the force by an increment of about 0.4-0.5 $\times 10^{-3}$ per 100 mN during the application of a sinusoidal signal of 25 V-, 37 V- and 50 V-peak drive voltages and 50 Hz frequency.

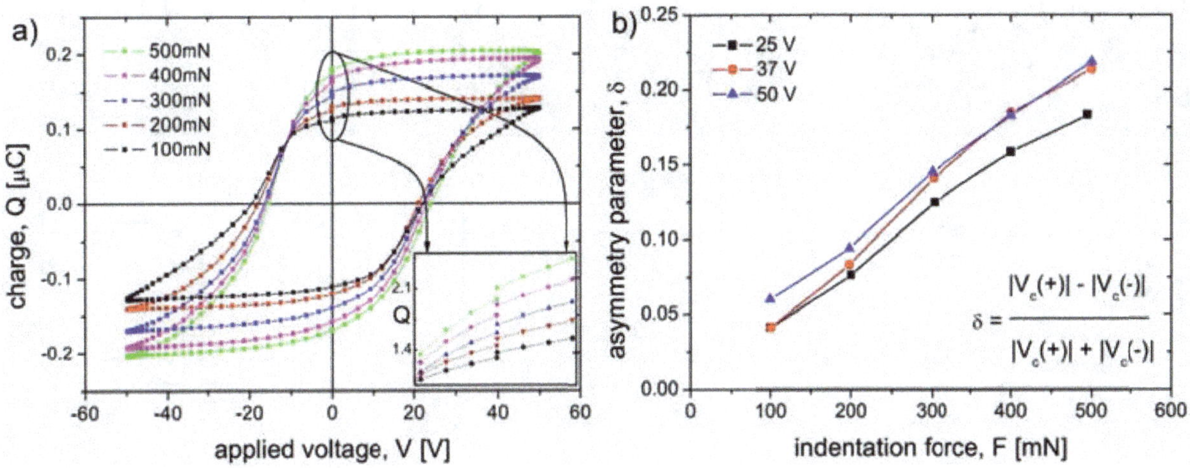


**Figure 10.** (a) The voltage shift and (b) the asymmetry parameter of the Q-V hysteresis loop of the Mn-doped PZT thin film under nanoindentation. Reproduced with permission from Koval *et al.* [62]. Copyright 2005, AIP Publishing LLC.

These results are interpreted in Ref. 62 by two mechanisms that may act in parallel and contribute to the indentation-driven voltage shift of the Q-V hysteresis loops: i) defect-dipole realignment and concurrent space charge rearrangement, and ii) reduction of the clamping effect on domains controlled by a variation of the internal residual stress. The later mechanism results in the enhanced polarization state of the film, the so-called interfacial poling, and potentially may lead to more space charge trapping at the bottom electrode due to liberation of defect-stabilized domain walls. Koval *et al.* in their early work [63] reported on an increase of the effective piezoelectric coefficient upon nanoindentation in the PZT film capacitors and attributed it to the stress-enhanced irreversible movement of ferroelastic domain walls. This scenario is consistent with the theoretical and experimental works of other researchers.[64, 65] Pertsev and Emelyanov [66] have shown that the residual stresses can change significantly when a $90^0$ domain wall is shifted from its equilibrium position by an external field.

Stress-driven effects in ferroelectric thin films indicate that reliability, performance and lifetime of ferroelectric capacitors can be significantly affected by mechanical force. Gruverman *et al.*

[67] have reported on the stress-induced poling and imprinting of (111)-oriented PZT-based capacitors. By using piezoresponse force microscopy (PFM) they observed that the application of either compressive or tensile stress can change the in-plane polarization component of the film via the flexoelectric effect, and thus produce FeRAM (Ferroelectric Random Access Memory) capacitors in a heavily imprinted state characterized by a strongly shifted hysteresis loop (Fig. 11). Recently, the possibility to write ferroelectric memory bits using pure mechanical force instead of electrical voltage in data storage devices has been proposed by Lu *et al.* [68].

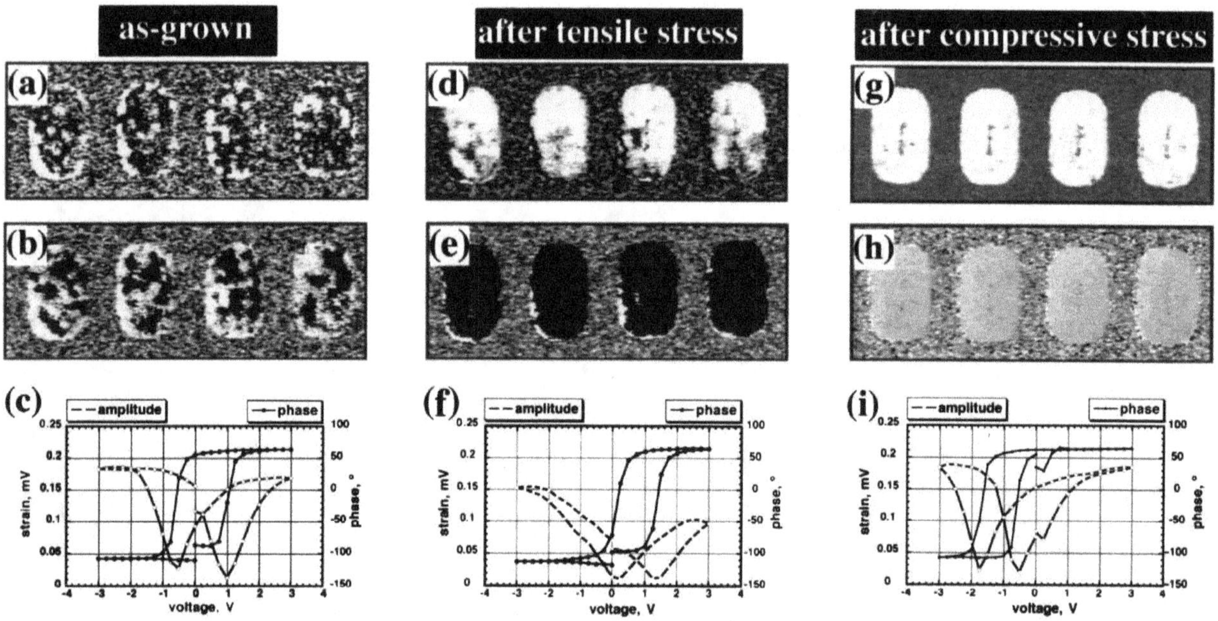


**Figure 11.** Stress-induced polarization reversal and switching characteristics of the (111)-oriented PZT capacitors. (a), (d) and (e) PFM amplitude images; (b), (e) and (h) PFM phase images; (c), (f) and (i) local hysteresis loops measured for as-grown capacitor, after tensile and compressive stress application, respectively. Reproduced with permission from Gruverman *et al.* [67]. Copyright 2003, AIP Publishing LLC.

## 3. Ferroelastic materials

### 3.1. Hysteresis and biasing effects in ferroelastic materials

Ferroelastic materials undergo a phase transition from a high temperature paraelastic phase to a low temperature ferroelastic phase at the Curie temperature, $T_c$. In the ferroelastic phase, there are two or more equilibrium and switchable orientation states, characterized by a different spontaneous strain. A shift from one state to another can occur during the application of a mechanical stress, and in most cases it gives rise to hysteretic stress-strain curves (Fig. 12).[69] In tension-compression loading, the stress at which the strain is zero is called *the coercive stress* (Fig. 12a). The intersection of the two branches of the hysteresis curve is defined as the "*critical stress*", where the system experiences significant changes in the domain pattern.

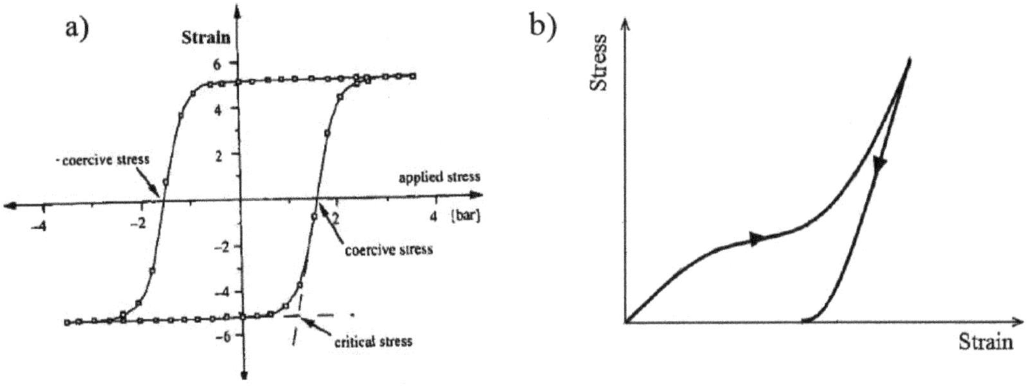


**Figure 12.** (a) Tension-compression stress-strain curve of a ferroelastic material. Reproduced with permission from Salje [67]. Copyright 1993, Cambridge University Press. (b) Compressive stress–strain curve of a ferroelastic material.

A number of complex oxides with a perovskite structure, and shape memory alloys exhibiting the martensitic transformation represent two important families of ferroelastic materials. In ferroelastic ceramics, it is difficult to generate the tension-compression hysteresis loops due to their brittleness, and thus ferroelasticity in these systems is preferably investigated by studying compression test curves.[70] A typical compressive stress–strain curve of a ferroelastic material is shown in Fig. 12b. At the beginning of loading, the strain increases almost linearly with stress. Above a critical stress, the stress-strain curve shows a clear nonlinearity due to ferroelastic domain switching. When most of the switchable domains have been reoriented, the strain will again increase linearly with the stress. The coercive stress is commonly identified with the stress corresponding to the minimum value of the local slope of the stress-strain curve. It should be mentioned that in polar ferroelastic systems, where the spontaneous strain reorientation is also accompanied by polarization changes, the mechanical response depends on the electrical boundary conditions. In particular, higher stiffness is usually observed in open circuit conditions.

### 3.2. Perovskite oxides

Perovskite oxides may exhibit asymmetric deformation under external tensile and compressive mechanical loading. Fett *et al.* [71] performed bending tests on unpoled and poled soft PZT ceramics and reported that in both cases the region of the samples under tension and compression exhibit different deformation behaviour (Fig. 13).

The compression stress-strain curves of soft and hard PZT ceramics show large differences. Acceptor-doped PZTs, which behave as "hard" ferroelectrics, present an analogous "hard" behaviour under mechanical loading, characterized by a low elastic compliance, a high coercive stress and a large mechanical factor of quality. Cao *et al.* [72] systematically studied the strain and polarization variations of soft and hard PZT ceramics upon unipolar compressive stress cycles applied parallel and perpendicularly to the poling direction. They found that both hard and soft PZT exhibit a nonlinear stress-strain curve when the stress exceeds a critical value at which ferroelastic domain switching begins to occur (Fig. 14). From the figure, one can see that the coercive stress of hard PZT is larger than that of the soft composition. In

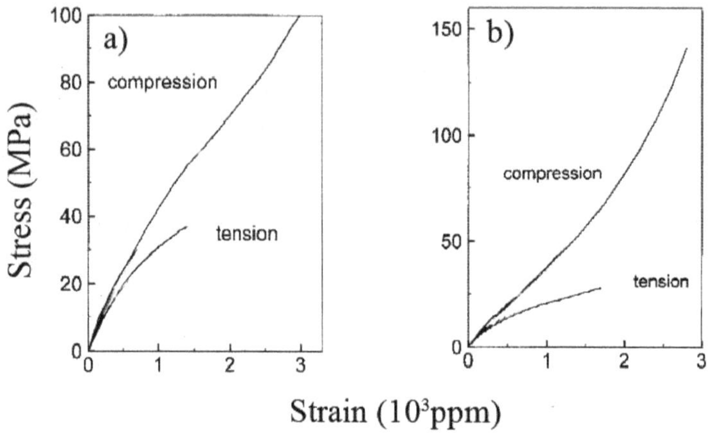


**Figure 13.** Stress–strain curves for the (a) unpoled and (b) poled PZT ceramics. After [71].

addition, hard PZT can recover most of their nonlinear strain upon unloading, whereas the nonlinear strain of soft PZT is mostly unrecoverable (Fig. 14a). Similar behaviour can be also observed in the stress-depolarization curves (Fig. 14b). It is generally accepted that the main variations of both strain and polarization in ferroelectric/ferroelastic materials under mechanical stress result from the switching of 90° domains. When the stress is perpendicular to the poling direction, the stress-strain curves show lower strain values (Fig. 14c) due to a smaller amount of domain reorientation compared to the parallel case. Hard PZT has a more stable domain structure and thus the strain and the polarization changes, induced by the applied mechanical stress, are smaller than those of the soft PZT ceramics (Fig. 14c). Marsilius *et al.* [73] studied the stress-strain behaviour of the soft and hard ferroelectric/ferroelastic ceramics at different temperatures and found that hard ceramics have a larger coercive stress and a larger coercive field than soft ceramics. In addition, they demonstrated that the difference between the coercive mechanical stresses of soft and hard materials is much larger than the difference between their coercive electric fields. This suggests that the effect of doping on stress-induced ferroelastic switching is greater than the effect of doping on the domain switching induced by an external electrical field. For mechanical stresses above 200 MPa, both the soft and hard ceramics showed similar mechanical behaviour, indicating that above a certain threshold the hardening mechanisms of doping can be eliminated by the switching of the defect dipoles in the direction perpendicular to the loading direction. Similar hardening effects were also observed in the ferroelastic non-polar $LaCoO_3$ ceramics after a partial substitution of La with the acceptor ions of Ca, which determined a suppression of domain wall movement within the sub-coercive stress region [74] and an increase of the coercive stress [75].

Picht *et al.* [76] measured the stress-strain curves of unpoled $BaTiO_3$ in short circuit conditions at different temperatures. As shown in Fig. 15, at room temperature $BaTiO_3$ displays a compressive stress-strain curve typical of a ferroelastic material. The remanent strain and the hysteresis area decrease with increasing temperature. When the stress-strain curves were measured at the Curie point (T = 127 °C) and at a slightly higher temperature (e.g., T = 130 °C), a double stress-strain loop was observed (Fig. 15). The hysteresis disappeared when the

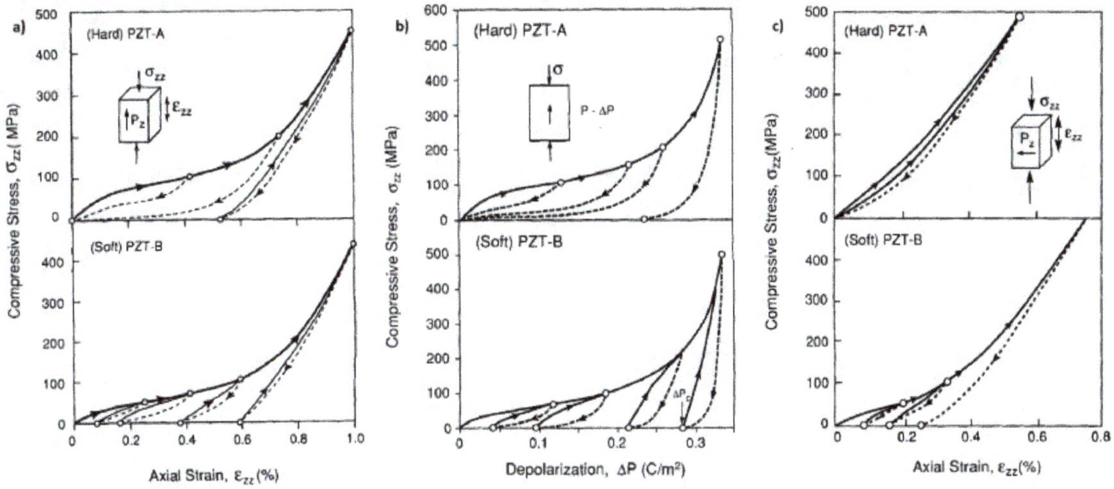

**Figure 14.** (a) Stress-strain curves of the soft and hard PZT ceramics under compression with the applied stress parallel to the polarization. (b) Depolarization vs. compressive stress curves of the hard and soft PZT ceramics. (c) Stress-strain curves of the soft and hard PZT ceramics under compression with the stress applied perpendicular to the polarization. Reproduced with permission from Cao and Evans [72]. Copyright 1993, Willey & Sons, Inc.

temperature was increased above 138 °C. The presence of the double loop above the Curie point was ascribed to a stress-induced paraelastic-to-ferroelastic phase transition.

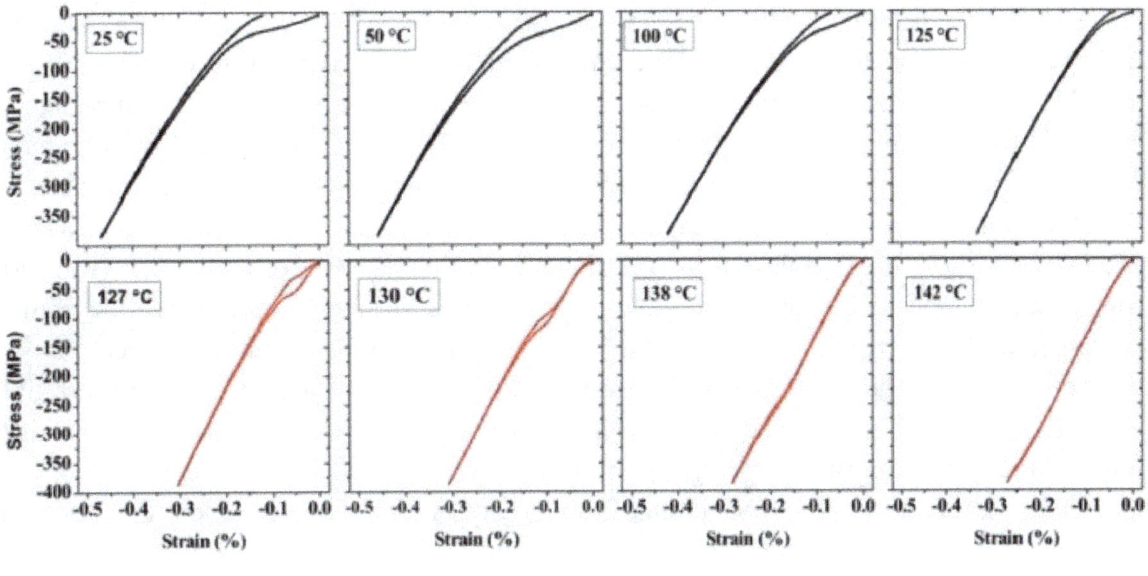

**Figure 15.** Stress-strain curves of $BaTiO_3$ ceramics at different temperatures. Reproduced with permission from Picht *et al.* [76]. Copyright 2012, AIP Publishing LLC.

## 3.3. Shape memory alloys

Shape memory effect refers to a shape recovery of ferroelastics during heating at high temperatures after a deformation process induced at a lower temperature. The most important class of materials exhibiting a strong shape memory effect is represented by Shape Memory

Alloys (SMAs). Plietsch *et al.* [77] have studied the stress-strain curves of the NiTi alloys subjected to either tension or compressive stress at different temperatures. They found that the pseudoelastic and martensitic NiTi phases show a pronounced asymmetry in stress-strain behaviour (Fig. 16). The asymmetry in the martensitic phase was ascribed to the different strength under tension and compression, which is likely caused by the presence of internal cracks, residual stresses or due to the Bauschinger effect.

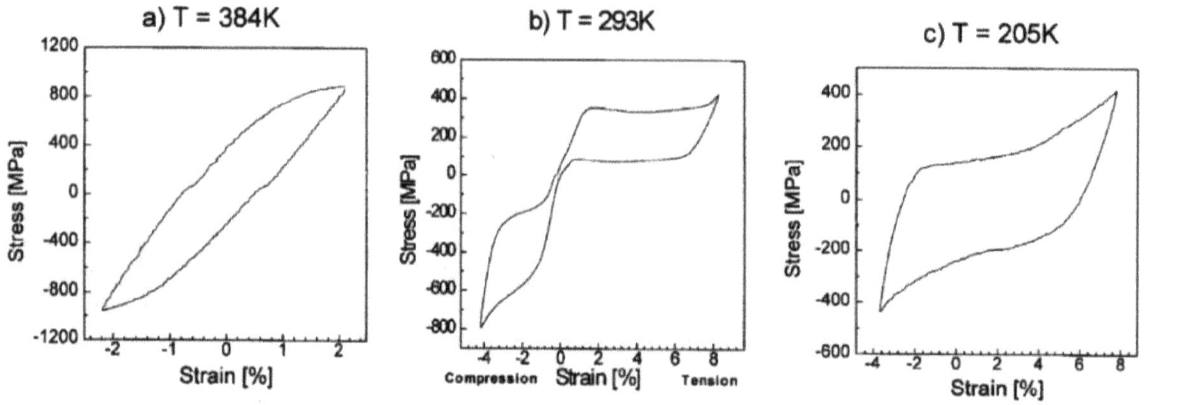


**Figure 16.** Tension/compression hysteresis for NiTi: (a) austenitic, (b) pseudoelastic, and (c) martensitic alloy. These images were published in Plietsch and Ehrlich [77]. Copyright 1997, Elsevier.

Liu *et al.* [78] have systematically studied tension-compression behaviour of martensitic NiTi under both monotonic and cyclic tensile/compressive loading. They found that the deformation mechanisms of the martensite phase are different under tension and compression, as evidenced by the asymmetric stress-strain curves (Fig. 17). Transmission electron microscopy (TEM) observations revealed that in the non-deformed specimen the martensite variants are well self-accommodated through the martensite twinning. Under tension, the interfaces between two variants are mobile and can migrate under stress. On the other hand, no migration of the junction planes between the neighbouring martensite plates was observed in samples compressed up to 4% strain. Instead, a high density of lattice defects, mainly dislocations, was observed inside both the martensite twin bands and twin boundaries. It was suggested that in the martensite phase the deformation mechanism under tension up to 4% may be predominantly related to the migration of variant interfaces, while under compression up to 4%, the strain is likely caused by the generation and movement of lattice defects, mostly dislocations.

### 3.4. The Bauschinger effect

The Bauschinger effect refers to the phenomenon by which the yield stress of metals reduces in the direction opposite to that of the very first stress applied.[79] This effect is believed to play a dominant role in the asymmetrical strain-stress behaviour of both monotonically and cyclic deformed metals. The effect is closely related to the presence of a long range internal stress (LRIS). The concept of LRIS assumes that the variations of the local stresses due to the

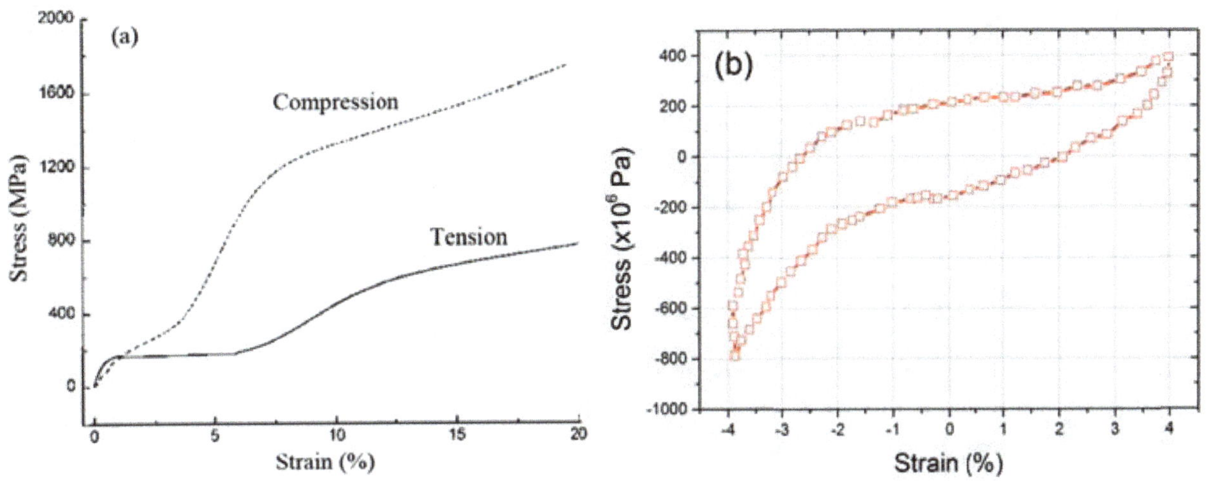


**Figure 17.** Stress-strain curves of a NiTi SMAs under (a) a monotonic tension and compression and (b) tension-compression cyclic loading within 4% strain. After [78].

applied stress occur over the long length scales. After plastic deformation have been produced, high dislocation density regions, such as sub-grain boundaries, and low dislocation density regions, such as cell interiors, characterized by a different yield stress are formed in the system. Under mechanical loading, the internal stress of the high-dislocation density regions is higher than that in low-dislocation density regions. Upon unloading, the average internal stress is zero; at the same time, the stress in the high dislocation density regions is positive, while it is negative in the low dislocation density regions. As a consequence, during stress reversal (applied stress changes sign), plasticity occurs in the low dislocation density regions at a lower stress and gives rise to the observed lower yield stress.

### 3.5. Microscopic mechanisms

In SMAs, the defects that might be present in the system, such as vacancies and solute atoms, tend to distribute following a short-range order symmetry, which will comply with the crystal symmetry after aging treatments.[80] In the aged martensite (Fig. 18a), the short-range order symmetry of defects conforms with the crystal symmetry of the martensite phase. During heating, the short-range ordering of defects does not change abruptly into a cubic symmetry at temperatures above the diffusionless martensitic transformation, because the diffusion of defects is a slow process (Fig. 18b). After aging, the short-range order symmetry of defects becomes conformed to the cubic crystal symmetry of the austenite parent phase (Fig. 18c). This phenomenon is referred to as the symmetry-conforming property of point defects in ferroelastic systems.[80] It should be mentioned that the short-range order of defects with martensitic symmetry can slightly distort the cubic lattice towards the martensitic symmetry, when the martensite is quickly heated up to the cubic phase. Such a lattice difference will produce a short range order-induced domain pattern in the cubic phase identical to the martensitic domain pattern. Thus, the symmetry property of point defects can give rise to the aging-induced two-way shape memory effect, namely the one observed in the aged martensitic phase of $Au_{51}Cd_{49}$ SMA.[80]

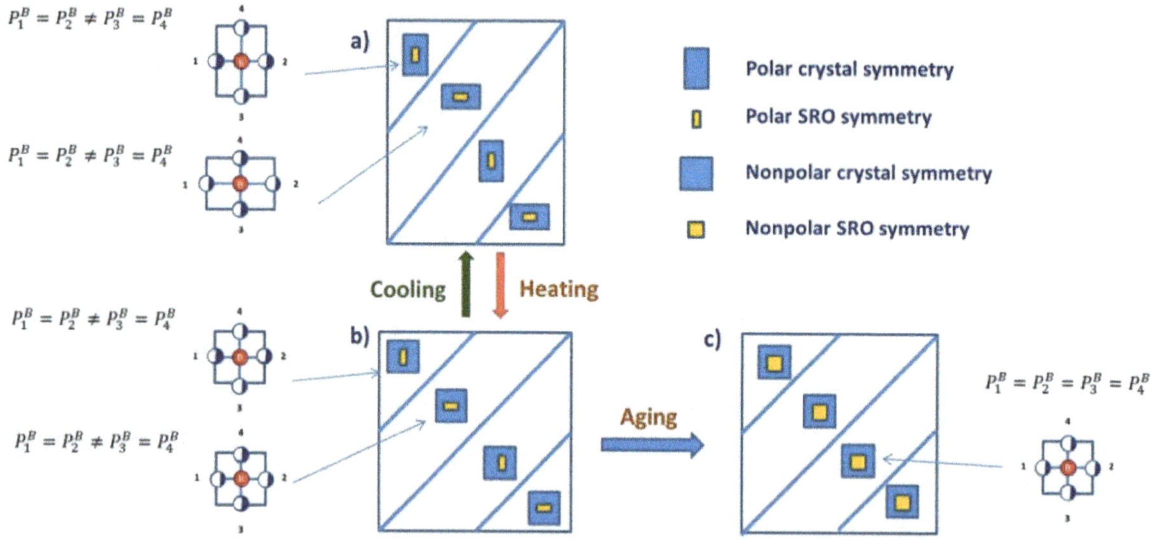


**Figure 18.** Predicted microstructure and short range order changes during reverse martensitic transformation from symmetry conforming short range order principle. (a) Aged martensite in which short range order symmetry conforms to the crystal symmetry of martensite. (b) Parent phase immediately after diffusionless transformation from (a). (c) Parent phase after aging that allows for the short range order symmetry of defects to conform to the cubic symmetry of the parent phase. After [80].

# 4. Ferromagnetic materials

## 4.1. Distortions of M-H hysteresis loops

In ferromagnetic materials, there is an intense direct exchange interaction between the magnetic moments of adjacent atoms due to the overlapping of atomic orbitals. The spins tend to align parallel each other and induce a spontaneous magnetization. The application of an alternating magnetic field produces a hysteretic magnetization-magnetic field (M-H) loop (Fig. 19). During the application of the very first magnetic field cycle, the ferromagnetic material becomes magnetized to the saturated state with a saturated magnetization ($M_s$). By decreasing the magnetic field to zero, the magnetization decreases and shows a remanent magnetization ($M_r$) at zero field. The magnetic field corresponding to zero magnetization represents the coercive magnetic field ($H_c$).

In absence of biasing effects, the M-H hysteresis loops of ferromagnetic materials display certain symmetry characteristics with respect to both H- and M-axis. However, these can be lost consequently to the development of different type of biasing processes that give rise to the deformed and shifted M-H loops along both axes. In this context, two main mechanisms will be discussed; the *exchange bias effect* and the *coexistence of amorphous and crystalline phases.*

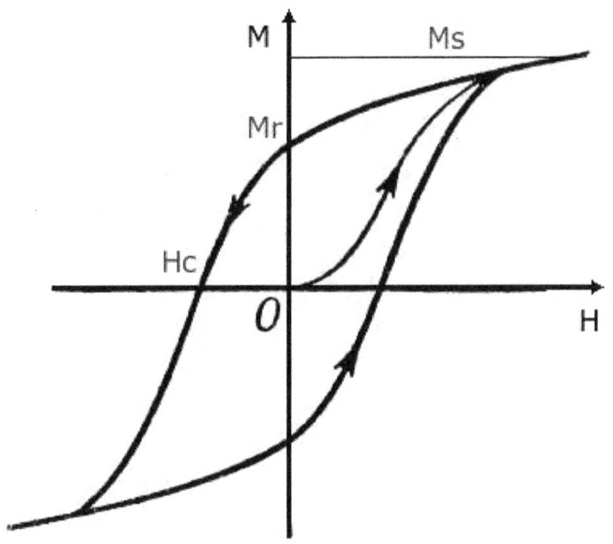


**Figure 19.** Magnetic M-H hysteresis loops of a ferromagnetic.

## 4.2. Exchange bias effect

### *4.2.1. Shift of the M-H loop along the magnetic field axis*

The exchange bias effect can be observed in systems containing interfaces between a ferromagnetic (FM) and an antiferromagnetic (AFM) phase, in which the Curie temperature ($T_c$) of the FM is higher than the Neel temperature ($T_N$) of the AFM.[81] When the FM-AFM system is cooled through the $T_N$ during the application of an external magnetic field (referred to as the cooling field $H_{cf}$), a shift of the M-H hysteresis loop of the FM-AFM system along the magnetic field axis may appear. This defines the exchange bias field ($H_E$).[82] The exchange bias phenomenon was firstly discovered by Meiklejohn and Bean in Co (ferromagnetic)-CoO (antiferromagnetic) system.[83] Later, the exchange bias has been observed in many other systems, including nanoparticles, inhomogeneous materials, single crystals and thin films containing FM-AFM, ferrimagnetic-FM and ferrimagnetic-AFM interfaces. The first intuitive mechanism proposed by Nogues *et al.* [84] to explain the exchange bias effect is schematically drawn in Fig. 20. In the range $T_N < T < T_C$, the cooling field induces the alignment of the FM spins along its direction (Fig. 20i), while at $T < T_N$, the spins of the AFM phase arrange in an antiferromagnetic configuration (Fig. 20ii). Along the FM-AFM interface, the AFM spins tend to align ferromagnetically as they are influenced by those of the FM due to the FM exchange interaction (Fig. 20ii). When the magnetic field is reversed, the FM spins tend to reorient following the applied field, while the AFM spins remain unchanged due to the large AFM anisotropy (Figs. 20iii - 20v). The ferromagnetic interaction at the FM-AFM interface provides a strong restoring force on the FM spins reorientations, and thus a shift of the M-H loop is produced. Generally, when the interaction at the interface is ferromagnetic and the cooling field $H_{cf}$ is applied along the positive direction, the exchange bias field $H_E$ is characterized as a negative exchange bias field and a shift of the M-H loop towards the negative direction is observed (Fig. 20).

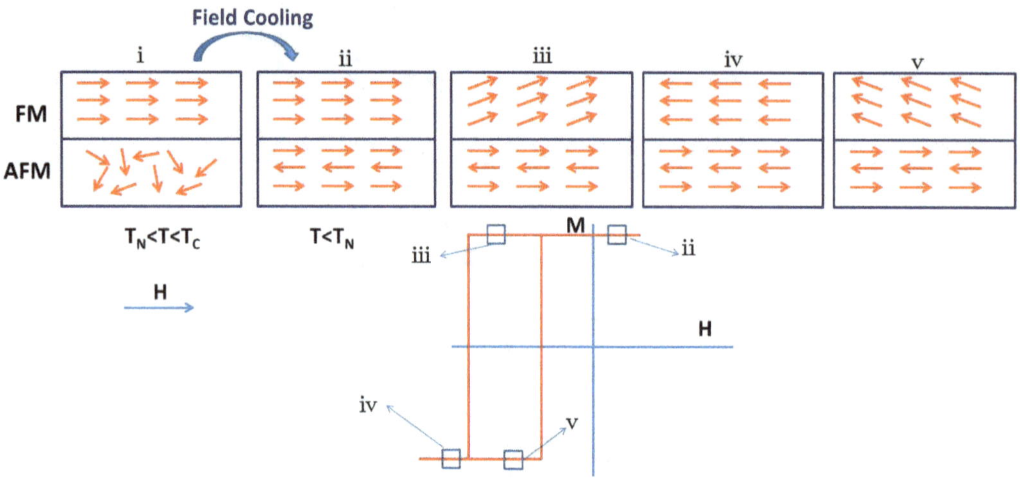


**Figure 20.** Schematic of the exchange bias mechanisms in case of ferromagnetic interaction between the FM and AFM spins at the interface. After [82].

However, it has been also understood that in some magnetic systems the interaction between the spins at the interfaces can be antiferromagnetic, meaning that below $T_N$ the AFM spins in the layer close to the interface are aligned antiparallel to the FM spins. In this case, the direction of the M-H loop shift depends on the strength of the cooling field. Nogues *et al.* [84] have found that in Fe (FM)-$FeF_2$ (AFM) bilayers, the M-H hysteresis loops can shift along the direction of the cooling field, after a sufficiently large cooling field is applied (Fig. 21). It was proposed that in case of the antiferromagnetic interaction, when the system is cooled through $T_N$, the spins of AFM at the interface tend to align antiparallel to those of FM. By applying a low cooling field along the positive direction, the FM spins will align along the positive direction, while the AFM spins will not switch. When the applied magnetic field is reversed, the FM spins will try to follow the field, while the AFM spins will oppose to that, and they will force the FM spins to be antiparallel to them, leading to a shift of the M-H loop towards left. However, when the positive cooling field is large enough, the AFM surface spins are forced to be parallel to the FM spins along the cooling field direction, and a positive exchange bias is developed. Therefore, when the applied field is reversed, the magnetization reversal is facilitated by the antiferromagnetic coupling of the spins at the interface, and a positive shift of the M-H loop is observed (Fig. 21).

In order to quantitatively describe the exchange bias effect, several phenomenological models have been proposed. The model developed by Meiklejohn [83] assumes that both the FM and the AFM are in single domain state, the FM-AFM interface is perfectly smooth, and the energy per unit interface area can be expressed as:

$$E = -HM_{FM}t_{FM}\cos(\theta-\beta) + K_{FM}t_{FM}sin^2(\beta) + K_{AFM}t_{AFM}sin^2(\alpha) - J_{INT}\cos(\beta-\alpha) \quad (8)$$

where $H$ is the applied field, $M_{FM}$ is the saturation magnetization, $t_{FM}$ and $t_{AFM}$ are the thickness of FM and AFM layers respectively, $K_{FM}$ and $K_{AFM}$ are the anisotropy of FM and AFM layers

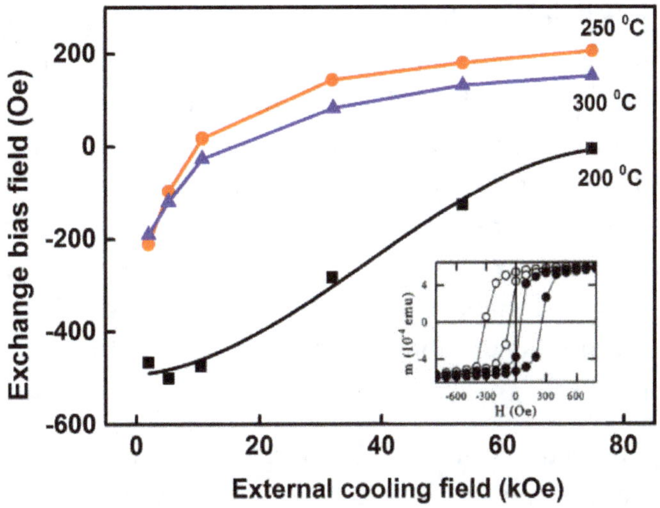


**Figure 21.** The exchange bias field as a function of the cooling field for $FeF_2$-Fe bilayers with the $FeF_2$ grown at different temperatures, square: 200°C, triangle: 250°C, and circle: 300°C. In-set: magnetization loops of the sample with $FeF_2$ grown at 300°C for low (hollow circle) and high (solid circle) cooling field. After [84].

respectively, and $J_{INT}$ is the interface coupling constant. The terms $\alpha$, $\beta$ and $\theta$ represent the angles between $K_{AFM}$ and $M_{AFM}$, $K_{FM}$, and $M_{FM}$, $K_{FM}$ and H, respectively.

By neglecting the FM anisotropy, which is much smaller than that of AFM, and by minimizing the energy with respect to $\alpha$ and $\beta$, the exchange bias is obtained as:

$$H_E = \frac{J_{INT}}{M_{FM} t_{FM}} \tag{9}$$

However, the exchange bias $H_E$ calculated using the Eq. 9 is usually several orders of magnitude larger than the value observed in the experiments. Malozemoff [85] proposed an exchange bias effect model based on the assumption of rough FM-AFM interfaces. A microscopically random exchange field at the interface due to the defects, roughness or lattice mismatch can give rise to a random field which produces a number of uncompensated spins at the interface, leading to the loop shift. It was assumed that the FM is in a single domain state, therefore, due to the presence of a random field the AFM system will split into domains in order to minimize the unidirectional anisotropy (i.e., one single stable configuration of FM spins). The model gives the following expression for the exchange bias field [85]:

$$H_E = \frac{2z}{\pi^2 M_{FM} t_{FM}} \sqrt{\frac{J_{INT} K_{AF}}{a}} \tag{10}$$

where $z$ is a constant in the order of unity related to the randomness degree of the interface and $a$ is the lattice parameter of the FM lattice which was considered cubic. The exchange bias values estimated by this model are consistent with the experiments. However, the main drawback of the model is that the calculated bias depends on the defect concentration at the interface. Mauri *et. al.* [86] developed a model assuming that the domain walls develop in the AFM layer, which sets an upper limit on the exchange coupling energy in such a manner that it ultimately gives rise to a significantly smaller exchange bias than that provided by the Meiklejohn's model calculations [83]. Using a micromagnetic approach, Koon [87] has stated that the orientation of the FM spins should be perpendicular to the AFM magnetic easy axes in the ground state. Although this model can explain the coercivity enhancement in some systems, it fails to provide reasonable estimations of the exchange bias. Therefore, the development of more general models to more accurately quantify the exchange bias in different systems is still ongoing.

Another kind of asymmetry characterized by a sharp step in the upper branch of the hysteresis loop was reported in the exchange biased Co-CoO dot array system.[88] The hysteresis loops of the Co-CoO dot arrays with four different dot sizes are shown in Fig. 22. Together with a shift of the hysteresis loops along the H-axis due to the exchange bias, a peculiar anomaly can be observed in the upper branch of the M-H hysteresis loop. This loop distortion is considered to originate from the presence of an intermediate magnetization saturated state. Both loop features showed a strong dependence on the size of dots. The H-shift of the M-H loop reduces with decreasing the dot size due to a reduction of the exchange bias effect. The deformation in the upper branch of the M-H loop was attributed to a magnetostatic interaction between the dots. The interaction determines an intermediate saturation of the magnetization (indicated in Fig. 22 by an arrow), which hinders the magnetization reversal and lowers its rate. The stability of the intermediate magnetization is larger in the array with bigger dots (compare Figs. 22a-d). For the dots with smaller size, the inter-dot magnetostatic interaction is weaker and the magnetization switching of each dot is less influenced by those of the surrounding dots. This determines a faster magnetization switching rate and an almost complete disappearance of the intermediate saturation magnetization (Fig. 22d).

#### *4.2.2. Shift of the M-H loop along the magnetization axis*

In some magnetic systems containing the FM-AFM interfaces, the M-H hysteresis loop can also shift along the magnetization axis. Generally, the M-shift of the M-H loop is caused by the presence of a pinned magnetization which cannot be switched by the applied external magnetic field. The pinned magnetization can be developed through different mechanisms, including: i) an incomplete spin reversal of the FM spins; ii) the presence of spin glass-like phases in FM material nearby the interface region; iii) the presence of uncompensated spins in the AFM layer, and iv) the spin canting in the AFM layer. An example of systems where a spin glass-like arrangement develops in the FM phase is represented by the composite made up of granular ferrimagnetic $NiFe_2O_4$ nanoparticles embedded in an antiferromagnetic NiO matrix.[89] Fig. 23a shows the M-H hysteresis loops of the system measured at 10 K after both the zero-field cooling (ZFC) and field cooling (FC) processes. In the latter process, the M-H hysteresis loop shows a field offset as well as a magnetization shift with the exchange bias linearly increasing with the vertical magnetization offset. The underlying mechanism of the switching asymmetry in these materials was suggested to originate from the spin glass-like

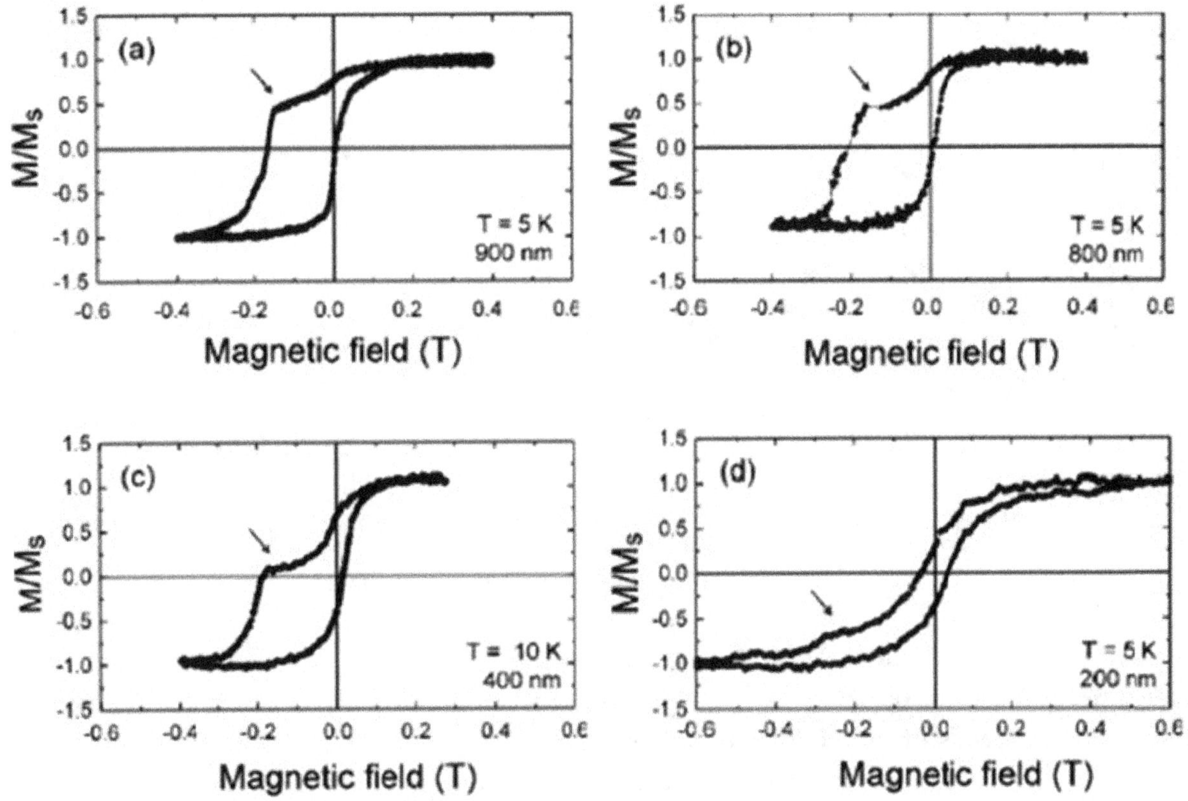


**Figure 22.** Hysteresis loops of the Co-CoO square dot arrays with four different dot sizes. Reproduced with permission from Girgis *et al.* [88]. Copyright 2003, APS.

phase formed in $NiFe_2O_4$ due to the fine particle size and the consequent structural disorder at the interface regions. During cooling under a magnetic field in the temperature range $T_{irr} < T < T_C$ ($T_{irr}$ is the temperature above which the exchange bias disappears), the FM spins align along the applied magnetic field, while the spins of the glass-like phase remain randomly oriented. As the temperature is lowered below $T_{irr}$, some of the net uncompensated spins of the glass-like phase also line-up along the field-cooling direction and produce a pinned magnetization state. During the field reversal, the latter does not follow the field, which is probably the reason for the vertical shift of the loop. At the same time, the frozen spins in the glass-like phase try to keep the spins of the ferrimagnetic $NiFe_2O_4$ along their original direction leading to a negative shift of M-H loop along the H-axis.

The vertical shift of M-H hysteresis loops due to spin canting in the AFM layer was reported by Yuan *et al.* [90] for the Co/$Ca_2Ru_{0.98}Fe_{0.02}FeO_4$ (FM/AFM) heterostructure, where the Co layer was sputtered on the $Ca_2Ru_{0.98}Fe_{0.02}FeO_4$ single crystal. When the FM/AFM system was cooled down to 10 K in a magnetic field, a horizontal and a vertical shift of the M-H loop was observed. In particular, the shift along the H-axis is negative for a positive cooling field and positive for a negative cooling field. On the other hand, the shift along the M-axis is positive for a positive cooling field and negative for a negative cooling field (Fig. 23b). The $Ca_2Ru_{0.98}Fe_{0.02}FeO_4$ single crystal does not have uncompensated AFM spins, however, the magnetic moments of the Ru (Fe) ions in the B-site of the AFM oxide are canted and give rise to the net ferromagnetic

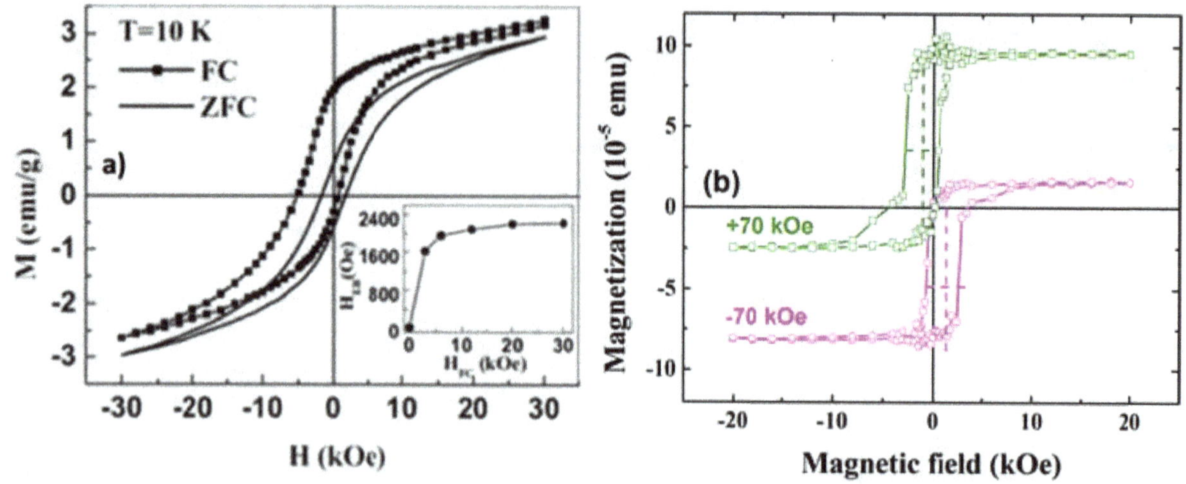


**Figure 23.** (a) The M-H hysteresis loops of $NiFe_2O_4$ nanoparticles embedded in an antiferromagnetic NiO matrix measured at 10 K. Reproduced with permission from Tian *et al.* [89]. Copyright 2008, AIP Publishing LLC. (b) The M-H hysteresis loops of $Co/Ca_2Ru_{0.98}Fe_{0.02}FeO_4$ heterostructure. Courtesy of S. J. Yuan, after [90].

moments. In the first layer near the Co phase, these moments align parallel to those of the Co layer, but they cannot be readily switched when the magnetic field is reversed. As a consequence, the vertical shift of the M-H hysteresis loops is produced (Fig. 23b). Nogues *et al.* [91] studied the dependence of the exchange bias and magnetization shift on cooling field in the Fe-$FeF_2$-Al system. They observed that for low cooling fields the vertical shift can be opposite to the cooling field suggesting that an antiferromagnetic coupling exists at the FM-AFM interface. Additionally, at low cooling fields, the vertical shift was negative when the thickness of the $FeF_2$ layer was 200 nm and it was positive for the 100 nm layer. This behaviour was attributed to the roughness of the interface; smooth interface tends to induce a negative vertical shift, while rough interface usually leads to a positive shift. For large cooling fields, a positive vertical shift was observed in all cases.

The vertical offsets of M-H hysteresis loops were also observed in magnetic systems without exchange bias. Watanabe *et al.* [92] reported on the shift of the magnetization hysteresis loop along the magnetization axis in the antiferromagnetic $LaFeO_3$ when cooled in a magnetic field below the Neél temperature. The shift phenomenon was explained by the fact that the field-cooled $LaFeO_3$ always shows a weak ferromagnetism with a small remanent magnetization, which cannot be reversed by the external field. A vertically shifted M-H hysteresis loop was also observed in $Co_2VO_4$ at 4.2 K when cooled in a small magnetic field.[93] Due to the high anisotropy and magnetic hardness of cobalt in spinel lattices, the spins of cobalt ions align along the cooling field direction when cooled through the Curie temperature and, at low temperatures, they preserve the orientation producing the shift of M-H loop along the magnetization axis.

### 4.3. Effect of the amorphous-crystalline phases coexistence

A distinct shift of the M-H hysteresis loop was observed in magnetic systems subjected to a conventional annealing at a temperature below their crystallization temperature. Ohta *et al.*

[94] studied the $Fe_5Co_{70}Si_{10}B_{15}$ amorphous alloys and reported that the as-quenched sample shows a symmetric hysteresis loop (Fig. 24a). After annealing at 420 °C, the hysteresis loop becomes pinched (Fig. 24b) and annealing under a magnetic field of 10 kOe leads to a significant shift of the loop along the H-axis (Fig. 24c).

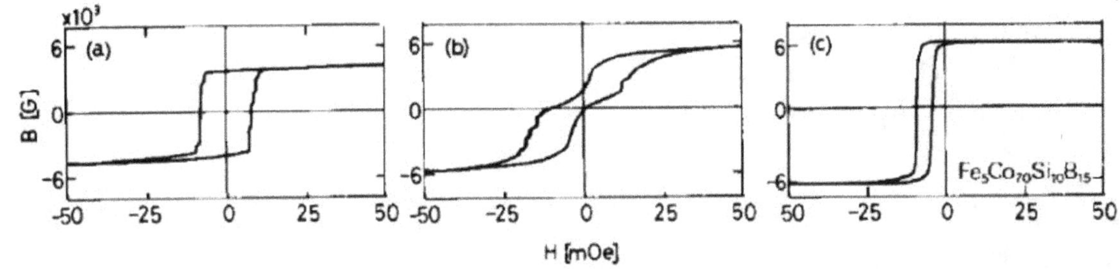


**Figure 24.** The magnetic induction (B) vs. magnetic field (H) hysteresis loops of amorphous alloys $Fe_5Co_{70}Si_{10}B_{15}$: (a) as-quenched, (b) annealed and (c) annealed in a magnetic field. These images were published in Ohta *et al.* [94]. Copyright 1980, Elsevier.

Kohmoto *et al.* [95] have reported on the loop shift of the $Fe_5Co_{70}Si_{10}B_{15}$ amorphous alloys annealed at 180 °C for different time. The shift along the field axis was found to increase with the increase of annealing time. Rivas *et al.* [96] observed the shifted M-H loops in the metallic-glass $Co_{66}Si_{16}B_{12}Fe_4Mo_2$ ribbons annealed at 510 °C and "pre-magnetized" under a magnetic field of 400 kA/m at room temperature. A large negative field offset was observed when the initial applied field was parallel to the pre-magnetizing field, and the field shift was positive as the initial applied field was antiparallel to the pre-magnetizing field. However, no shift of M-H hysteresis loops was observed when the magnetic field was applied perpendicular to the direction of the pre-magnetizing field. The results suggest that the shift of M-H hysteresis loops in the annealed pre-magnetized ribbons originate from the coexistence of the amorphous and crystalline phases. After annealing, the amorphous matrix contains partially crystallized particles which are much magnetically harder if compared to the amorphous matrix. When subjected to a relatively high pre-magnetizing field, the magnetic moments of the particles tend to align along the applied field direction. Due to the large magnetic anisotropy, the magnetic spins of the hard-magnetic particles remain unchanged under the ac applied magnetic field used for the reorientation of the magnetic moments in the soft matrix. Therefore, a strong unidirectional dipolar magnetic field is formed, which can exert a strong restoring force on the reorientations of magnetic moments of the soft matrix, and thus produces a shift of M-H hysteresis loops.

## 4.4. Magnetic aging

The term "magnetic aging" indicates the time-dependent changes in the magnetization of ferromagnetic materials. These changes are also commonly referred to as "magnetic viscosity" or "magnetic after-effects". The typical relaxation time spans several orders of magnitude; it ranges from less than a second in superparamagnetic particles to millions of years in magnetic rocks.[97] Various mechanisms that contribute to the magnetic aging have been proposed, including i) thermal fluctuations, ii) the diffusion after-effect of ferrous ions and vacancies, iii) chemical modifications.[98] The magnetic aging originated from thermal fluctuations is driven by the thermally activated jumps over energy barriers of domain walls. This process determines the time-, temperature- and magnetic field-dependence of the remanent magnetization. The diffusion of point defects, impurity atoms or electrons towards domain walls can cause the departure of the domain wall energy from a minimum which determines the displacement of the domain walls in a new equilibrium configuration. This type of magnetic aging mechanism is usually defined as "diffusion after-effect". It was first observed in iron containing carbon or nitrogen impurities.[98]

## 5. Summary

Biasing effects in different ferroic systems show a highly complex phenomenology which depends on several variables whose significance and contribution vary from case to case. Biased P-E and S-E ferroelectric loops are mainly generated by aging and fatigue processes, which are generally caused by the arrangement of charged defects and charge carriers in a certain configuration. A common cause of these two processes can be attributed to the development of an internal bias, which induces preferential polarization and strain configurations. In aging, the presence of an internal bias gives rise to pinched and asymmetric/shifted loops, while in fatigued samples the bias results in asymmetric and shifted loops. These features are mostly reversible upon opportune thermal treatments or electrical cycling, but fatigue shows also irreversible effects consisting, for instance, of mechanical damage such as microcracks that cannot be recovered. Aging effects are homogeneous since the controlling mechanisms are active down to the unit cell scale according to the symmetry-conforming property of point defects. Fatigue features instead can be highly inhomogeneous due to the screening retardation of the locally varying depolarizing field during electric field cycles. Under certain circumstances, aging of ferroelectrics can be explained within the concept of the symmetry-conforming property of point defects, which seems to be able to satisfactorily explain also biasing effects in non-polar ferroelastic materials. According to this principle, short-range ordered defects in ferroelastics tend to arrange themselves over time following the crystal symmetry of the hosting unit cell. The time-dependent change in the symmetry of point defects induces memory effects of equilibrium configurations, determining biased stress-strain hysteresis loops. In ferroelectric/ferroelasic perovskites, a partial replacement of A- and B-site cations with acceptor and donor species, respectively, produces analogous hardening and softening effects on ferroelectric and ferroelastic hysteresis loops under electric field and mechanical stress. Biasing effects in single phase ferromagnetic systems are usually generated by spatially inhomogeneous arrangement of spins on the unit cell scale. In magnetic heterostructures and composites characterized by the simultaneous existence of ferromagnetic and antiferromagnetic ordering within the material, biasing effects are due to the competitive contributions from the coexisting soft and hard magnetic phases. The exchange interaction at the interfaces between dissimilar phases, the presence of the uncompensated spins, the formation of spin glass-like phases and spin canting effects, together with pinned preferential magnetization states are the most important possible mechanisms of biased M-H hysteresis loops. The thermally activated magnetic aging and the existence of domain wall pinning/de-pinning and rejuvenation effects in ferromagnetic materials introduce common characteristics with aging phenomena in ferroelectric/ferroelastic systems.

## Acknowledgements

This work was done within the frame of the project no. 2/0057/14, which is supported by the Grant Agency of the Slovak Academy of Sciences.

## Author details

Vladimir Koval[1*], Giuseppe Viola[2,3] and Yongqiang Tan[4]

*Address all correspondence to: vkoval@imr.saske.sk

1 Institute of Materials Research, Slovak Academy of Sciences, Kosice, Slovakia

2 Department of Applied Science and Technology, Institute of Materials Physics and Engineering, Torino, Italy

3 School of Engineering and Materials Science, Queen Mary University of London, London, UK

4 School of Physics, Shandong University, China

## References

[1] Yan H, Inam F, Viola G, Ning H, Zhang H, Jiang Q, Zhang T, Gao Z, Reece MJ. The contribution of electrical conductivity, dielectric permittivity and domain switching in ferroelectric hysteresis loops. Journal of Advanced Dielectrics 2011;1(1) 107-118.

[2] Viola G, Saunders T, Wei X, Chong KB, Luo H, Reece MJ, Yan H. Contribution of piezoelectric effect, electrostriction and ferroelectric/ferroelastic switching to strain-electric field response of dielectrics. Journal of Advanced Dielectrics 2013;3(1) 1350007-11.

[3] Jin L, Li F, Zhang S. Decoding the Fingerprint of Ferroelectric Loops: Comprehension of the material properties and structures. Journal of American Ceramics Society 2014;97(1) 1–27.

[4] Genenko YA, Glaum J, Hoffmann MJ, Albe K. Mechanisms of aging and fatigue in ferroelectrics. Materials Science and Engineering B 2015;192 52-82.

[5] Tan Y, Zhang J, Wang C. Aging behaviours of CuO modified $BaTiO_3$ ceramics. Advances in Applied Ceramics 2014;113(4) 223-227.

[6] Arlt G, Neumann H. Internal bias in ferroelectric ceramics: Origin and time dependence. Ferroelectrics 1988;87(1) 109-120.

[7] Feng Z, Ren X. Striking similarity of ferroelectric aging effect in tetragonal, orthorhombic and rhombohedral crystal structures. Physical Review B 2008;77(13) 134115-6.

[8] Zhang L, Ren X. Aging behavior in single-domain Mn-doped BaTiO3 crystals: Implication for a unified microscopic explanation of ferroelectric aging. Physical Review B 2006;73(9) 094121-6.

[9] Tutuncu G, Damjanovic D, Chen J, Jones JL. Deaging and asymmetric energy landscapes in electrically biased ferroelectrics. Physical Review Letters 2012;108(17) 177601-5.

[10] Carl K, Hardtl KH. Electrical after-effects in Pb(Ti, Zr)$O_3$ ceramics. Ferroelectrics 1978;17(1) 413-486.

[11] Lupascu D, Genenko YA, Balke N. Aging in ferroelectrics. Journal of American Ceramic Society 2006;89(1) 224–229.

[12] Takahashi M. Space charge effect in lead zirconate titanate ceramics caused by the addition of impurities. Japan Journal of Applied Physics 1970;9(10) 1236-1246.

[13] Morozov MI, Damjanovic D. Charge migration in Pb(Zr, Ti)$O_3$ ceramics and its relation to ageing, hardening and softening. Journal of Applied Physics 2010;107(3) 034106-10.

[14] Tan Q, Li J, Viehland D. Role of lower valent substituent-oxygen vacancy complexes in polarization pinning in potassium-modified lead zirconate titanate. Applied Physics Letters 1999;75(3) 418-420.

[15] Balke N, Lupascu D, Granzow T, Rödel J. Fatigue of lead zirconate titanate ceramics. I: Unipolar and DC loading. Journal of American Ceramics Society 2007;90(4) 1081-1087.

[16] Luo Z, Glaum J, Granzow T, Jo W, Dittmer R, Hoffman M, Rödel J. Bipolar and unipolar fatigue of ferroelectric BNT-based lead-free piezoceramics. Journal of American Ceramics Society 2011;94(2) 529-535.

[17] Viola G, Tan Y, Saunders T, Reece MJ. Unpublished data.

[18] Jiang QY, Subbarao EC, Cross LE. Fatigue in PLZT: acoustic emission as a discriminator between microcracking and domain switching. Ferroelectrics 1994, 154(1) 113–118.

[19] Nuffer J, Lupascu D, Glazounov A, Kleebe HJ, Rödel J. Microstructural modifications of ferroelectric lead zirconate titanate ceramics due to bipolar electric fatigue. Journal of the European Ceramic Society 2002;22(13) 2133-2142.

[20] Priya S, Kim HW, Ryu J, Uchino K., Viehland D. Fractal cluster modeling of the fatigue behavior of lead zirconate titanate. Applied Physics Letters 2002;80(4) 1625-1627.

[21] Chen J, Harmer MP, Smyth DM. Compositional control of ferroelectric fatigue in perovskite ferroelectric ceramics and thin films. Journal of Applied Physics 1994;76(9) 5394-5398.

[22] Zuo R, Xu Z, Li L. Dielectric and piezoelectric properties of $Fe_2O_3$-doped $(Na_{0.5}K_{0.5})_{0.96}Li_{0.04}Nb_{0.86}Ta_{0.1}Sb_{0.04}O_3$ lead-free ceramics. Journal of Physics and Chemistry of Solids 2008;69(7) 1728-1732.

[23] Glaum J, Hoffman M. Electric Fatigue of Lead-Free Piezoelectric Materials. Journal of the American Ceramics Society 2014;97(3) 665-680.

[24] Viola G, Ning H, Reece MJ, Wilson R, Correia TM, Weaver P, Cain M, Yan H. Reversibility in electric field-induced transitions and energy storage properties of bismuth-based perovskite ceramics. Journal of Physics D: Applied Physics 2012;45(35) 355302-7.

[25] Viola G, Ning H, Wei X, Deluca M, Adomkevicius A, Khaliq J, Reece MJ, Yan H. Dielectric relaxation, lattice dynamics and polarization mechanisms in $Bi_{0.5}Na_{0.5}TiO_3$-based lead-free ceramics. Journal of Applied Physics 2013;114(1) 014107-9.

[26] Viola G, McKinnon R, Koval V, Adomkevicius A, Dunn S, Yan H. Lithium-induced phase transitions in lead-free $Bi_{0.5}Na_{0.5}TiO_3$ based ceramics. Journal of Physical Chemistry C 2014;118(16) 8564-8570.

[27] Luo Z, Granzow T, Glaum J, Jo W, Rödel J, Hoffman M. Effect of ferroelectric long-range order on the unipolar and bipolar electric fatigue in $Bi_{1/2}Na_{1/2}TiO_3$-based lead-free piezoceramics. Journal of the American Ceramics Society 2011;94(11) 3927-3933.

[28] Simons H, Glaum J, Daniels JE, Studer AJ, Liess A, Rödel J, Hoffman M. Domain fragmentation during cyclic fatigue in 94% $(Bi_{1/2}Na_{1/2})TiO_3$-6% $BaTiO_3$. Journal of Applied Physics 2012;112(4) 044101-5.

[29] Shur VY. Kinetics of polarization reversal in normal and relaxor ferroelectrics: relaxation effects. Phase Transitions 1998;65(1-4) 49-72.

[30] Balke N, Granzow T, Rödel J. Degradation of lead-zirconate-titanate ceramics under different dc loads. Journal of Applied Physics 2009;105(10) 104105-7.

[31] Shur VY, Nikolaeva EV, Shishkin EI, Baturin IS, Shur AG, Utschig T, Schlegel T, Lupascu D. Deaging in $Gd_2(MoO_4)_3$ by cyclic motion of a single planar domain wall. Journal of Applied Physics 2005;98(7) 074106-8.

[32] Shur VY, Rumyantsev EL, Nikolaeva EV, Shishkin EI, Baturin IS, Lupascu D, Nuffer J, Randall C, Ozgul M. Fatigue effect in bulk ferroelectrics. In: Lynch CS. (ed.) SPIE Proceedings on Smart Structures and Materials: Active Materials Structure and Mechanics 4699, July 15, 2002, San Diego, USA; 40-50. doi:10.1117/12.475015.

[33] Setter N, Damjanovic D, Eng L, Fox G, Gevorgian S, Hong S, Kingon A, Kohlstedt H, Park NY, Stephenson GB, Stolitchnov I, Taganstev AK, Taylor DV, Yamada T, Streiffer S. Ferroelectric thin films: Review of materials, properties, and applications. Journal of Applied Physics 2006;100(5) 051606.

[34] Ramesh R, Aggarwal S, Auciello O. Science and technology of ferroelectric films and heterostructures for non-volatile ferroelectric memories. Materials Science and Engineering: R: Reports 2001;32(6) 191-236.

[35] Koval V, Bharadwaja SSN, Trolier-McKinstry. Mist deposited lead zirconate titanate films. Ferroelectrics 2011;421(1) 23-29.

[36] Polcawich RG, Trolier-McKinstry S. Piezoelectric and dielectric realibility of lead zirconate titanate thin films. Journal of Materials Research 2000;15(1) 2505-2513.

[37] Warren WL, Tuttle BA, Dimos D, Pike GE, Al-Shareef HN, Ramesh R, Evans JT. Imprint in ferroelectric capacitors. Japanese Journal of Applied Physics 1996;35(2S) 1521-1524.

[38] Grossmann M, Lohse O, Bolten D, Boettger U, Schneller T. The interface screening model as origin of imprint in $PbZr_xTi_z$-$xO_3$ thin films. I. Dopant, illumination, and bias dependence. Journal of Applied Physics 2002;92(5) 2680-2687.

[39] Warren WL, Dimos D, Pike GE, Vanheusden K, Ramesh R.. Alignment of defect dipoles in polycrystalline ferroelectrics. Applied Physics Letters 1995;67(12) 1689-1691.

[40] Kim SH, Hong JG, Streiffer SK, Kingon AI. Imprint and fatigue properties of chemical solution derived $Pb_{1-x}La_x(Zr_yTi_{1-y})(1-x/4)O_3$ thin films. Journal of Materials Research 1999;14(4) 1371-1377.

[41] Folkman CM, Baek SH, Nelson CT, Jang HW, Tybell T, Pan XQ, Eom CB. Study of defect-dipoles in an epitaxial ferroelectric thin film. Applied Physics Letters 2010;96(5) 052903.

[42] Dimos D, Warren WL, Sinclair MB, Tuttle BA, Schwartz RW. Photoinduced hysteresis changes and optical storage in (Pb, La)(Zr, Ti)$O_3$ thin films and ceramics. Journal of Applied Physics1994;76(7) 4305-4315.

[43] Sze SM. Physics of Semiconductor Devices. Section 7.3.4, New York: Wiley&Sons; 1969.

[44] O'Dwyer J. The Theory of Dielectric Breakdown in Solids. Oxford: Clarendon Press; 1964.

[45] Kim WY, Lee YS, Lee HCh, Ka DY, Kim SY. Analysis of static imprint phenomenon

in ferroelectric VDF-TrFE copolymer film for nonvolatile memory devices. Journal of the Korean Physical Society 2010;57(6) 1690-1694.

[46] Pertsev NA, Contreras JR, Kukhar VG, Hermanns B, Kohlstedt H, Waser R. Coercive field of ultrathin $Pb(Zr_{0.52}Ti_{0.48})O_3$ epitaxial films. Applied Physics Letters 2003;83(16) 3356-3358.

[47] Tagantsev AK, Stolichnov I, Setter N, Cross JS. Nature of nonlinear imprint in ferroelectric films and long-term prediction of polarization loss in ferroelectric memories. Journal of Applied Physics 2004;96(11) 6616-6623.

[48] Abe K, Yanase N, Yasumoto T, Kawakubo T. Voltage shift phenomena in a heteroepitaxial $BaTiO_3$ thin film capacitor. Journal of Applied Physics 2002;91(1) 323-330.

[49] Le Rhun G, Bouregba R, Poullain G. Polarization loop deformations of an oxygen deficient $Pb(Zr_{0.25}Ti_{0.75})O_3$ ferroelectric thin film. Journal of Applied Physics 2004;96(10) 5712-5721.

[50] He JQ, Vasco E, Jia CL, Wang RH. Direct observation of a fully strained dead layer at $Ba_{0.7}Sr_{0.3}TiO_3/SrRuO_3$ interface. Applied Physics Letters 2005, 87(6) 062901.

[51] Cillessen J, Prins M, Wolf R. Thickness dependence of the switching voltage in all-oxide ferroelectric thin-film capacitors prepared by pulsed laser deposition. Journal of Applied Physics 1997;81(6) 2777-2783.

[52] Tagantsev AK, Landivar M, Colla E, Setter N. Identification of passive layer in ferroelectric thin films from their switching parameters. Journal of Applied Physics 1995;78(4) 2623-2630.

[53] Hesse D, Alexe M. Ferroelectric nanostructures for future high-density non-volatile memory applications – preparation methods, microstructure, and physical properties, Autumn School, Berlin, 7. 10. 2005, http://crysta.physik.hu-berlin.de/as2005/pdf/as2005_talk_17_Hesse.pdf.

[54] Zubko P, Jung D J, Scott J F. Space charge effects in ferroelectric thin films. Journal of Applied Physics 2006;100(11) 114113.

[55] Fredrickson KD, Ponath P, Posadas AB, McCartney MR, Aoki T, Smith DJ, Demkov AA. Atomic and electronic structure of the ferroelectric $BaTiO_3$/Ge(001) interface. Applied Physics Letters 2014;104(24) 242908.

[56] Spaepen F. Interfaces and Stresses in Thin Films. Acta Materialia 2000;48 31-42.

[57] Wu WB, Wong KH, Choy CL. Interface-oxygen-loss-controlled voltage offsets in epitaxial $Pb(Zr_{0.52}Ti_{0.48})O_3$ thin-film capacitors with $La_{0.7}Sr_{0.3}MnO_3$ electrodes. Applied Physics Letters 2004;85(21) 5013-5015.

[58] Lian L, Sottos NR. Stress effects in sol-gel derived ferroelectric thin films. Journal of Applied Physics 2004;95(2) 629-634.

[59] Ma W, Cross LE. Strain-gradient-induced electrical polarization in lead zirconate titanate ceramics. Applied Physics Letters 2003;82(19) 3293-3295.

[60] Tuttle BA, Voigt JA, Garino TJ, Goodnow DC, Schwartz RW, Lamppa DL, Headley TJ, Eatough MO. Chemically prepared Pb(Zr, Ti)$O_3$ thin films: The effects of orientation and stress. In: Liu M, Safari A, Kingon A, Haertling G. (eds.) Proceedings Of the Eight International Symposium On Applications of Ferroelectrics, Greenwille USA, August 30 – September 2; 1992. p344-348.

[61] Alpay SP, Misirlioglu IB, Nagarajan V, Ramesh R. Can interface dislocations degrade ferroelectric properties? Applied Physics Letters 2004;85(11) 2044-2046.

[62] Koval V, Reece MJ, Bushby AJ. Enhanced ferroelectric loop asymmetry of lead zirconate titanate thin films under nanoindentation. Journal of Applied Physics 2007;101(2) 024113.

[63] Koval V, Reece MJ, Bushby AJ. Ferroelectric/ferroelastic behavior and piezoelectric response of lead zirconate titanate thin films under nanoindentation. Journal of Applied Physics 2005;97(7) 074301.

[64] Emelyanov AYu, Pertsev AN, Kholkin AL. Effect of external stress on ferroelectricity in epitaxial thin films. Physical Review B 2002;66(21) 214108.

[65] Kelman MB, McIntyre P, Hendrix BC, Bilodeau SM, Roeder JF. Effect of applied mechanical strain on the ferroelectric and dielectric properties of $Pb(Zr_{0.35}Ti_{0.65})O_3$ thin films. Journal of Applied Physics 2003;93(11) 9231-9236.

[66] Pertsev NA, Emelyanov A Yu. Domain-wall contribution to the piezoelectric response of epitaxial ferroelectric thin films. Applied Physics Letters 1997;71(25) 3646-3648.

[67] Gruverman A, Rodriguez BJ, Kingon AI, Nemanich RJ, Tagantsev AK, Cross JS, Tsukada M. Mechanical stress effect on imprint behavior of integrated ferroelectric capacitors. Applied Physics Letters 2003;83(4) 728-730.

[68] Lu H, Bark CW, Ojos D, Alcala J, Eom CB, Catalan G, Gruverman A. Mechanical writing of ferroelectric polarization. Science 2012;336 59-61.

[69] Salje EKH. Phase transitions in ferroelastic and co-elastic crystals. Cambridge University Press; 1993.

[70] Araki W, Malzbender J. Ferroelastic deformation of $La_{0.58}Sr_{0.4}Co_{0.2}Fe_{0.8}O_{3-\delta}$ under uniaxial compressive loading. Journal of the European Ceramic Society 2013;33(4) 805-812.

[71] Fett T, Munz D, Thun G. Nonsymmetric Deformation behavior of lead zirconate titanate determined in bending tests. Journal of the American Ceramic Society 1998;81(1) 269-272.

[72] Cao H, Evans AG. Nonlinear deformation of ferroelectric ceramics. Journal of the

American Ceramic Society 1993;76(4) 890-896.

[73] Marsilius M, Webber KG, Aulbach E, Granzow T. Comparison of the temperature-dependent ferroelastic behavior of hard and soft lead zirconate titanate ceramics. Journal of the American Ceramic Society 2010;93(9) 2850-2856.

[74] Viola G, Verbylo D, Orlovskaya N, Reece MJ. Effect of composition on rate dependence of ferroelastic/ferroelectric switching in perovskite ceramics. Materials Science and Technology 2009;25(11) 1312-1315.

[75] Lugovy M, Slyunyayev V, Orlovskaya N, Verbylo D, Reece MJ. Room-temperature creep of $LaCoO_3$-based perovskites: Equilibrium strain under compression. Physical Review B 2008;78(2) 024107-9.

[76] Picht G, Webber KG., Zhang Y, Kungl H, Damjanovic D, Hoffmann MJ. Critical mechanical and electrical transition behavior of $BaTiO_3$: The observation of mechanical double loop behaviour. Journal of Applied Physics 2012;112(12) 124101.

[77] Plietsch R. Ehrlich K. Strength differential effect in pseudoelastic NiTi shape memory alloys. Acta Materialia 1997;45(6) 2417-2424.

[78] Xie Z, Liu Y, Humbeeck JV. Microstructure of NiTi shape memory alloy due to tension–compression cyclic deformation. Acta Materialia 1998;46(6) 1989-2000.

[79] Kassner ME, Geantil P, Levine LE, Larson BC. Backstress, the Bauschinger effect and cyclic deformation. Materials Science Forum 2009;604-605 39-51.

[80] Ren XB, Otsuka K. Universal symmetry property of point defects in crystals. Physical Review Letters 2000;85 1016-1019.

[81] Koval V, Skorvanek I, Yan H. Low-temperature magnetic and dielectric anomalies in rare-earth-substituted $BiFeO_3$ ceramics. Journal of the American Ceramic Society 2014;97(12) 3729-3732.

[82] Nogues J, Schuller Ivan K. Exchange bias. Journal of Magnetism and Magnetic Materials 1999;192(2) 203-232.

[83] Meiklejohn WH. Exchange anisotropy-A review. Journal of Applied Physics 1962;33(3) 1328-1335.

[84] Nogues J, Lederman D, Moran TJ, Schuller Ivan K. Positive Exchange bias in $FeF_2$-Fe bilayers. Physical Review Letters 1996;76(24) 4624-4627.

[85] Malozemoff AP. Random-field model of exchange anisotropy at rough ferromagnetic-antiferromagnetic interfaces. Physical Review B 1987;35(7) 3679-3682.

[86] Mauri D, Siegmann HC, Bagus PS, Kay E. Simple model for thin ferromagnetic films exchange coupled to an antiferromagnetic substrate. Journal of Applied Physics 1987;62(7) 3047-3049.

[87] Koon NC. Calculations of exchange bias in thin films with ferromagnetic/antiferromagnetic interfaces. Physical Review Letters. 1997;78(25) 4865-4868.

[88] Girgis E, Portugal RD, Loosvelt H, M. Van Bael J, Gordon I, Malfait M, Temst K, Van Haesendonck C. Enhanced asymmetric magnetization reversal in nanoscale Co/CoO arrays: Competition between exchange bias and magnetostatic coupling. Physical Review Letters 2003;91(18) 187202-4.

[89] Tian ZM, Yuan SL, Yin SY, Liu L, He JH, Duan HN, Li P, Wang CH. Exchange bias effect in a granular system of $NiFe_2O_4$ nanoparticles embedded in an antiferromagnetic NiO matrix. Applied Physics Letters 2008;93(22) 222505-3.

[90] Yuan SJ, Li L, Qi TF, DeLong LE, Cao G. Giant vertical magnetization shift induced by spin canting in a $Co/Ca_2Ru_{0.98}Fe_{0.02}FeO_4$ heterostructure. Physical Review B 2013;88(2) 024413-5.

[91] Nogues J, Leighton C, Schuller Ivan K. Correlation between antiferromagnetic interface coupling and positive exchange bias. Physical Review B 2000;61(2) 1315-1317.

[92] Watanabe H. Thermo-remanent magnetism in the systems $La_2O_3$-$Fe_2O3$ and $Nd_2O_3$-$Fe_2O_3$. Journal of the Physical Society of Japan 1959;14(4) 511-516.

[93] Menyuk N, Dwight K, Wickham DG. Magnetization reversal and asymmetry in cobalt vanadate (IV). Physical Review Letters 1960;4(3) 119-120.

[94] Ohta K, Matsuyama T, Kajiura M, Teranishi T. Disaccommodation of magnetic permeability and induced anisotropy in amorphous Fe-Co alloys. Journal of Magnetism and Magnetic Materials 1980;19(1-3) 165-167.

[95] Kohmoto O, Yamaguchi N, Ohya K, Fujishima H. Change of the magnetic property in some amorphous alloys by low temperature annealing. Japanese Journal of Applied Physics 1978;17(1) 257-258.

[96] Rivas M, García JA, Tejedor M, Bertrán E, Céspedes JG. Influence of the dipolar interactions in the magnetization reversal asymmetry of hard–soft magnetic ribbons. Journal of Applied Physics 2005;97(2) 023903-5.

[97] Skomski R, Zhou J, Kirby R, Sellmyer DJ. Magnetic aging. In: Chipara M, Puglisi O, Skomski R, Jones F, Hsiao B. (eds.) Proceedings of the Materials Research Society 2006, Symposium 887; 133-138.

[98] Moskowitz BM. Magnetic viscosity, diffusion after-effect, and disaccommodation in natural and synthetic samples. Geophysical Journal International 1985;82(2) 143-161.

5

# Perovskite-Based Mesostructures and Related Composites—Influence Exerted by Morphology and Interface

Guilhermina F. Teixeira, Rafael A. Ciola,
Walter K. Sakamoto and Maria A. Zaghete

**Abstract**

The present work seeks to report about the properties of lead-free $NaNbO_3$ particles with different morphology (cubic-like and fiber-like particles) and their application in composites with poly(vinylidenefluoride) (PVDF) polymer. The composites are obtained using a range of varying volume fractions of $NaNbO_3$ particles (30%, 40%, 50% and 60%). The best conditions for obtaining the $NaNbO_3$ particles as well as the composites have been thoroughly studied. It was observed that the highest volume fraction of $NaNbO_3$ particles undermined the flexibility of the composites. The transition percolation phenomenon, commonly known as percolation threshold, was calculated as a function of the dielectric constant and conductivity of the composite. The composites exhibit piezoelectric and ferroelectric properties and both are found to improve by anisotropy of the NaNbO3 particles.

**Keywords:** lead-free composites, piezoelectricity, ferroelectricity, anisotropy, percolation threshold

## 1. Introduction

The ongoing technological developments lead undoubtedly to the quest for anisotropic ceramic material which presents a specific property or a combination of them. To this end, it is necessary to control the growth parameter of nanostructures controlling the morphology of the particles.

Research on unidimensional nanostructures (1D) arouses multidisciplinary interest owing to their properties which allow the use of these nanostructures in self-assembly devices applicable in optical, electronic, photonic and biological systems [1,2]. Currently, researchers have been trying to develop efficient techniques for obtaining 1D ordered particles in nanoscale, controlling the size and shape of the particles since this is the key to improving the use of existing materials while allowing room for new multifunctional devices.

Ferroelectric and piezoelectric materials are much influenced by structural control, because some materials are found to exhibit piezoelectricity only when they are grown with 1D morphology. Another interesting property of 1D material that is worth mentioning is the photoluminescence (PL). Photoluminescence spectroscopy provides us with important information about electronic and optical structures of the materials such as the intermediate states between the valence band (VB) and the conductive band (CB) known as the band gap [3]. When a material combines the photoluminescence and piezoelectric properties, it can be applicable in piezophotonic devices [4].

Lead zirconate titanate (PZT)-based materials present excellent optical and electronic properties; oddly enough though, despite their good properties, materials like PZT are prone to cause environmental damage which may eventually cause the need for their substitution for lead-free materials.

Alkali niobates, such as sodium niobate ($NaNbO_3$), are promising materials for substituting materials like PZT. $NaNbO_3$ is a lead-free piezoelectric perovskite semiconductor with different phase transitions, which has attracted increasing attention among researchers owing to its capacity to form the basis of the class of environmentally friendly materials [5,6].

The efficiency of these materials is said to be directly related to their crystalline and morphological structures. One possibility of obtaining particles with crystalline and morphological well-defined characteristics lies in hydrothermal synthesis, which is a variation of solvothermal synthesis.

The heating in hydrothermal synthesis can be carried out by microwave-assisted technique and when this happens the synthesis is named microwave hydrothermal synthesis. This process, in particular, is truly a low-temperature method for the preparation of monophasic materials of different sizes and shapes. This method is found to save energy and is environmentally friendly in that the reactions take place in closed system conditions. Unlike the conventional heating which requires a long time of reaction, microwave-assisted heating is a greener approach toward the synthesis of materials in a shorter time (several minutes to hours) with lower power consumption as a result of the direct and uniform heating of the contents [7].

Although some ceramics are known to have high ferroelectric and piezoelectric properties, their poor mechanical properties and other properties including the mismatch of the acoustic impedance with water and human tissue are found to restrict their application. On the other hand, ferroelectric polymers possess excellent mechanical properties such as flexibility and deformation, but their piezoelectric activity is low. To circumvent these problems, composite materials made with ferroelectric ceramic and polymer have been investigated as an alternative material which combines the electric properties of ceramic and mechanical properties of polymer [8].

A core parameter to be considered when discussing about composites lies in the connectivity patterns, which establish the arrangement of the phases comprising the composites. The first concept of connectivity was developed by Skinner et al [9] and Newnham et al [10], where the manner in which the individual phases are self-connected are described. There are 10 types of arrangements in which the two components, the matrix phase and the dispersed phase, can be connected forming the composite, ranging from unconnected 0-3 pattern to a 3-3 pattern in which both phases are three-dimensionally self-connected. The first number in the notation represents the dimension of connectivity for the piezoelectric active phase, while the second number refers to the electromechanically inactive polymer phase [11,12]. The composites with 0-3 connectivity possess the merits of being highly flexible and having a relatively high piezoelectric coefficient, though it is very difficult to obtain 0-3 composites with a high ceramic content. The high ceramic content provides a mixed connectivity in 0-3 composites due to the percolation of the particles, besides that the high concentration of big ceramic particles is found to reduce the flexibility of the composite [8]. The 3-3 composites when compared with the 0-3 composites are found to present a relatively higher piezoelectric coefficient [12] though they are quite more rigid and difficult to manufacture than the latter (0-3). Studies conducted on lead-free composites using KNN particles (sodium potassium niobate) and poly(vinylidene-fluoride) (PVDF) as polymer concluded that the piezoelectric activity tends to increase given an increase in the ceramic content. However, a decrease in the polymeric phase of the composite weakens the mechanical properties. The piezoelectric properties of the KNN/PVDF composite were compatible with the PZT-based materials, implying that it is a good option to be used in place of PZT-based materials [13].

A wide array of studies on piezoelectric composites can be found in the literature. Dargahi et al [14] developed a composite to be used in the medical area as a sensor in minimally invasive surgeries without damage to the body. Fuzari Jr et al [15] studied a promising material to use as an acoustic emission sensor. A $NaNbO_3$/PVDF composite used in energy harvest was obtained by Mendoza et al. and Srinivas [16,17], where they compared the $NaNbO_3$/PVDF and lead-based ceramic/PVDF composite in order to demonstrate the feasibility of replacing lead-containing materials in high-energy-density dielectric capacitors.

Despite the range of advantages attributable to composite materials, the poling process still poses a challenge owing to the fact that the effective electric field for polarizing the ceramic particles is much lower compared to the applied electric field [15,18,19]. The poling efficiency of ceramic particles dispersed in a polymer can be enhanced inserting a third phase into the composite. This phase creates an electrical flux path between the ceramics particles, and to promote this effect a conducting polymer, polyaniline (PAni), for instance, can be included in the composite to control the electrical conductivity [15].

In light of that, the main purpose of this study is to discuss the preparation and characterization of $NaNbO_3$ ceramic particles with different morphology obtained by microwave-assisted hydrothermal synthesis. The influence exerted by the synthesis parameter in obtaining the particles with the best characteristics as well as the use of these particles in the fabrication of composite films with 0-3 connectivity, using PVDF as polymer matrix, are all presented and discussed.

## 2. Experimental processing

### 2.1. $NaNbO_3$ synthesis

$NaNbO_3$ powders were obtained by microwave hydrothermal synthesis which was carried out beginning with NaOH (p. a. Quemis) and $Nb_2O_5$ (Alfa Aesar, 99%). In the reaction to form $NaNbO_3$, the NaOH acts as a mineralizing agent and sodium source. The reaction was carried out in a Teflon vessel model XP-1500 (CEM Corp.), in a MARS-5 (CEM Corp.) microwave oven. The precursor suspensions were prepared with 0.696 g of $Nb_2O_5$ added to 30 mL of 8 mol.$L^{-1}$ NaOH water solution. The suspension was then transferred into Teflon vessels which were placed inside a microwave furnace. The synthesis was carried out at 180°C with different conditions as given in Table 1. The obtained powders were thoroughly washed with distilled water by centrifugation and finally dried at room temperature.

| Microwave Power (W) | Synthesis Time (minutes) |
|---|---|
| 300 | 30 |
| | 60 |
| | 120 |
| | 180 |
| | 240 |
| 600 | 30 |
| | 60 |
| 800 | 30 |
| 1000 | 30 |

**Table 1.** Microwave hydrothermal synthesis conditions for obtaining $NaNbO_3$ particles.

### 2.2. Reaction Mechanism to form $Na_2Nb_2O_6.H_2O$ and $NaNbO_3$

The $NbO_7$ decahedra and $NbO_6$ octahedra resulted from the breaking up of the $Nb_2O_5$. The $NbO_7$ and $NbO_6$ are connected by corner forming $Nb_6O_{19}^{8-}$ hexaniobate *Lindqvist* ion, which can provide different niobate compositions. These compositions are related to the different synthesis conditions used to obtain the materials.

In solution, the $Nb_6O_{19}^{8-}$ ions incorporate $Na^+$ ions forming $Na_2Nb_2O_6.nH_2O$ microfibers, followed by structural rearrangement, releasing water molecules forming $NaNbO_3$ structures [20,21,22].

The reaction mechanism between $Nb_2O_5$ and NaOH to form $NaNbO_3$ can be described by the equations below:

$$3\ Nb_2O_5 + 8\ OH^- \rightarrow Nb_6O_{19}^{8-} + 4\ H_2O \quad (1)$$

$$Nb_6O_{19}^{8-} + 34\ OH^- \rightarrow 6\ NbO_6^{7-} + 17\ H_2O \quad (2)$$

$$2\ NbO_6^{7-} + 2\ Na^+ + 7\ H_2O \rightarrow Na_2Nb_2O_6.H_2O + 12\ OH^- \quad (3)$$

$$Na_2Nb_2O_6.H_2O \rightarrow 2\ NaNbO_3 + H_2O \quad (4)$$

### 2.3. $NaNbO_3$ particles characterization

The obtained powders were characterized by X-ray powder diffraction using a rotatory anode diffractometer (Rigaku-RINT 2000) with divergent cleft of 0.25°, soller cleft of 2.5° of divergence in the 2θ range from 20° to 80° with 0.2 graus.min$^{-1}$. The morphology of as-prepared samples was observed using a high-resolution field-emission gun scanning electron microscopy FE-SEM (GERMA JEOL JSM 7500F Field Emission Scanning Electron Microscopy). PL spectra were collected with a Thermal Jarrel-Ash Monospec 27 monochromator and a Hamamatsu R446 photomultiplier. The 350 nm exciting wavelength of a krypton ion laser (Coherent Innova) was used with the nominal output power of the laser kept at 550 mW. All measurements were performed at room temperature.

Figure 1 shows the diffraction powder patterns of X-ray obtained for the powder synthesis carried out at 300 W. Based on the results, it is observed that a pure orthorhombic crystalline structure of $NaNbO_3$ is favored by a rise in the synthesis time period, which is in accordance with the JCPDS file nº 33.1270. For the 30 and 60 minute synthesis times, the obtained product consisted of $Na_2Nb_2O_6.nH_2O$. When the synthesis times are increased to 120 and 180 minutes, the products are a mixture of $Na_2Nb_2O_6.nH_2O$ and orthorhombic $NaNbO_3$ and a diffractogram peak attesting to the decreasing of $Na_2Nb_2O_6.nH_2O$ until it disappears following 240 minutes of synthesis indicating the formation of orthorhombic $NaNbO_3$ pure phase.

The presence of microwave enables the obtaining of $NaNbO_3$ crystalline phase in a short synthesis time and this time can be even smaller when the power of microwave is increased. Figure 2 shows the X-ray diffraction of products obtained using a higher power above 300 W. Using 600 W for 30 minutes, peaks related to $Na_2Nb_2O_6.nH_2O$ were observed and a pure orthorhombic $NaNbO_3$ was seen using 600 W for 60 minutes. By increasing the power to 800 or 1000 W, it is worth noting that the time for obtaining the orthorhombic $NaNbO_3$ pure phase was found to decrease to 30 minutes.

According to Zhu et al [21], $Na_2Nb_2O_6.nH_2O$ is a microporous metastable phase with fiber morphology resulting from $NaNbO_3$ crystallization. The $Na_2Nb_2O_6.nH_2O$ phase presents monoclinic crystalline structure belonging to the $C2/c$ space group with lattice parameters $a$ = 17.0511 Å, $b$ = 5.0293 Å, $c$ = 16.4921 Å e β=113.942 Å and are applied in SOMS, Sandia Octahedral Molecular Sieves [20]. The dehydration of $Na_2Nb_2O_6.nH_2O$ promotes the formation of $NaNbO_3$ perovskite phase which is energetically more stable [20].

Under hydrothermal conditions, $NaNbO_3$ formation occurs as a result of the dissolution of $Nb_2O_5$ in NaOH environment, which is favored by the rise in synthesis time. While the $Nb_2O_5$ is dissolved, $OH^-$ ions are adsorbed on the $Nb_2O_5$ surface, thereby promoting the affinity between Nb=O and O-H bonds forming Nb=O···OH···O=Nb [23].

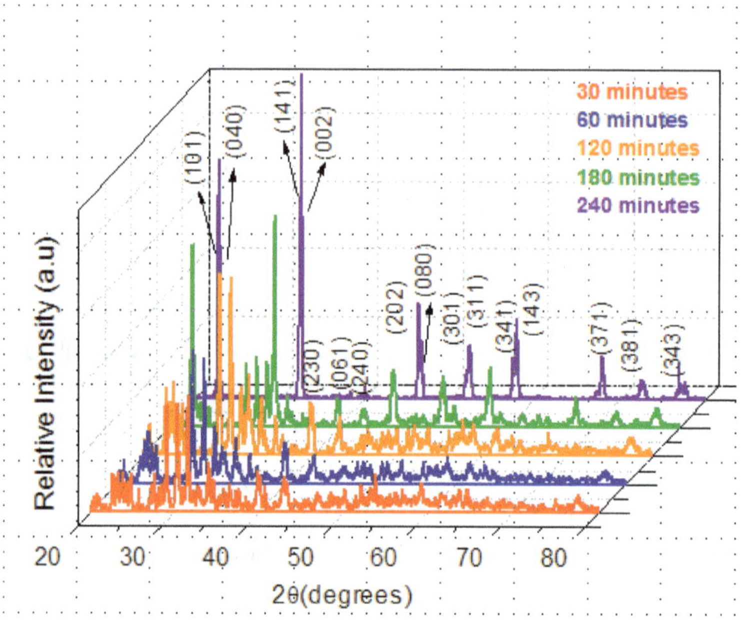


**Figure 1.** XRD patterns of particles obtained by microwave hydrothermal synthesis using 300 W at 180°C in different synthesis time.

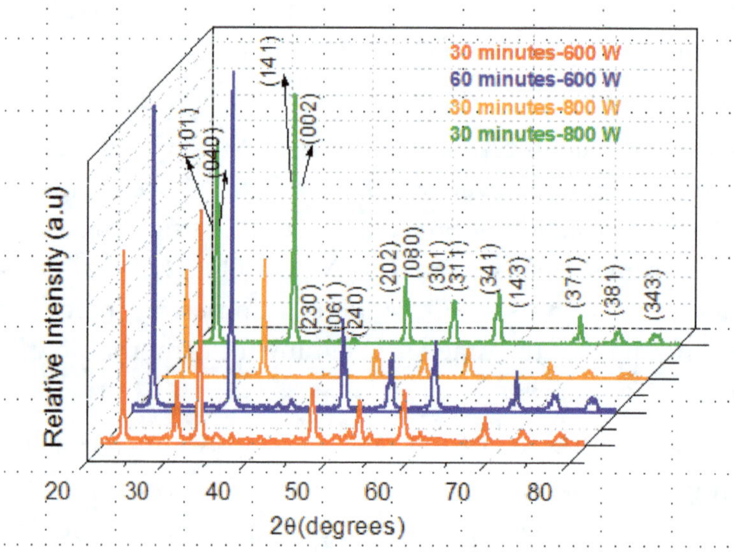


**Figure 2.** XRD patterns of particles obtained by microwave hydrothermal synthesis using 600 W, 800 W and 1000 W at 180°C.

The morphological changes regarding the process are shown in Figure 3, where one can observe that the products presented different morphologies. It is noteworthy that for the 30 and 60 minutes synthesis times with 300 W of power, the samples were composed of $Na_2Nb_2O_6.nH_2O$ structures with fiber-like shape (Figures 3 A–B). For the 120 and 180 minutes synthesis times, there is a mix of fiber and plate-like structures (Figure 3C), and when the synthesis time is increased to 240 minutes, only particles with cube-like morphology are observed (Figure 3D). These results confirm the XRD results (Figure 1) as it is possible to observe that given an increase in the synthesis time, the fiber morphology corresponds to the $Na_2Nb_2O_6.nH_2O$ until the formation of orthorhombic $NaNbO_3$ with a cubic shape [24,25].

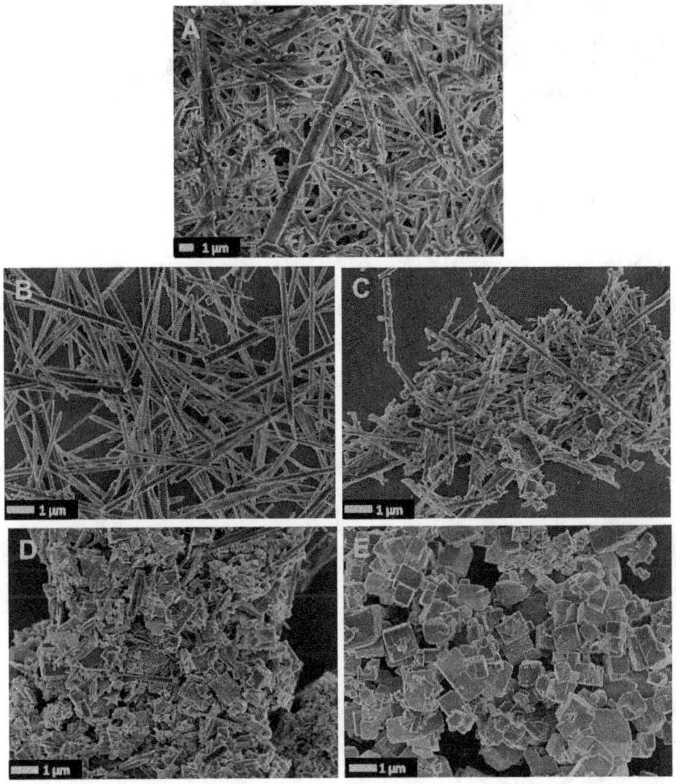


**Figure 3.** FE–SEM of particles obtained by microwave hydrothermal synthesis at 300 W. (A) 30 minutes; (B) 60 minutes; (C) 120 minutes; (D) 180 minutes; (E) 240 minutes.

Using 600 W power, the particles obtained in the shortest time – 30 minutes – also presented fiber morphology, while presenting a cubic morphology after 60 minutes of synthesis (Figure 4 A–B). By increasing the power, the time to obtain the $NaNbO_3$ cubic morphology decreases to 30 minutes, though the particles are clustered as can be seen in Figure 4 C–D.

The morphological evolution of $NaNbO_3$ obtained by hydrothermal synthesis was extensively studied by Zhu et al [21]. NaOH breaks the octahedra corner sharing formed by Nb–O from $Nb_2O_5$ forming intermediaries octahedral edge sharing, which presents low crystallinity. The dissolution of intermediate gives $NbO_6$ which promotes the growth of metastable compound with fiber shape. Finally, increasing the synthesis time, the fibers are dissolved and the $NbO_6$ octahedral takes the form of $NaNbO_3$ with a cubic shape.

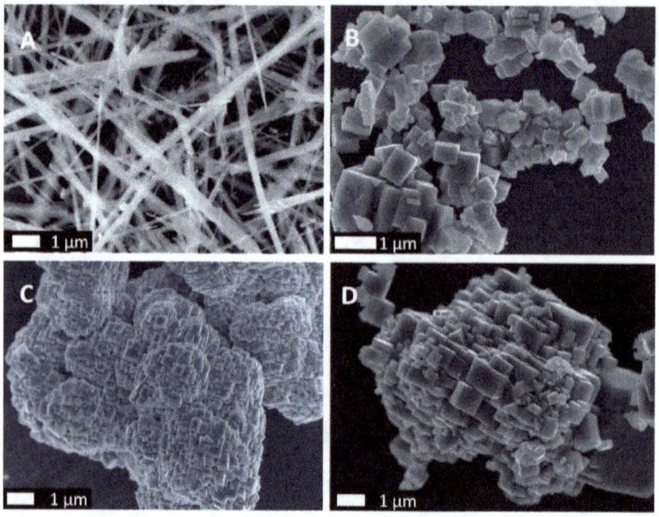


**Figure 4.** FE–SEM of particles obtained by microwave hydrothermal synthesis using different powers.(A) 30 minutes and 600 W; (B) 60 minutes and 600 W; (C) 30 minutes and 800 W; (D) 30 minutes and 1000 W.

By means of the conventional hydrothermal synthesis using the Pluronic P123 copolymer (EO20PO70EO20, BASF, USA), Shi et al [26] obtained orthorhombic $NaNbO_3$ with a fiber shape. The copolymer assists the particles crystallization during the reaction between $Nb_2O_5$ and NaOH at 200°C for 24 hours. Replacing the P123 with PEG-1000 (polyethylene glycol), cubic-like particles with a narrow size distribution is obtained. Ke et al [27] report the change of $NaNbO_3$ morphology in line with the NaOH concentration used in the synthesis. The materials obtained from $NaNbO_3$ with 1D morphology including fibers and wires can present good piezoelectric property [4]. A process used to obtain this morphology is the thermal treatment of $Na_2Nb_2O_5.H_2O$ at temperatures around 300–550° C. The thermal treatment changes the crystalline structure of the material maintaining the 1D shape. Although piezo-electricity is an anisotropic property, the $NaNbO_3$ with cubic morphology hardly presents good piezoelectricity due to the high symmetry of the cubes.

In this work, the thermal treatment of the product was carried out for a period of 4 hours at 550°C by microwave hydrothermal synthesis with 300 W at 180°C for 30 minutes; the meta-stable phase of $Na_2Nb_2O_5.nH_2O$ is transformed into the orthorhombic structure of $NaNbO_3$ maintaining the fiber-like morphology as shown in Figure 5. The thermal treatment promotes the dehydration of $Na_2Nb_2O_6.nH_2O$, forming $NaNbO_3$ perovskite phase, which is found to be energetically more stable than $Na_2Nb_2O_6.nH_2O$.

In a previous work, we reported about the PL properties at room temperature of $NaNbO_3$ particles and film as a function of anisotropic morphology; the $NaNbO_3$ particles in question are fiber and cubic-like morphologically structured, being that the fibers are originated from the thermal treatment of $Na_2Nb_2O_6.nH_2O$ fibers. The particles which presented major PL intensity are $Na_2Nb_2O_6.nH_2O$ fibers prior to thermal treatment, and the smaller PL emission is found to be presented by $NaNbO_3$ fiber-like particles [22].

The order–disorder of the materials including surface defects, cation or anion vacancies, lattice distortions and the recombination process in semiconductors can be evaluated based on the

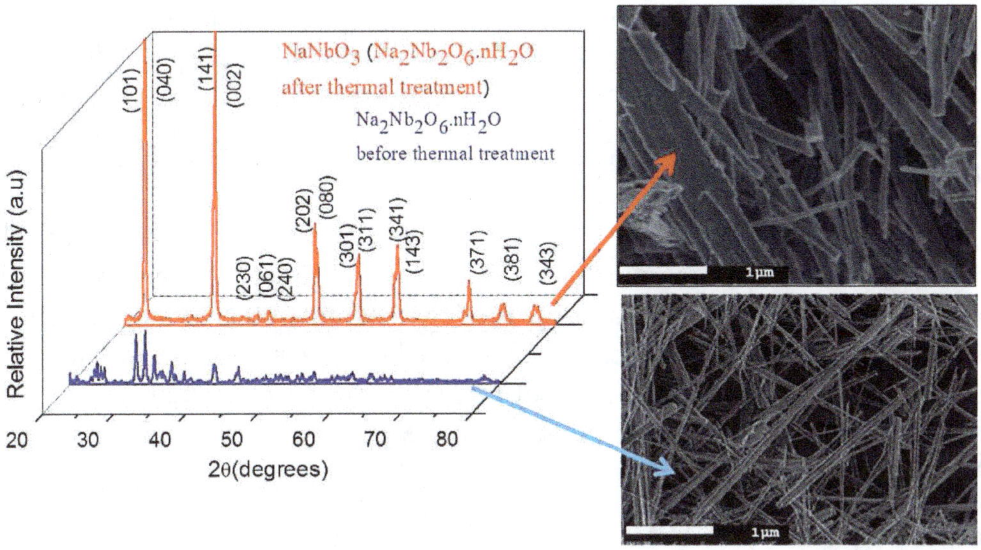


**Figure 5.** XRD patterns and FE–SEM of $Na_2Nb_2O_5.nH_2O$ prior to following thermal treatment at 500°C for 4 hours to form $NaNbO_3$ particles.

PL process. PL is an efficient method for investigating the electronic structure and optical properties of materials. PL properties are a great probe for investigating changes around the site environment at short-range (2–5 Å) and medium-range order (5–20 Å) of clusters where the degree of local order is pronounced [28]. The PL emission region is associated to the defects of the materials. When the material presents deep-level defects, it generates intermediary energy levels near the conduction band which are represented by the PL emission at low energy (yellow/green region of electromagnetic spectrum). Shallow levels generate states next to the valence band where they exhibit PL emission at a higher energy (violet/blue region) [28].

Prior to thermal treatment, the fibers show PL emission around 550 nm, which is then found to decrease following thermal treatment. Furthermore, the band is seen to shift toward a higher energy region. Like the $NaNbO_3$ fiber, the cubic particles also emit higher energy (450 nm) (cf. Figure 4 Ref 22), and this displacement indicates that $NaNbO_3$ crystalline structure is more stable compared to $Na_2Nb_2O_6.nH_2O$ [22].

At low temperature, Almeida et al [29] obtained two PL bands centered at approximately 430 nm and 540 nm in glass-ceramic-containing $NaNbO_3$ crystal. The emission at 430 nm was attributed to the direct exciton recombination, and the low quantum efficiency is likely to indicate that the emission at 540 nm emanates from trap states. At room temperature, the PL band is centered only in the region around 430 nm, next to $NaNbO_3$ PL emission region as we have reported in a previous work [22].

In a previously published paper [22], we proposed that the PL bands can be attributed to the bulk energy levels consistent with their attribution to the transitions between $O^{2-}$ and $Nb^{5+}$, based on the local structure of $NbO_6$ octahedra. By similar structure analyses, $NbO_6$ assumes the form of a regular octahedron, which comes from the strong covalent bonding as a result of the orbital hybridization between Nb **d** and O **2p**. Without the presence of structural defects capable of changing the clusters organization, this crystalline structure is characterized by a high symmetry. However, the order–disordered $NaNbO_3$ presents a symmetry rupture along the O–Nb–O bonds that result in complex clusters with different coordination numbers ($[NbO_6]$–$[NbO_5]$) or distortions in the $[NbO_6]$–$[NbO_6]$ octahedral clusters. The wide-band visible emission observed is a characteristic property of practically all self-activated $ABO_3$ perovskites [22,30].

Some materials present a change in the PL emission according to the morphology [31]. However, as far as our investigation is concerned, the PL results are seen not to be related to the morphology of the $NaNbO_3$ but rather to the organization of the crystalline structure. This observation is backed by the results observed in [22] where the orthorhombic $NaNbO_3$ presents fiber-like and cubic-like particles, where both have PL emission in the same region. The PL emission region observed here is not the same for $Na_2Nb_2O_6.nH_2O$ which presents fiber-like particles to boot [22].

The microwave hydrothermal process is capable of producing $NaNbO_3$ with a more disorganized structure when compared to $NaNbO_3$ obtained by thermal treatment of $Na_2Nb_2O_6.nH_2O$. This can be associated to the heating of $Na_2Nb_2O_6.nH_2O$ so as to yield $NaNbO_3$ with fiber-like morphology which tends to promote a better self-assembly of the crystalline structure than in cubic-like particles. This assembly decreases the structural defects as well as the intermediary levels within the band gap, resulting in a lower PL emission [22].

## 3. Composite processing

### 3.1. Polymer matrix: PVDF

PVDF is a piezoelectric polymer ($d_{33}$=–33 pC/N) that is able to produce variation in the surface charge when subjected to mechanical stress without requiring additional energy sources or electrodes for the generation of electrical signal. This polymer consists of a carbon-based chain with alternating hydrogen and fluorine units ($(–CH_2–CF_2–)n$), and its molecular weight is around 105 $g.mol^{-1}$. PVDF presents polymorphism and can be found in four structural phases ($\alpha$, $\beta$, $\gamma$ and $\delta$) where the piezoelectric phase is said to be the $\beta$ phase once it is a polar phase [8,32].

### 3.2. Obtaining $NaNbO_3$ 1D nanostructures modified by polyaniline (PAni)

PAni (polyaniline) is a conducting polymer in which the electrical conductivity can be modified by the protonation process controlling the reaction pH [15], and its use to cover ceramic particles has the ability to improve the conductivity of particles.

To obtain PAni, the monomer aniline ($C_6H_5NH_2$) (Sigma-Aldrich) was used after vacuum distillation to remove photoxidized molecules. The oxidant ammonium persulfate was employed for the polymerization process of aniline (MERCK).

To obtain $NaNbO_3$ fiber particles coated with PAni, the fibers were incorporated into a solution of aniline, cloridric acid water solution, 1 $mol.L^{-1}$, and ammonium persulfate under stirring at temperature around 2°C for approximately 1 hour. This suspension displayed a green color, a feature of protonated PAni (conductive). The deprotonation process was performed by washing the green particles with ammonium hydroxide water solution, 0.1 mol.L$^{-1}$; in this condition the particles presented a blue color (not conductive). Finally, the particles were protonated again through the use of a solution with pH = 3.7 containing cloridric acid (1 mol.L$^{-1}$) and ammonium hydroxide (0.1 mol.L$^{-1}$). At the end of the process, the particles displayed a green color again (conductivity controlled).

Observing the FE–SEM image depicted in Figure 6, it is possible to note that the covering did not change the morphology of the particles.

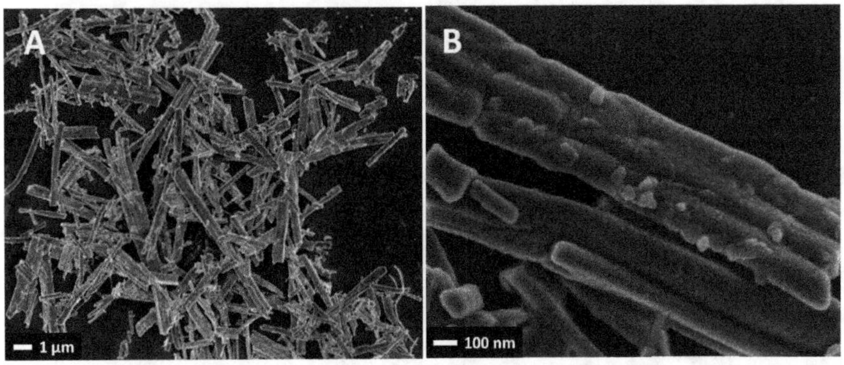


**Figure 6.** FE–SEM image of (A) $NaNbO_3$ fiber-like particles covered with reprotonated PAni and (B) larger image.

### 3.3. Fabrication of composite

$\alpha$-PVDF in powder form was mixed in a mortar with pure $NaNbO_3$ particles with both morphologies (fiber and cubic particles). In addition, the $NaNbO_3$ fiber-like particles coated with PAni were mixed in a mortar with PVDF until a homogeneous mixture was formed. The mixtures were then placed between sheets of Kapton and hot pressed at 190°C for 5 minutes with a pressure of 5 MPa. The thickness of the films is in the range of 190 to 500 µm depending on the ratio of ceramic/polymer content. The composite films from pure $NaNbO_3$ particles were obtained with volumetric fractions of ceramic (30%, 40%, 50% and 60%), while those from $NaNbO_3$ fibers modified with PAni were obtained with 30% and 40% of volume fraction of ceramic. The composites obtained from fiber-like particles without PAni will be denoted by FbNN while the particles coated with PAni will be denoted by PbNN-PAni$_{rep}$, and composites from cubic-like particles will be represented by CbNN.

The volume fraction of ceramic was calculated using the equation 5 [8,33].

$$m_c = \frac{m_p \rho_c}{\rho_p} \frac{\Phi_c}{1-\Phi_c} \tag{5}$$

In the equation $m$ is the mass and $\varrho$ is the density. The subscript $c$ and $p$ are related to ceramic and polymer, respectively. $\Phi_c$ is the volume fraction of ceramic.

Figure 7, 8 and 9 show the FE–SEM images of the composite sample. The distribution of the ceramic particles in the polymer matrix likewise the difference between the samples surface according to the increase in the ceramic particles can be observed. The inhomogeneity of the sample with 60% of ceramic particles led to the lowest flexibility of the composite owing to the high concentration of particles and the change in connectivity.

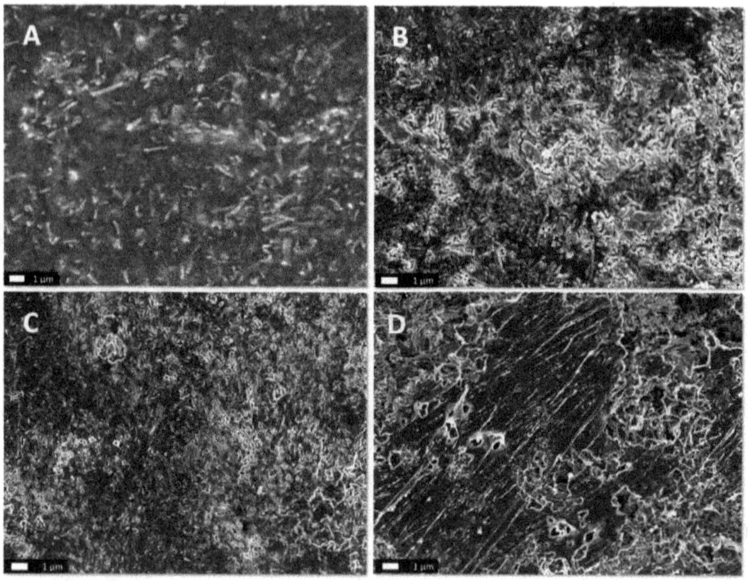

**Figure 7.** FE–SEM of surface of FbNN composites with the respective volume fraction of $NaNbO_3$ particles: (A) 30%; (B) 40%; (C) 50%; (D) 60%.

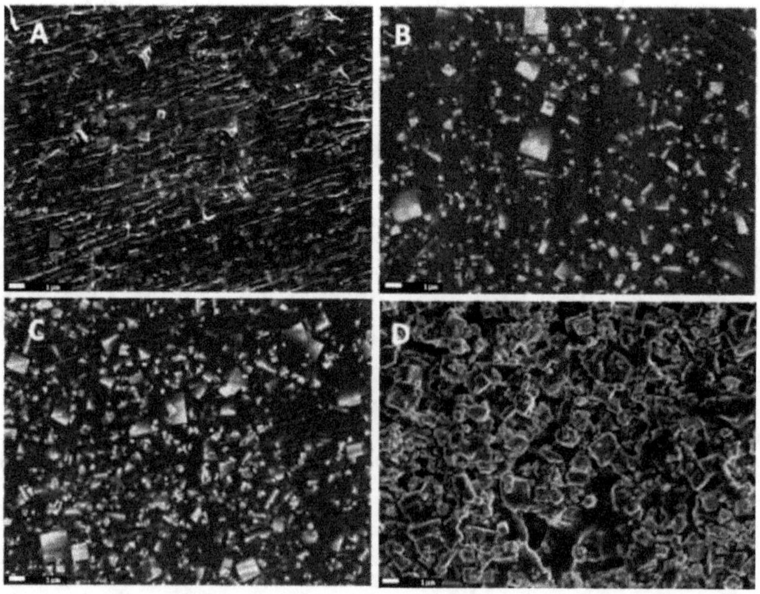

**Figure 8.** FE–SEM of surface of CbNN composites with the respective volume fraction of $NaNbO_3$ particles: (A) 30%; (B) 40%; (C) 50%; (D) 60%.

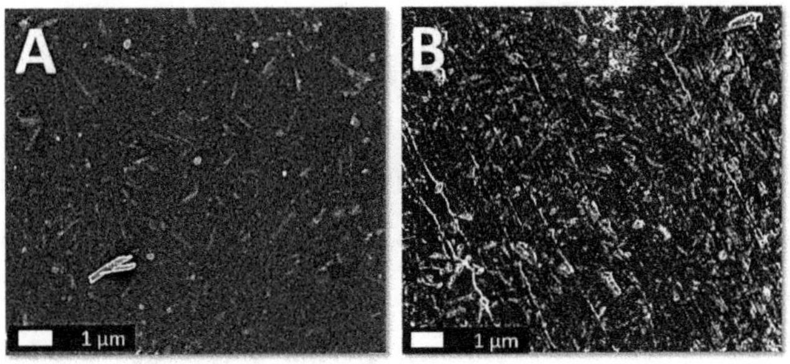


**Figure 9.** FE–SEM of surface of FbNN–PAni$_{rep}$ composites with the respective volume fraction of $NaNbO_3$ particles of: (A) 30%; (B) 40%.

### 3.4. Composite characterization

To carry out the electric measurements, a contact was deposited onto both sides of the samples; gold electrodes with 1.0 cm of diameter were vacuum evaporated. The composite films were poled with an electric field of 5 MV/m at 90°C during 60 minutes in silicone oil.

A TREK high-voltage power supply was used for the poling process. By measuring the longitudinal piezoelectric coefficient $d_{33}$, the piezoelectric activity of the composite was studied. In order to acquire the longitudinal piezoelectric coefficient $d_{33}$ Pennebaker Model 8000Piezo $d_{33}$ Tester was used (American Piezo Ceramics Inc) coupled to a multimeter 34401A (Hewlett Packard).

To avoid problems with lack of uniformity of the composites, the measurements were made at least in 10 different points for each sample, where the average value of these points was taken as the coefficient $d_{33}$.The sample with 50% and 60% of ceramic particles could not be poled because it got ruptured during the polarization process due to the high electric field applied. The improvement in the piezoelectric coefficient is proportional to the anisotropy of the $NaNbO_3$ particles and the volumetric ratio of the ceramic particles was found to scatter in the polymeric matrix. The values of $d_{33}$ obtained for the composites containing $NaNbO_3$ particles are listed in Table 2.

| Sample | Volumetric fraction of ceramic (%) | *d33* (pC/N) |
|---|---|---|
| FbNN | 30 | 2.5 |
| FbNN | 40 | 7.8 |
| CbNN | 30 | 0.9 |
| CbNN | 40 | 1.1 |
| FbNN–PAni$_{rep}$ | 30 | 5.7 |
| FbNN–PAni$_{rep}$ | 40 | 9.0 |

**Table 2.** The values of $d_{33}$ found for the composite with $NaNbO_3$ obtained in different conditions.

Table 3 shows the values of $d_{33}$ piezoelectric constant for some ceramic/polymer composites. It is possible to observe that for the lowest volumetric ratio of ceramic in the composites, the piezoelectric response for the composite with the ceramic particle covered with PAni (FbNN–PAni rep/PVDF composite) was found to be better than almost all the composites with the same volumetric fraction of ceramic. This value is only smaller than that of PZT/PAni/PVDF composite.

| Composites | Volumetric fraction of ceramic (%) | *d33* (pC/N) |
|---|---|---|
| PZT/PHB [34] | 30 | 1.0 to ~3.2* |
| PZT/PHB [34] | 30 | ~1.0 to 4.0** |
| PZT/PAni/PVDF [15] | 30 | 16 |
| PZT/PVDF [35] | 30 | 4.5 |
| BT/PVDF [36] | 30 | 4.2 |
| BT/PVDF [36] | 50 | 5.5 |

*Sample poled at different temperatures; **Sample in different electric field

**Table 3.** The values of $d_{33}$ found for some composites in the literature.

Malmonge et al verified the increase in $d_{33}$ as a consequence of the increase in temperature and electric field used for the poled PHB–PZT composite (PHB = poly-β-hydroxybutyrate), indicating the possible application of the composite in areas related to sensors [34].

Hysteresis loop provides information about the remnant polarization ($P_r$) of the composites. The $Pr$ values were obtained using a Radiant Technologies Inc RT 6000 HVS High Voltage Test System. The measurements were carried out at room temperature using varying electric fields (100 and 600 V) and frequencies (2.5 Hz, 5 Hz, 20 Hz and 40 Hz). The FbNN and CbNN (containing 30% and 60% volumetric fraction of ceramic) and FbNN–$PAni_{rep}$ with 30% are graphically represented by hysteresis loop, Figures 10, 11 and 12.

The increase in voltage applied provides higher values of electric field; and for the samples with a greater thickness, the electric field value is found to decrease. The increase observed in the $P_r$ is proportional to the increase of voltage applied and to the decrease in frequency. Low frequency promotes the increasing of $P_r$ value and the rounding of hysteresis loop to the same conditions. This happens as a result of the longer cycle times and the longer relaxation time for different charge carriers and dipoles. This can be explained by the fact that some charges are able to follow the electric field whereas other charges are not. There are charges that need longer relaxation time. The increase in the volumetric fraction of the ceramic leads to a higher $P_r$ value, though with the demerit of having a lower flexibility of composite. In the FbNN–$PAni_{rep}$, the presence of a conducting polymer is found to improve the $P_r$ value. Another factor that improves the $P_r$ value is the anisotropy of the particles because it is possible to observe that for the FbNN composites all the $P_r$ values are better than the ones for the CbNN composites.

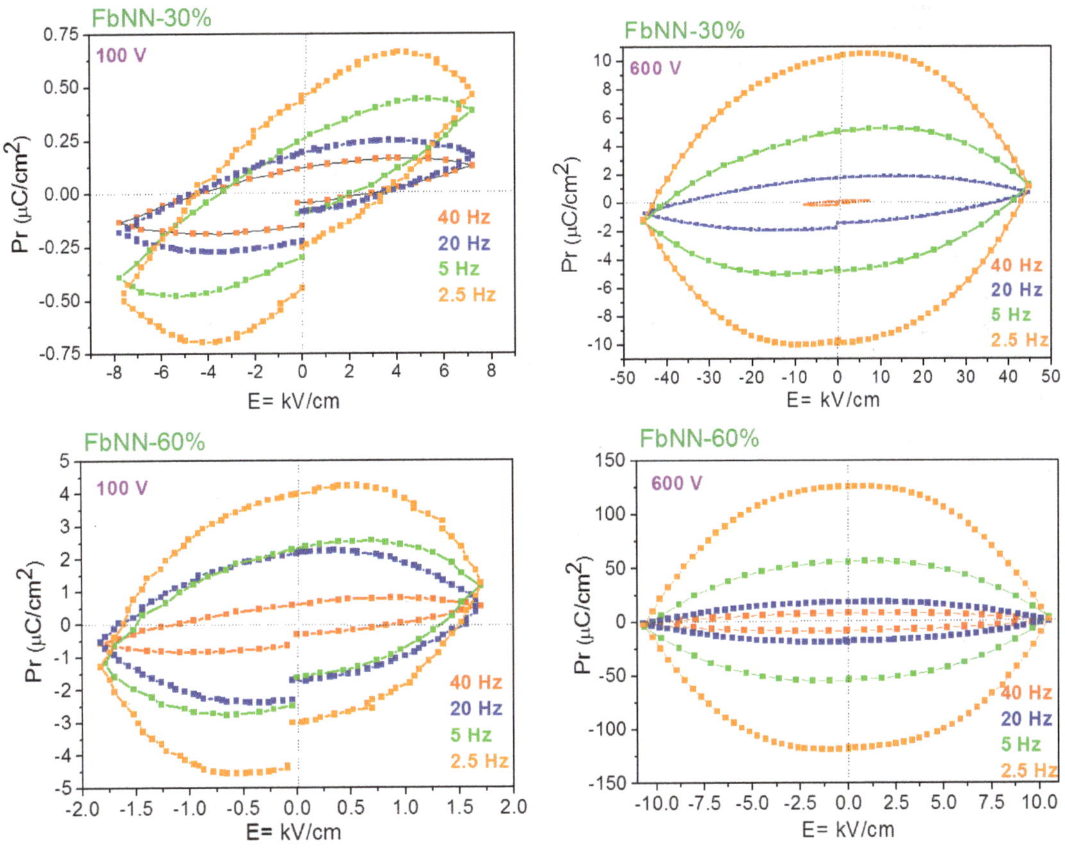


**Figure 10.** *P-E* loops for FbNN composites. (Applied voltage: 100 and 600 V)

The solid materials are classified according to their ease of conducting electrical current. When they have low conductivity in the range of $10^{-14}$ to $10^{-10}$ Ω-m, they are classified as insulators; semiconductors have conductivity between $10^{-9}$ and $10^{-1}$ Ω-m; besides, a conductive material has conductivity higher than $10^{2}$ Ω-m [37].

The characterization of composite materials indicates that conductivity is of essential relevance owing to the fact that there are multimodal microstructures and each phase has intrinsic properties which determine the properties as well as the applications of the composite.

Considering that the union of the different phases occurs only by physical contact of the surfaces, known as interfaces, no chemical bond formation is observed. Therefore, the layers that form the composite (matrix/disperse phase) can be said to retain their distinct conductivity.

The heterogeneity of microstructure consists of single phase regions which are statistically distributed in the volume of composite, and the macroscopic properties of the composite that are directly connected to this distribution [38]. In the composite preparation method and in the volumetric fraction of each phase there are parameters which directly influence the resulting properties. The influence of each phase in the composite characteristics is associated with the intrinsic characteristics. The volume fraction of the disperse phase (one perovskite oxide) and the composites matrix which is an insulating polymer, is dependent on the

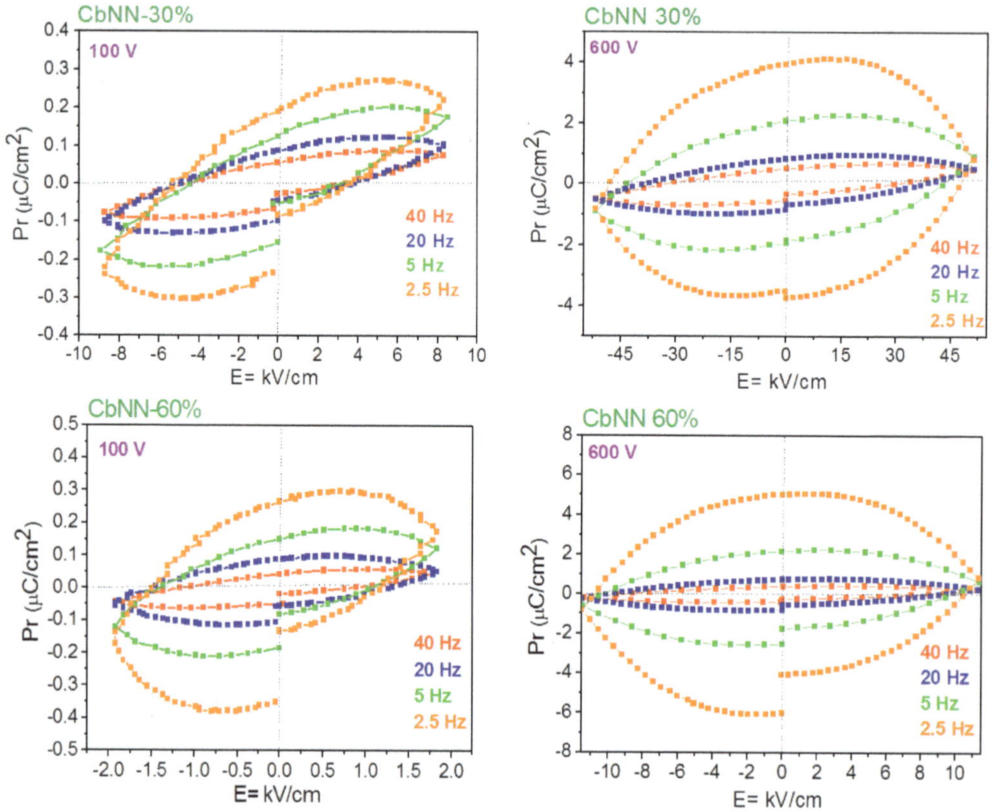


**Figure 11.** *P-E* loops for CbNN composites. (Applied voltage: 100 and 600 V)

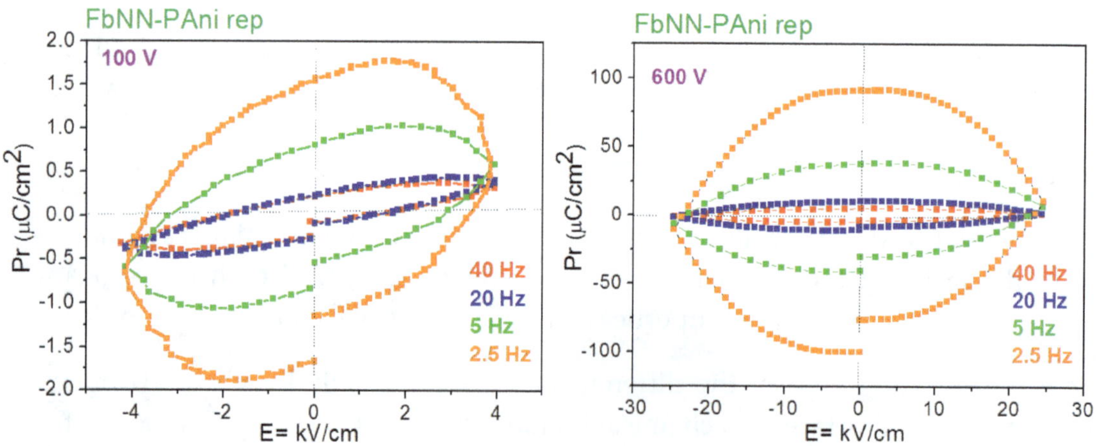


**Figure 12.** *P-E* loops for FbNN–PAni$_{rep}$ composite. (Applied voltage: 100 and 600 V)

embedded oxide fraction in the polymer matrix. Depending on the process, it can be isolated or connected to the polymer. The percolation transition phenomenon, commonly known as the percolation threshold, occurs during the transition from the isolated oxide behavior to the interconnected oxide behavior. The percolation threshold is a mathematical term related to the percolation theory which involves long-range bond formation in a random system. Below the limit, the component or system as a whole is not connected, and over the limit there is a greatness component that is bigger than the system itself and which exerts influence on the overall behavior of the composite. The presence of a metal as a dispersed phase in the polymer matrix paves the way for the occurrence of an insulator/conductor interface transition [39].

Close to the percolation threshold, the value of the electrical conductivity and the dielectric constant of the composite increases abruptly in several orders of magnitude. However, the composite type insulator/semiconductor presents a remarkable increase in the dielectric constant with high values and a relatively low conductivity range, close to the percolation threshold [40].

The actual conductivity of the material is directly proportional to the interaction between the dispersed phase and the matrix phase expressed through the power law (a functional relationship between two quantities, where one varies as potency of the other). Mathematically, the effective conductivity (σ) or the effective dielectric constant (ε) of the composite is described by the following equations [38,40].

$$\sigma_{eff} = \sigma_f \left(f_f - f_c\right)^t, \text{ to } f_f > f_c \quad (6)$$

$$\sigma_{eff} = \sigma_{PVDF} \left(f_c - f_f\right)^{-s}, \text{ to } f_f < f_c \quad (7)$$

$$\varepsilon_{eff} = \varepsilon_f \left(f_f - f_c\right)^{t'}, \text{ to } f_f > f_c \quad (8)$$

$$\varepsilon_{eff} = \varepsilon_{PVDF} \left(f_c - f_f\right)^{-s'}, \text{ to } f_f < f_c \quad (9)$$

where $\mathbf{f_f}$ is the fraction of the disperse phase; $\mathbf{f_c}$ is the percolation threshold; $\boldsymbol{\sigma_f}$ the conductivity of the dispersed phase; $\boldsymbol{\sigma_{PVDF}}$ and phase conductivity matrix (PVDF); $\boldsymbol{\varepsilon_f}$ the dielectric constant of the disperse phase and $\boldsymbol{\varepsilon_{PVDF}}$ the dielectric constant of the matrix phase. The **t** e **s** exponents are the critical values of the conductive region and the insulating region. Some articles that have already been published reported that the percolation threshold is influenced by the size, morphology and constitution of particles that form the dispersed phase [41,42]. The composites analyzed in this study consist of PVDF, as polymer matrix phase and $NaNbO_3$ as dispersed phase. The chemical composition and crystal structure of the dispersed phase are fixed, though we have two distinct particle morphology – cubes and fibers.

To calculate the percolation threshold of the composites, the measurements of electrical conductivity and dielectric constant at 1 kHz for the various rates (matrix / dispersed phase) were carried out.

Figure 13 shows the curves of dielectric constant and conductivity measurements for the CbNN/PVDF composite, where one can observe that there is an abrupt increase in both dielectric constant and conductivity as the ratio is higher than 50% (w/w) for the cubes morphology. Equation 9 jointly with the best linear fit (in the illustrated set) with the value of s = 0.31 +/– 0.08 allow us to indicate that the percolation mechanism (percolation threshold) in this composite and for these processing conditions occur at the rate of 54.6% (w/w). In Figure 13, the threshold value is close to the abrupt increase in conductivity. The composites when subjected to an electric field responded to the field applied accumulating charges on the surface of particles; this charge interferes in the characteristics of the particles–matrix interfaces. The composites formed by the concentrations of morphologically different particles, where the dispersed phase needs different rates (in the matrix phase) so as to occur an interconnection between them. A minimum distance required between the particles prior to the occurrence of percolation is known as interconnection [43]. When the morphology of the particle is in the form of cubes, the distance must be very small in order to promote the contact between them since polarization is homogeneous in the cube–matrix interfaces. Thus, for cubes-like particles a high density of ceramics in the array is required to achieve the interfacial polarization needed to promote the percolation threshold for 54.6% (w/w) of the cubes in the matrix. The variation of dielectric constant and conductivity values prior to the percolation threshold, as can be observed at 40% (w/w), can be linked to the interfacial polarization and the percolation paths in the matrix [44].

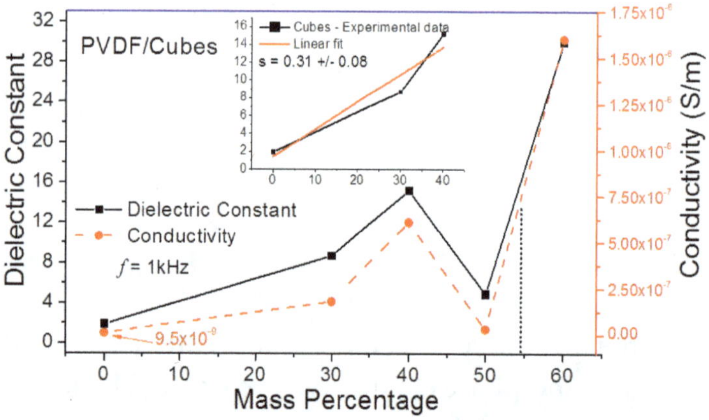


**Figure 13.** Dielectric constant and conductivity for the various fractions mass of cubes dispersed phase in PVDF matrix. Inset for the best linear fit of the percolation threshold.

The dielectric constant and conductivity values for the FbNN/PVDF composite are illustrated in Figure 14; in this case, there is an abrupt increase in conductivity after 30% (w/w) of fibers. Applying the best linear fit (shown in the inset) with s = 0.90 +/– 0.05 in equation 9 gives you a concentration of 38.0% (w/w) for the percolation threshold. Therefore, the morphology of the fibers, nanoparticles 1D, allows the polarization to reach a long distance over the length of the fiber. A lower concentration of particles enables the interconnection between them so the percolation threshold occurs at only 38% (w/w).

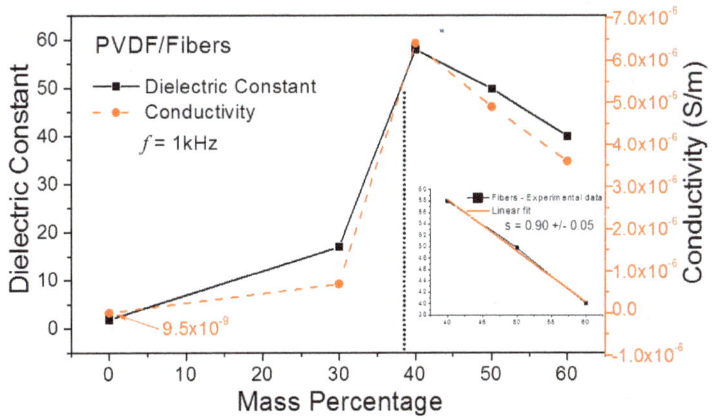


**Figure 14.** Dielectric constant and conductivity for the various fractions mass of fibers dispersed phase in PVDF matrix. Inset for the best linear fit of the percolation threshold.

Figure 15 presents the data obtained by plotting the dielectric constant and conductivity for the composite FbNN–PAni$_{rep}$/PVDF. The abrupt increase in the dielectric constant and conductivity values is observed from the first analyzed concentration of 30% (w/w). Because of this, it is believed that the percolation threshold is set to a lower concentration. Comparing the three types of dispersed phase of this review, this is the one with the lowest value of percolation threshold, which can be explained by the surface characteristics. The particle morphology 1D indicates a behavior similar to that seen in Figure 14; however, the covering of the fibers with polyaniline which promotes conductivity and the connectivity allowing the percolation threshold is reached with lower concentrations of the dispersed phase. Therefore, to establish the percolation threshold of the composites, a study involving the minor fractions dispersed phase is required. There is also an anomalous behavior observed in Figure 15, for the concentration of 50% (w/w) of dispersed particles. At this point, the results of conductivity and dielectric constant are $1.1x10^{-8}$ and 4.0, respectively. These values are close to those of pure PVDF matrix. This anomalous behavior for the concentrations above the percolation threshold has been observed in some papers that have been published already [41,45,46]. It is suggested that there is a formation of agglomerates with dispersed phase which are found to be isolated functioning as microcapacitors, enveloped by an insulating film that prevents both the load passage and a continued polarization of the network, so a decrease in the dielectric constant and conductivity is verified at this concentration.

The dielectric constant and conductivity values for the dispersed phases and for the pure PVDF are plotted in Figure 16. These values were measured in pellet of these materials. The values of electric conductivity are low contrary to the dielectric constant values which are found to be high. The dielectric constant and conductivity increase in the following order PVDF< cubes< fibers< fibers–PAni$_{rep}$, which can be explained by the fact that this polymer is an insulating material with a rigid structure linked by strong covalent bonds, which can generate a polarization though with difficulty resulting in very low dielectric constant and conductivity. Interestingly, the dispersed phase, which are perovskite oxides, have dipoles in their crystalline structure, making them capable of being oriented quite more easily.

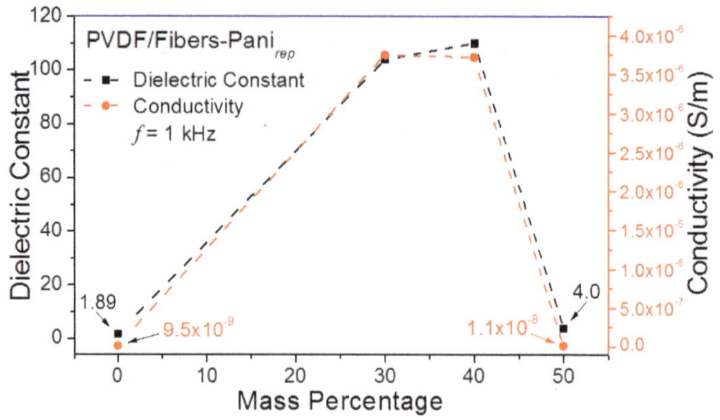


**Figure 15.** Dielectric constant and conductivity for the various fractions mass of fibers–PAni$_{rep}$ dispersed phase in PVDF matrix. Inset for the best linear fit of the percolation threshold.

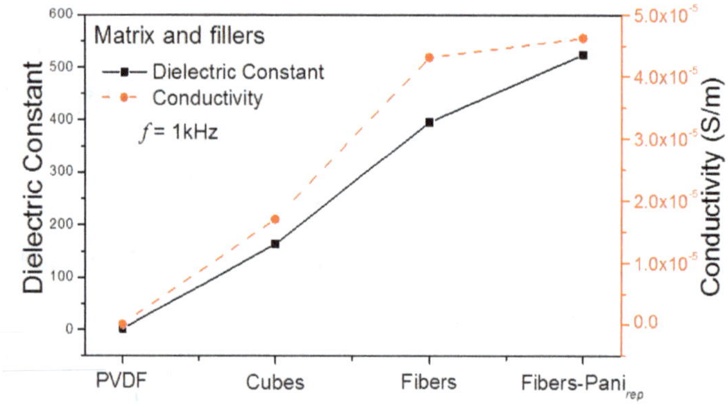


**Figure 16.** Dielectric constant and conductivity for pure PVDF matrix and the dispersed phase, $NaNbO_3$ with different morphology (cubes, fiber and fiber–PAni$_{rep}$).

The morphology of the dispersed phase directly influences this behavior; the fibers compared with cubes as shown in Figure 16 allow for better charge distribution and thus a higher conductivity. The surface-modified fibers with a conductive polymer, in this case PAni, promote conductivity and permittivity (dielectric constant). The values found for the dielectric constant and conductivity decrease in the order fibers–PAni$_{rep}$, fiber, cubes. The same result is observed for the percolation threshold values, when these materials are used as the dispersed phase in the PVDF matrix, indicating that these factors are interrelated.

## 4. Conclusion

By microwave hydrothermal synthesis, it is possible to obtain $Na_2Nb_2O_6.nH_2O$ and $NaNbO_3$ orthorhombic crystalline structure particles. The $Na_2Nb_2O_6.nH_2O$ is obtained in fiber-like morphology while $NaNbO_3$ presented cubic-like morphology. The results are associated to synthesis conditions including synthesis time and microwave power. Using 300 W, the increase in synthesis time favors the formation of $NaNbO_3$; and by 800 and 1000 W the $NaNbO_3$ is obtained within a relatively shorter time. The $Na_2Nb_2O_6.nH_2O$ is used as a precursor to get a fiber-shaped $NaNbO_3$. The thermal treatment of $Na_2Nb_2O_6.nH_2O$ promotes the dehydration and the formation of 1D $NaNbO_3$. The characteristics of the particles exert influence on the material characteristics. Both morphologies of $NaNbO_3$ particles can be mixed with PVDF for the production of flexible composites. An increase in the amount of the ceramics particles leads to the loss of flexibility in the composite owing to the change of connectivity of the ceramic particles and polymer. The anisotropy of ceramic particles improves the composites characteristics, such as $d_{33}$, Pr, conductivity and dielectric constant values. By inserting a third phase into the composite, these values tend to be higher. The characterization of composite materials indicates the importance of conductivity due to the fact that these materials have multimodal microstructure and each phase has intrinsic properties that determine the properties and applications of the composite formed. These preliminary results demonstrate that the $NaNbO_3$ composite can be used with piezoelectric and ferroelectric properties depending on the material.

## Acknowledgements

The authors would like to express their sincerest gratitude and indebtedness to Maximo Siu Li of São Carlos Physical Institute for providing the PL measurements, the LMA-IQ for FEG–SEM facilities and the Brazilian research funding agencies CNPq and FAPESP-CEPID/CDMF 2013/07296-2 for granting the financial support for this research project.

## Author details

Guilhermina F. Teixeira[1], Rafael A. Ciola[1], Walter K. Sakamoto[2] and Maria A. Zaghete[1*]

*Address all correspondence to: zaghete@iq.unesp.br

1 Department of Biochemistry and Chemistry Technology, Chemistry Institute, São Paulo State University – UNESP, Araraquara, Brazil

2 Department of Physics and Chemistry, School of Engineering, São Paulo State University, UNESP, Ilha Solteira, Brazil

## References

[1] Wu J, Xue D. In situ precursor-template route to semi-ordered $NaNbO_3$ nanobelt arrays. Nanosc Res Lett 2011;6(1):14.

[2] Lue P, Xue D. Growth of one-dimensional MnO2 nanostructure. Mod Phys Lett B 2009;23(31):3835–41.

[3] Schroder DK. Semiconductor Materials and Device Characterization. New York: John Wiley & Sons; 1990.

[4] Wang ZL. Piezopotential gated nanowire devices: piezotronics and piezo-phototronics. Nano Today 2010;5(1):540–52.

[5] Chaiyoa N, Muanghlua R, Niemcharoen S, Boonchom B, Vittayakorn N. Solution combustion synthesis and characterization of lead free piezoelectric sodium niobate ($NaNbO_3$) powders. J Alloy Comp 2011;509(5):2445–9.

[6] Shiratori Y, Magrz A, Ficher W, Pithan C, Waser R. Temperature-induced phase transitions in micro, submicro, and nanocrystalline $NaNbO_3$. J Phys Chem C 2007;111(1): 18493–502.

[7] Boris IK, Oxana VK, Ubaldo OM. Microwave hydrothermal and solvothermal processing of materials and compounds. The development and application of microwave heating. Rijeka: InTech, 2011; pp. 107–40. Available from: http://www.intechopen.com/books/the-development-and-application-of-microwave-heating/microwave-hydrothermal-and-solvothermal-processing-of-materials-and-compounds (accessed 14 January 2015).

[8] Sakamoto WK, Fuzari Jr GC, Zaghete MA, Freitas RLB. Lead titanate-based nanocomposite: fabrication, characterization and application and energy conversion evaluation, Ferroelectrics – material aspects, Rijeka: InTech 2011; pp. 251–77. Available from:http://www.intechopen.com/books/ferroelectrics-material-aspects/lead-titanate-based-nanocomposite-fabrication-characterization-and-application-and-energy-conversion (accessed 14 January 2015).

[9] Skinner DP, Newnham RE, Cross LE. Flexible composite transducers. Mat Res Bull 1978;13(6):599–607.

[10] Newnham RE. Composite electroceramics. Ferroelectrics 1986;68(1):1–32.

[11] Akdogan EK, Allahverdi M, Safari A. Piezoelectric composites for sensor and actuator applications. IEEE Trans Ultrason Ferr 2005;52(5):746–75.

[12] Lee HJ, Zhang S, Bar-Cohen Y, Sherrit S. High temperature, high power piezoelectric composite transducer. Sensors (Basel) 2014;14(8):14526–52.

[13] Seol J-H, Lee JS, Ji H-N, Ok Y-P, Kong GP, Kim K-S, Kim CY, Tai W-P. Piezoelectric and dielectric properties of $(K_{0.44}Na_{0.52}Li_{0.04})(Nb_{0.86}Ta_{0.10}Sb_{0.04})O_3$–PVDF composites. Ceram Int 2012;38(1):S263–6.

[14] Dargahi J. An endoscopic and robotic tooth-like compliance and roughness tactile sensor. J Mech Des 2002;124(3):576–82.

[15] Fuzari Jr G de C, Arlindo EPS, Zaghete MA, Longo E, Sakamoto WK. Poled polyaniline coated piezo composite using low electric field and reduced poling time: a functional material. J Mat Sci Eng B. 2014;4(4):109–15.

[16] Mendoza M, Khan MAR, Shuvo MAI, Guerrero A, Lin Y. Development of lead-free nanowire composites for energy storage applications. ISRN Nanomaterials 2012; doi: 10.5402/2012/151748.

[17] Srinivas K. Possible lead-free nanocomposite polymer dielectrics for high energy storage applications. Int J Adv Eng Nano Technol (IJAENT) 2014;1(9):9–14.

[18] Furukawa T, Ishida K, Fukada E, Piezoelectric properties in the composite systems of polymers and PZT ceramics. J Appl Phys 1979;50(7):4904–12.

[19] Liu XF, Xiong CX, Sun HJ, Dong LJ, Li R, Liu Y. Piezoelectric and dielectric properties of PZT/PVC and graphite doped with PZT/PVC. Compos Mater Sci Eng B Solid State Mater Adv Technol 2006;127(2–3):261–6.

[20] Xu H, Nyman M, Nenoff TM, Navrotsky A. Prototype sandia octahedral molecular sieve (SOMS) $Na_2Nb_2O_6.H_2O$: synthesis, structure and thermodynamic stability. Chem Mater 2004;16(10):2034–40.

[21] Zhu H, Zheng Z, Gao X, Huang Y, Yan Z, Zou J, Yin H, Zou Q, Kable SH, Zhao J, Xi Y, Martens WN, Frost RL. Structural evolution in a hydrothermal reaction between $Nb_2O_5$ and NaOH solution: from $Nb_2O_5$ grains to micropourous $Na_2Nb_2O_{6\cdot 2/3}H\ _2O$ fibers and $NaNbO_3$ cubes. J Am Chem Soc 2006;128(7):2373–84.

[22] Teixeira GF, Wright TR, Manfroi DC, Longo E, Varela JA, Zaghete MA. Photoluminescence in $NaNbO_3$ particles and films. Mater Lett 2015;139(1):443–6.

[23] Wu SY, Liu XQ, Chen XM. Hydrothermal synthesis of $NaNbO_3$ with low NaOH concentration. Ceram Int 2010;36(1):871–7.

[24] Yu A, Liu JQL, Pan H, Zhou X. Surface sprouting growth of $Na_2Nb_2O_6$ $H_2O$ nanowires and fabrication of $NaNbO_3$ nanostructures with controlled morphologies. Appl Surf Sci 2012;258(8):3490–6.

[25] Teixeira GF, Zaghete MA, Varela JA, Longo E. Synthesis and characterization of $NaNbO_3$ mesostructure by a microwave-assisted hydrothermal method. MRS Proc 2014;1675:145–50.

[26] Shi H, Li X, Wang D, Yuan Y, Zou Z, Ye J. $NaNbO_3$ nanostructures: facile synthesis, characterization and their photocatalytic properties. Catal Lett 2009;132(1):205–12.

[27] Ke T-Y, Chen H-A, Sheu H-S, Yeh J-W, Lin H-N, Lee C-Y, Chiu H-T. Sodium niobate nanowire and its piezoelectricity. J Phys Chem C 2008;112(26):8827–31.

[28] Junior ES, La Porta FA, Liu M, Andrés J, Varela JA, Longo E. A relationship among structural, electronic order-disorder effects and optical properties in crystalline $TiO_2$ nanomaterials. Dalton Trans 2015;44(7):3159–75.

[29] Almeida E, Menezes L de S, Araújo CB de, Lipovskii AA. Luminescence properties and optical dephasing in a glass-ceramic containing sodium-niobate nanocrystals. J Appl Phys 2011;109(11):113108–5.

[30] Teixeira GF, Zaghete MA, Gasparotto G, Costa MGS, Espinosa JWM, Longo E, Varela JA. Photoluminescence properties and synthesis of a PZT mesostructures obtained by the microwave-assisted hydrothermal method. J Alloy Comp 2012;512(1):124–7.

[31] Teixeira GF, Gasparotto G, Paris EC, Zaghete MA, Longo E, Varela JA. Photoluminescence properties of PZT 52/48 synthesized by microwave hydrothermal method using PVA with template. J Lumin 2012;132(1):46–50.

[32] Ribeiro C, Correia DM, Ribeiro S, Sencadas V, Botelho G, Lanceros-Méndez S. Piezoelectric poly(vinylidene fluoride) microstructure and poling state in active tissue engineering. Eng Life Sci 2014;00:1–6.

[33] Marin-Franch P, Martin T, Tunnicliffe DL, Das-Gupta DK. PTCa/PEKK piezo-composites for acoustic emission detection. Sens Actuators A: Physical 2002;99(3):236–43.

[34] Malmonge JA, Malmonge LF, Fuzari Jr GC, Malmonge SM, Sakamoto W.K. Piezo and dielectric properties of PHB–PZT composite. Polym Compos 2009;30(9):1333–7.

[35] Oliveira CA. Otimização do processamento para a obtenção de compósitos polímero/cerâmica com propriedades piezo e piroelétricas. Masters dissertation. UNESP; 2012.

[36] Macario LR. Preparo e caracterização de compósitos polímero/cerâmica com potencial aplicação como sensor multifuncional. PhD thesis. UNESP; 2013.

[37] Pang H, Xu L, Yan DX, Li ZM. Conductive polymer composites with segregated structures. Prog Polym Sci 2014;39(11):1908–33.

[38] Nan C-W. Physics of inhomogeneous inorganic materials. Prog Mater Sci 1993;37(1): 1–116.

[39] Shante VKS, Kirkpatrick S. An introduction to percolation theory. Adv Phys 1971;20(85):325–7.

[40] Dang ZM, Nan CW, Xie D, Zhang YH, Tjong S.C. Dielectric behavior and dependence of percolation threshold on the conductivity of fillers in polymer-semiconductor composites. Appl Phys Lett 2004;85(1):97–9.

[41] Deepa KS, Gopika MS, James J. Influence of matrix conductivity and Coulomb blockade effect on the percolation threshold of insulator-conductor composites. Compos Sci Technol 2013;78(1):18–23.

[42] Deepa KS, Sebastian MT, James J. Effect of interparticle distance and interfacial area on the properties of insulator-conductor composites. Appl Phys Lett 2007:91:202904.

[43] Hong JL, Schadler LS, Siegel RW. Rescaled electrical properties of ZnO/low density polyethylene nanocomposites. Appl Phys Lett 2003;82(12):1956–8.

[44] Tsangaris GM, Psarras GC, Kouloumbi N. Electric modulus and interfacial polarization in composite polymeric systems. J Mater Sci 1998;33(8):2027–37.

[45] Fan P, Wang L, Yang J, Chen F, Zhong M. Graphene/poly(vinylidene fluoride) composites with high dielectric constant and low percolation threshold. Nanotechnol 2012;23(36):365702.

[46] Li Q, Xue Q, Hao L, Gao X, Zheng Q. Large dielectric constant of the chemically functionalized carbon nanotube/polymer composites. Compos Sci Technol 2008;68(10–11):2290–6.

6

# Raman Spectra of Soft Modes in Ferroelectric Crystals

Takeshi Shigenari

**Abstract**

The interpretation of Raman spectra of soft modes in ferroelectric crystals is not easy because of its over-damped line shape and the complicated temperature dependence near the transition temperature. There exist not a few cases which are difficult to determine the transition type is either order-disorder or displacive type. The main purpose of this paper is to point out the characteristics of Raman spectra of polar modes and several cautions for the analysis of soft modes. We discuss on some general properties of the susceptibility functions and also show possible reasons of confusion which might lead to the incorrect conclusions in the specific cases of $KH_2PO_4$ (KDP), ferroelectric $SrTiO_3$ and proton–ordered Ice crystals.

**Keywords:** Soft mode, Raman spectroscopy, Phase transition, Polar mode

## 1. Introduction

Historically, phase transitions in ferroelectric crystals have been explained by the *order–disorder type* mechanism that is similar to magnetic crystals. In the1950s, however, the idea of soft mode for *displacive-type* structural phase transitions was proposed by Anderson and others [1]. Since the1960s, with the advent of laser technology, laser Raman spectroscopy became one of the most powerful experimental methods for the study of soft mode particularly in ferroelectric crystals, partly because soft modes below $T_c$ are always Raman-active [2]. An excellent review of soft mode spectroscopy in the early stage was given by Scott [3]. A theoretical textbook, *Scattering of Light by Crystals*, by Hayes and Loudon was published in 1978 [4].

A displacive-type transition is induced by the softening (decreasing frequency) of a particular lattice vibrational mode (phonon) toward the transition temperature $T_c$ and eventually the freezing of the mode pattern changes the symmetry of the crystal. The interpretation of a soft mode, however, is not easy because its Raman spectrum is often observed without any definite peak and becomes a broad quasi-elastic peak near $T_c$. Therefore, interpretations of Raman spectra of polar modes, such as soft modes in ferroelectric crystals must be carefully done. Otherwise, incorrect conclusions might be obtained. In the following sections, several examples related to this problem are shown for the cases of KDP, ferroelectric $SrTiO_3$, and proton-ordered Ice crystals.

## 2. Characteristics of polar phonons

### 2.1. General properties of Polar phonons

If the displacement of a phonon $Q$ produces an electric dipole moment $\vec{\mu} = e^* Q$ ($e^*$: effective charge), the mode is called a polar mode or polar phonon. A soft mode is not always a polar mode, for example, the soft mode in the $\alpha - \beta$ transition of quartz ($SiO_2$) is nonpolar. However, for ferroelectric transitions it is always a polar mode.

Due to its interaction with any kind of electric field, the mechanism of the scattering of light from polar modes is much more complicated than nonpolar modes [4]. The Raman spectrum of polar phonons has several characteristics as follows (Fig.1):

i. Frequency of a polar phonon $Q$ depends on the direction of its propagation vector $\vec{K}_p$ due to the depolarization field $\overleftarrow{E}_d \propto -\vec{\mu}/\varepsilon$. If $\vec{K}_p$ is perpendicular to $\vec{\mu}$, it is a transverse optic phonon (TO) and if $\vec{K}_p$ is parallel to $\vec{\mu}$, it is a longitudinal optic phonon (LO). Generally speaking, the frequency of LO is higher than TO because of the depolarization field. If the polar nature of $Q$ is strong (and much stronger than the anisotropy of the crystal), large LO/TO splitting is expected to take place (Fig.1).

ii. As shown in Fig. 1(b), TO interacts with infrared (IR) photon and propagates as the coupled mode "polaritons". In Raman spectra, polaritons can be observed only with a very small angle forward-scattering geometry [5].

iii. In the IR spectra, reflectivity between LO and TO frequency is high and almost flat, while in Raman spectra they are observed as separate peaks (Fig. 1(d, e)).

### 2.2. General properties of soft modes

i. Above $T_c$, the soft mode is the lowest frequency TO phonon. Below $T_c$, the soft mode becomes a totally symmetric mode ($A_1$). Therefore, it is Raman-active. In typical cases, its frequency follows the equation

$$\omega_{TO}^2(q=0) = a\left|T - T_c\right| \tag{1}$$

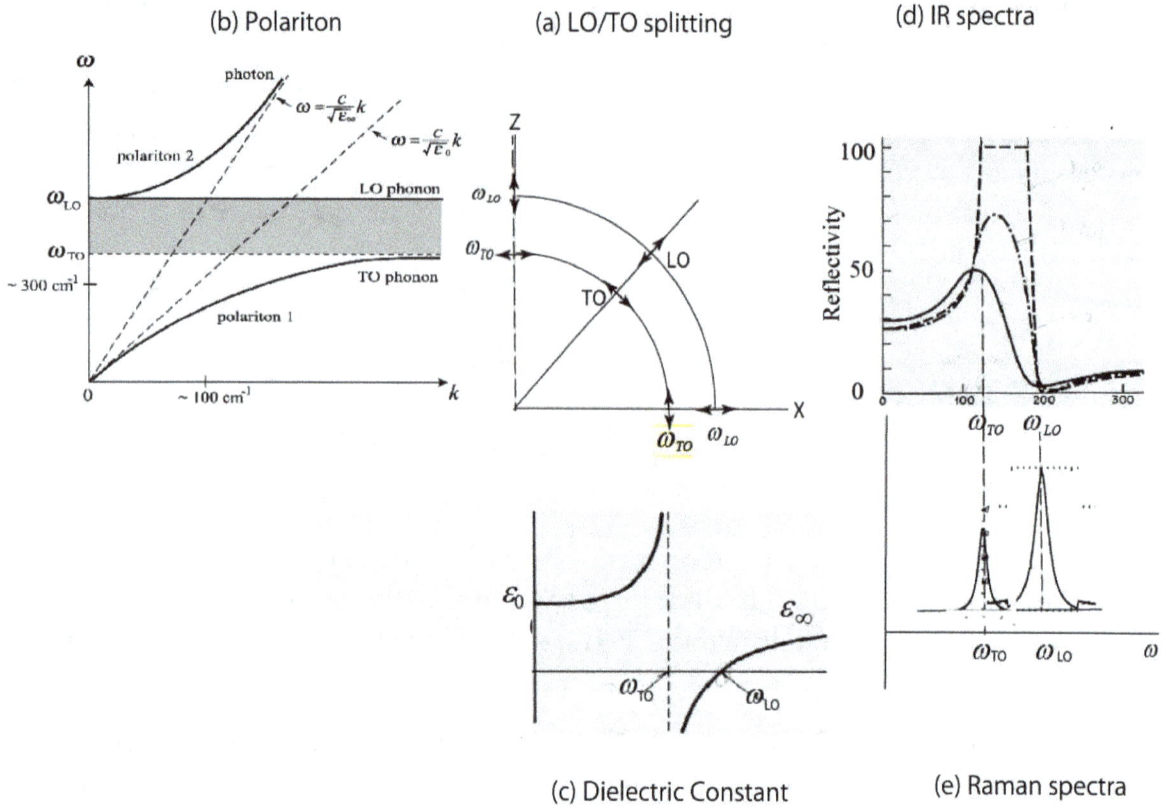


**Figure 1.** Various properties of polar LO and TO phonon in a cubic crystal. (a) LO/TO splitting in space, (b) dispersion of polariton, (c) dielectric constant, (d) IR reflection spectra, and (e) Raman spectrum.

**ii.** If the soft mode above $T_c$ is degenerate, the degeneracy is lifted below $T_c$ and one of them is totally symmetric ($A_1$) but $A_1$ is not necessarily the lowest frequency mode as in the $SrTiO_3$-18 case in section 5.2. Other split modes may or may not be Raman-active [6].

**iii.** Raman spectra of soft modes in the ferroelectric phase ($T < T_c$) depends on the angle between the spontaneous polarization $\vec{P}$ and the phonon propagation direction $\vec{K}_p$ because of the effect of the depolarization field $\overleftarrow{E}_d \propto -\vec{P}/\varepsilon$ as in the $SrTiO_3$ (section 5.2).

**iv.** Soft mode frequencies $\omega_{TO}$ and $\omega_{LO}$ are related to the dielectric constant via LST (Lyddayne–Sacks–Teller) relation.

$$(\omega_{TO}/\omega_{LO})^2 = \varepsilon_\infty/\varepsilon_0 \tag{2}$$

**iv.** Integrated intensity of the soft mode Raman spectra is related to the *real part* of the low frequency dielectric constant $\varepsilon(\omega)$ via Kramers–Krong relation and Eq. (4) in the next section,

$$\operatorname{Re}\varepsilon(\omega)-\varepsilon_{\infty}=\frac{2}{\pi}P\int_{0}^{\infty}\frac{\omega'}{\omega'^{2}-\omega^{2}}\left(\frac{I(\omega')}{n(\omega')+1}\right)d\omega' \quad (3)$$

An example of this relation was demonstrated in the case of KDP [7].

# 3. Susceptibility $\chi_Q(\omega, T)$ for soft mode

## 3.1. Intensity and line shape of Raman spectra

Temperature and frequency dependence of Raman spectrum is represented by the imaginary part of the susceptibility $\chi_Q(\omega, T)$, which is the response of a phonon $Q$ to its conjugate force $f_Q$.

$$I_S(\omega,T)\propto\left[n(\omega,T)+1\right]\operatorname{Im}\chi_Q(\omega,T) \quad (4)$$

or

$$\operatorname{Im}\chi_Q(\omega,T)=\frac{I_S(\omega,T)}{n(\omega,T)+1}=\frac{I_{AS}(\omega,T)}{n(\omega,T)} \quad (5)$$

where $I_S(\omega, T)$ and $I_{AS}(\omega, T)$ are intensity of Stokes and anti-Stokes side spectrum, respectively, and $n(\omega, T)=[\exp(\hbar\omega/k_BT)-1]^{-1}$ is the Bose factor. It should be noted that the approximations, $\hbar\omega/k_BT>>1$ or $\hbar\omega/k_BT<<1$, cannot be used for the soft mode.

In most cases, spectra are analyzed using either the so-called Damped Harmonic Oscillator (DHO) model for displacive-type transitions

$$\chi_Q(\omega)=\chi_Q(0)\frac{\omega_0^2}{\omega_0^2-\omega^2-i\omega\gamma} \quad (6)$$

or the Debye model (DB) for the relaxational behavior of order–disorder-type transitions,

$$\chi_Q(\omega)=\chi_Q(0)\frac{\gamma'}{\gamma'-i\omega} \quad (7)$$

where in ferroelectrics $\chi_Q(0)\approx C/(T-T_c)$ is proportional to the Curie constant $C$.

It seems rather strange, however, why these two formulas are exclusively used and what the meaning of the damping constant $\gamma$ and $\gamma'$ is. Since soft modes are often observed without any apparent peaks, it is important to recognize the difference between the various forms of the susceptibilities. Thus, we discuss the more general formula of $\chi(\omega)$.

### 3.2. Generalized susceptibility and the stability limit

Phase transitions occur when the system becomes unstable. In other words, as $T \to T_c$, at least one of the poles of the $\chi(\omega)$ approaches the origin in the complex $\omega$ plane. Typical traces are given in Fig. 2 for DHO, DB, and a more general case.

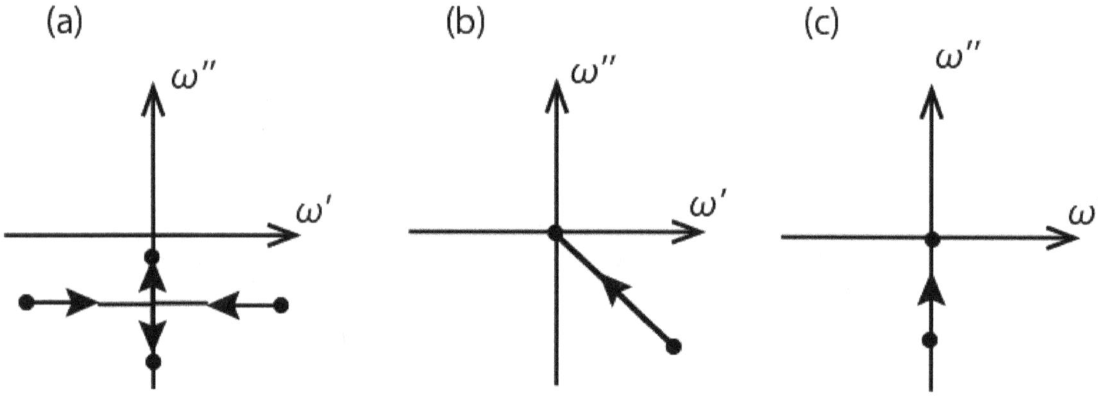


**Figure 2.** Motion of the pole of the susceptibility in the complex $\omega$-plane as stability limit is approached. (a) Damped harmonic oscillator, (b) general case, (c) relaxation type (Debye).

As Van Vleck and Weisskopf showed in 1945 [8] for the case of microwave spectroscopy of gas, DHO and Debye types are qualitatively different in nature and DB cannot be obtained as a limit of vanishing characteristic frequency $\omega_0$. It is because the $\gamma'$ in DB (Eq. (7)) means the inverse relaxation time $\tau$ of the order parameter, which is a stochastic variable. In contrast, the $\gamma$ in DHO (Eq. (6)) is a dynamical variable representing the damping proportional to the velocity of the order parameter. Nevertheless, in the field of Raman spectra, especially for an overdamped soft mode associated with structural phase transitions, it is often misleadingly stated that DHO smoothly reduced to DB by decreasing the ratio $\omega_0^2/\gamma \to 0$.

We note that the qualitative difference between DHO and DB originates from the fact that in DB the system is described for a finite time interval by neglecting the instantaneous change of the system. This "coarseness in the time domain" means that one cannot apply the Debye formula to the high frequency part of spectra. Therefore, the criteria, "high frequency tail of Raman spectra is proportional to $\omega^{-3}$ for DHO and $\omega^{-1}$ for DB" cannot be used to discriminate the displacive-type transitions from the order–disorder type.

Since a real system would not be represented by a single damping mechanism, we have proposed a more general form of susceptibility (GVWF) which includes two damping constants [9]. It is regarded as a generalization of the Van Vleck, Weisskopf, and Froehlich (VWF)-type susceptibility ($\gamma = \gamma'$).

$$\chi(\omega)=\chi(0)\frac{\omega_0^2+\gamma\gamma'-i\gamma'\omega}{\omega_0^2+\gamma\gamma'-i\left(\gamma+\gamma'\right)\omega-\omega^2} \tag{8}$$

In general, this form can be described by the two poles $\Omega_{1,2}$ in the complex $\omega$ plane as follows:

$$\chi(\omega)=\frac{\chi(0)}{2}\left\{\frac{\Pi_1}{\Omega_1-\omega}+\frac{\Pi_2}{\Omega_2-\omega}\right\} \tag{9}$$

where

$$\begin{aligned}\Omega_{1,2}&=-\frac{i\left(\gamma+\gamma'\right)}{2}\pm\sqrt{\omega_0^2-\left(\gamma-\gamma'\right)^2/4}, && \text{a}\\ \Pi_{1,2}&=-i\gamma'\pm\left[\omega_0^2+\gamma'(\gamma-\gamma')/2\right]\frac{1}{\sqrt{\omega_0^2-\left(\gamma-\gamma'\right)^2/4}} && \text{b}\end{aligned} \tag{10}$$

The behavior of the poles in the complex $\omega$ plane is shown in Fig. 3 for the case of the softening ($\omega_0 \to 0$) with constant $\gamma$ and $\gamma'$.

**a.** If $\gamma >> \gamma' = 0$, GVWF corresponds to DHO and two poles move parallel to $\omega'$ axis and reach the imaginary axis when $\omega_0 = \gamma/2$ and they separately move on the $\omega''$ axis, one toward the origin and another toward $-i\gamma$. In other words, if $\omega_0 < \gamma/2$, DHO cannot be discriminated from DB.

**b.** If $\gamma > \gamma' \neq 0$ (GVWF), two poles behave similar to the case (a) but even for $\omega_0 \to 0$ neither of them can reach the origin. They stop at $-i\gamma$ and $-i\gamma'$. The "intensity" of the two poles, $\Pi_{1,2}$, also changes; a mode toward $-i\gamma'$ increases, while the other pole gradually vanishes.

**c.** If $\gamma = \gamma'$ (VWF), two poles reach $\omega''$ axis only when $\omega_0 = 0$.

**d.** If $\gamma < \gamma'$, behaviors of two poles are similar to the case (b) except that the "stronger" pole $-i\gamma'$ lies further from the origin than the weaker pole $-i\gamma$.

Behavior of various $\chi(\omega)$ can be understood in terms of the three parameters ($\omega_0$, $\gamma$, $\gamma'$) of GVWF. Figure 4 represents schematically the paths of the pole on approaching $T_c$. There are many paths leading to the divergence of $\chi(\omega)$. The stability limit leading to the phase transition takes place when one of the poles moves toward the origin of complex $\omega$ plane ($\gamma$ axis in Fig.4).

The pole of DHO moves in the DHO-plane ($\omega_0$, $\gamma$, $\gamma'=0$) as an ideal softening in displacive-type transitions. Ideal relaxation in order–disorder-type transitions (DB) corresponds to

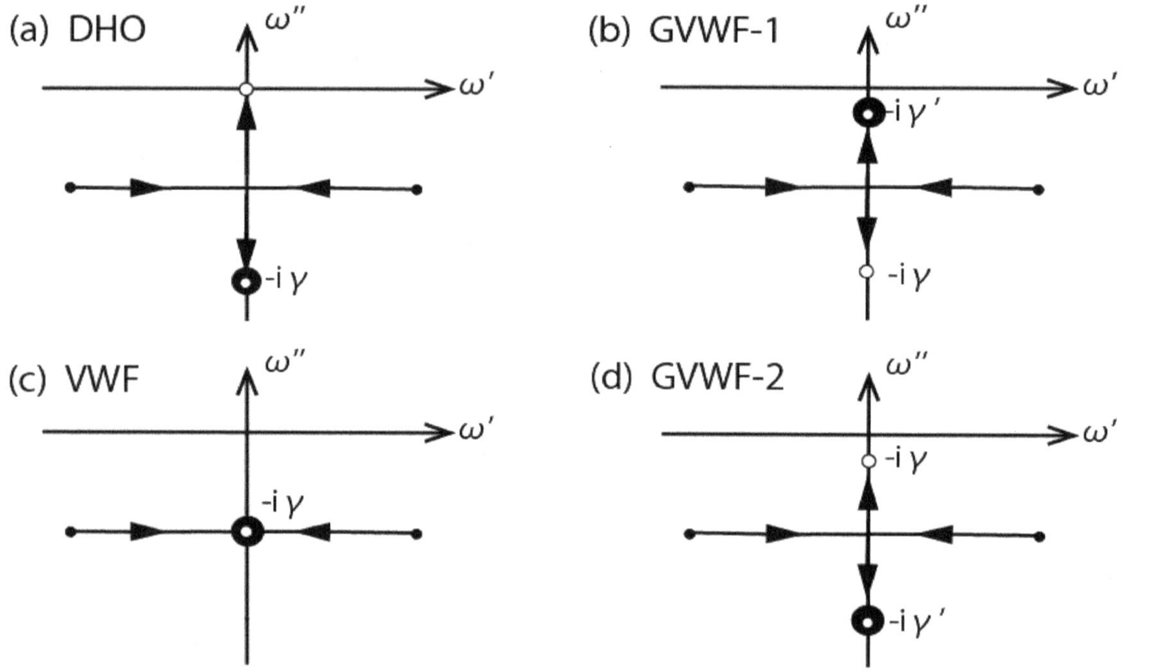


**Figure 3.** Behavior of the poles of the GVWF in the complex $\omega$ plane for $\omega_0 \to 0$ with fixed $\gamma$ and $\gamma'$. (a) $\gamma >> \gamma'=0$, DHO; (b) $\gamma > \gamma' \neq 0$, GVWF; (c) $\gamma = \gamma'$, VWF; (d) $\gamma < \gamma'$, GVWF. Small white circles mean the position of the pole for $\omega_0=0$ and the big black circles denote the DB relaxation mode.

$\gamma' \to 0$ regardless of $\omega_0$ and $\gamma$. The pole of VWF moves in the $(\omega_0, \gamma=\gamma')$ plane. The hatched plane indicates the position where the pole reaches the imaginary axis. Thus, GVWF is the most general approach to the instability, indicating that the application of DHO or DB is not trivial but based on the hidden assumptions.

## 4. Phase transition of KDP ($KH_2PO_4$)

### 4.1. Raman spectrum of soft mode in KDP

The first experimental study of the soft mode was done by Kaminov and Damen in 1968 on the ferroelectric phase transition of KDP ($KH_2PO_4$) [10]. The structure changes from tetragonal $D_{2d}$ ($I\bar{4}2d$) to orhorhombic $C_{2v}$ ($Fdd2$). KDP is a unique ferroelectric crystal because of its large isotope effect on $T_c$, being 122K and 212K for KDP and DKDP ($KD_2PO_4$), respectively. Its origin was first attributed by Blinc et al. from the IR spectra to the tunneling of a proton between two $PO_4$ ions [11]. The soft mode has $B_2$ symmetry, which is Raman-active in the (xy) polarized spectrum. Contrary to most high-frequency modes, the lowest strong mode is a quasi-elastic mode that shows no peak down to zero frequency. Kaminov et al. analyzed the spectra by DHO and concluded that the characteristic frequency $\omega_0^2$ goes to zero at $T_c$ and suggested that the results are consistent with either the soft mode model [12] or the collective-tunneling-mode model [13, 14]. Their conclusions, however, were obtained simply assuming that the damping

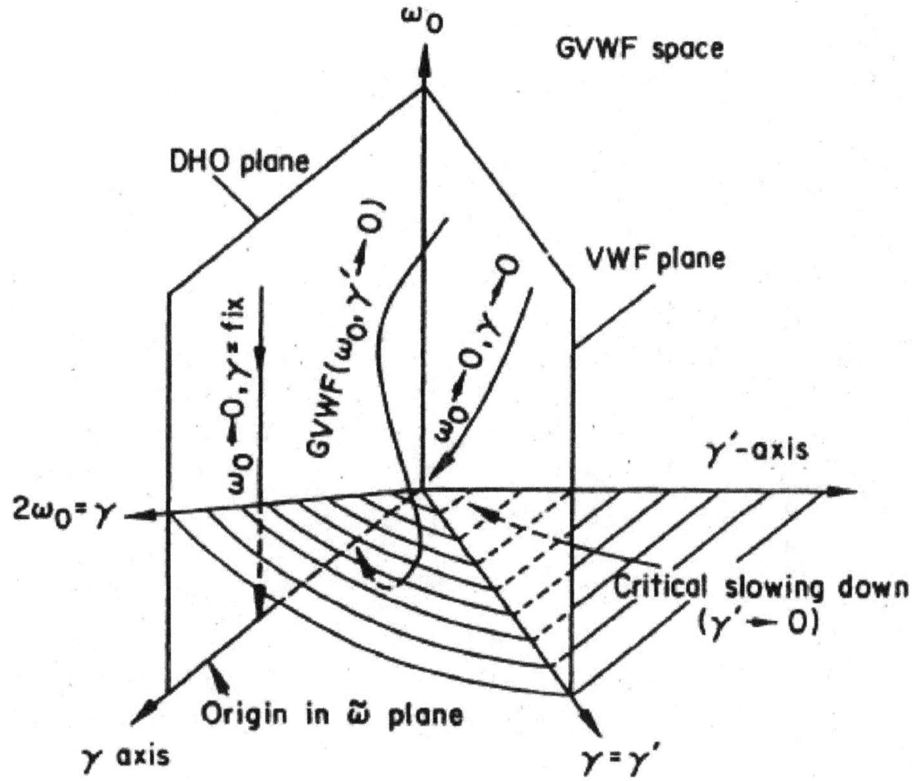


**Figure 4.** Behavior of the generalized susceptibility GVWF in $(\omega_0, \gamma, \gamma')$ space. Various paths of the pole to the origin of the complex $\omega$ plane are shown for DHO ($\omega_0 \rightarrow 0$, $\gamma$=const.), VWF ($\omega_0$, $\gamma=\gamma' \rightarrow 0$), GVWF ($\omega_0$, $\gamma' \rightarrow 0$), and DB ($\gamma' \rightarrow 0'$) cases.

constant $\gamma$ is a constant. Quasi-elastic spectra are often observed in other crystals. (We shall call it simply as an overdamped mode unless it is necessary to discriminate them explicitly.) As mentioned in the previous section, it is not easy to analyze such a spectrum. It might be an overdamped soft mode (DHO) or a Debye-type relaxational mode (DB).

Temperature dependence of Raman spectra of overdamped soft modes is often shown in the *log scale* instead of the *linear scale* because of the significant increase of intensity as shown in Fig. 5(a) in linear scale. In the pioneering data by Kaminov et al. [10] and also in the papers by Tominaga et al. [16] for KDP and by Takesada et al. [17] for $SrTiO_3$ (see section 5), spectra are shown in log scale. We note, however, that this might mislead its interpretation. In the log scale spectra, the product of the Bose factor $n(\omega)$ and $\mathrm{Im}\chi_Q$ is replaced by the sum of them. Then the low intensity part (i.e., the tail of the overdamped spectra) is apparently strengthened since the Bose factor is a monotonically increasing function toward infinity ($n\ (\omega=0)=\infty$ ), while $\mathrm{Im}\ \chi_Q$ is zero at $\omega=0$ ($\mathrm{Im}\chi_Q(\omega=0)=0$). As a consequence, the line shapes look quite different from that in the linear scale. For example, as in Fig. 5(b), the curvature of spectra changes from convex ($T<<T_c$) to concave ($T \approx T_c$). It looks as if there exist two components in the overdamped spectra. Figure 5(c) is the $\mathrm{Im}\chi_Q(\omega, T)$ spectra corresponding to the linear scale spectra by removing the Bose factor contribution [15]. Its temperature dependence is very smooth toward

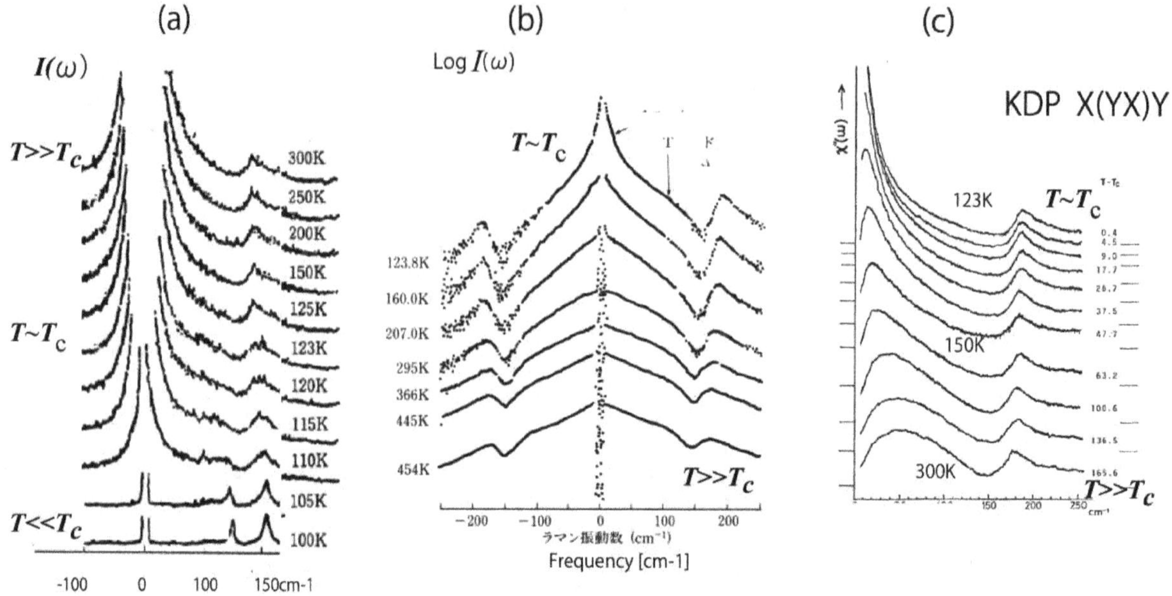


**Figure 5.** Temperature dependence of Raman spectra of KDP. (a) $I(\omega)$ in linear scale [15]. (b) Similar spectra $I(\omega)$ in log scale [16]. (c) Spectra of Im $\chi_Q = I(\omega)/n + 1$ in linear scale. (See the text for the difference between these spectra.)

$T_c$ without any qualitative change of the line shape. (The apparent peak in Fig. 5(c) is simply due to the fact $\mathrm{Im}\chi_Q(\omega=0)=0$.)

### 4.2. Displacive or order–disorder ?

Later studies by various groups however, revealed that the interpretation on the origin of the phase transition by Kaminov et al. [10] is not conclusive even from the Raman spectroscopic point of view. It still remains unclear whether it is a displacive type or an order–disorder type.

Difficulty of the interpretation of Raman spectra comes from the following reasons :

**i.** The low-frequency part of quasi-elastic spectra near $T_c$, can be fitted either by DHO as an overdamped soft mode or by DB as a relaxational mode. It can be fitted to any other function discussed in the previous section since from such a monotonic spectrum only a single parameter can be definitely determined, for example, $\omega_0^2/\gamma$ for the DHO model or $\gamma$ for the DB model.

**ii.** Above $T_c$, because of the coupling with a higher-frequency mode at about 230 cm$^{-1}$, shown in Fig. 5, the line shape is distorted and cannot be well fitted to any single-mode line shape function. Interactions with modes near the quasi-elastic peak require the so-called coupled-mode analysis [18].

**iii.** Below $T_c$, another mode at about 100 cm$^{-1}$ emerges from the overdamped mode observed by Kaminov et al., which suggests that the overdamped mode itself is not a single mode [19].

**iv.** Tominaga et al. insist that the phase transition of KDP is the order–disorder type [20]. The main evidence for their interpretation is based on the observation of extra Raman peaks above $T_c$ in the 500–1000 $cm^{-1}$ range, which should not be observed if the $PO_4$ units are the regular tetrahedra with $S_4$ symmetry. Thus, the existence of those peaks would indicate the $PO_4$ tetragonal unit is *not* regular but already distorted to $C_2$ symmetry above $T_c$. If it were true, the two different distortions of $PO_4$ (Fig. 6(b)) would cause the order–disorder transition. However, the observation of the extra peaks does not necessarily mean that $PO_4$ is statically distorted, because two protons vibrate coherently with $PO_4$ tetrahedron so that vibrational modes of the $H_2$-$PO_4$ system have $C_2$ symmetry and the extra peaks would be observed in Raman spectrum even if $PO_4$ is regular tetrahedra [21].

**v.** The role of protons in this transition is not clear in the order–disorder model. The coupled proton–$PO_4$ model proposed by Kobayashi [22] shown in Fig. 6(a) seems to be more realistic. As for the large difference of $T_c$ between KDP and DKDP, no direct evidence for the proton tunneling as the origin of the isotope effect has so far been obtained. It was also revealed that there exists some difference of structure between KDP and DKDP. Details were discussed in [21]. Thus, it still remains unclear whether it is a displacive type or an order–disorder type.

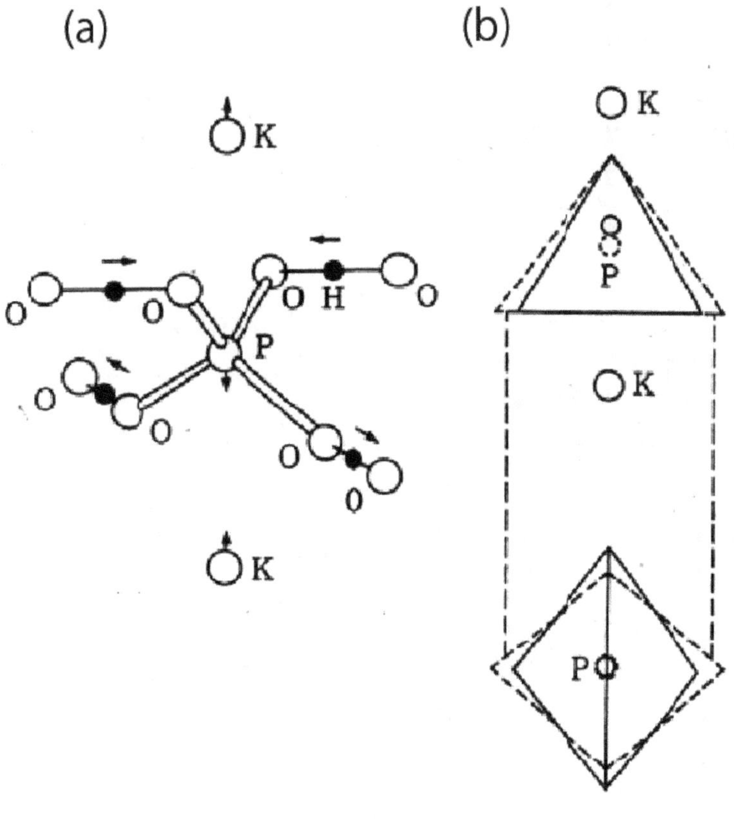


**Figure 6.** Two models for the ferroelectric transition in KDP. (a) Soft mode pattern of a proton–$PO_4$ coupled system proposed by Kobayashi [22]. (b) The order–disorder model of a deformed $PO_4$ [20].

### 4.3. Influence of the ferroelectric domain

Raman spectrum measures the macroscopic region of the sample. Therefore, in the Raman spectra of ferroelectric crystals the appearance of the domains with various scales and orientations in the sample should not be ignored. For example, a strange oscillatory behavior of the Raman intensity was observed in KDP. Raman intensity of $A_1$ mode, measured with the scattering geometry shown in Fig. 7(a), change oscillatory with temperature in the range $0<T_c-T<15K$ [23]. This behavior can be explained in terms of the interference of light reflected by the domain walls. Spontaneous polarizations $\pm\vec{P}$ in KDP are parallel to the c-axis and the thickness of 180 domain grows on cooling. The oscillatory behavior of Raman (Fig. 7(c)) is due to the periodical change of the polarization of light caused by the reflection from the domain walls and the birefringence between the neighbor domains as was confirmed by the similar oscillatory behavior the reflected light (Fig. 7(d)) [23].

The influences of domains in Raman spectra are also important in the case of more complicated domain structures in the case of STO18, discussed in section 5.2.

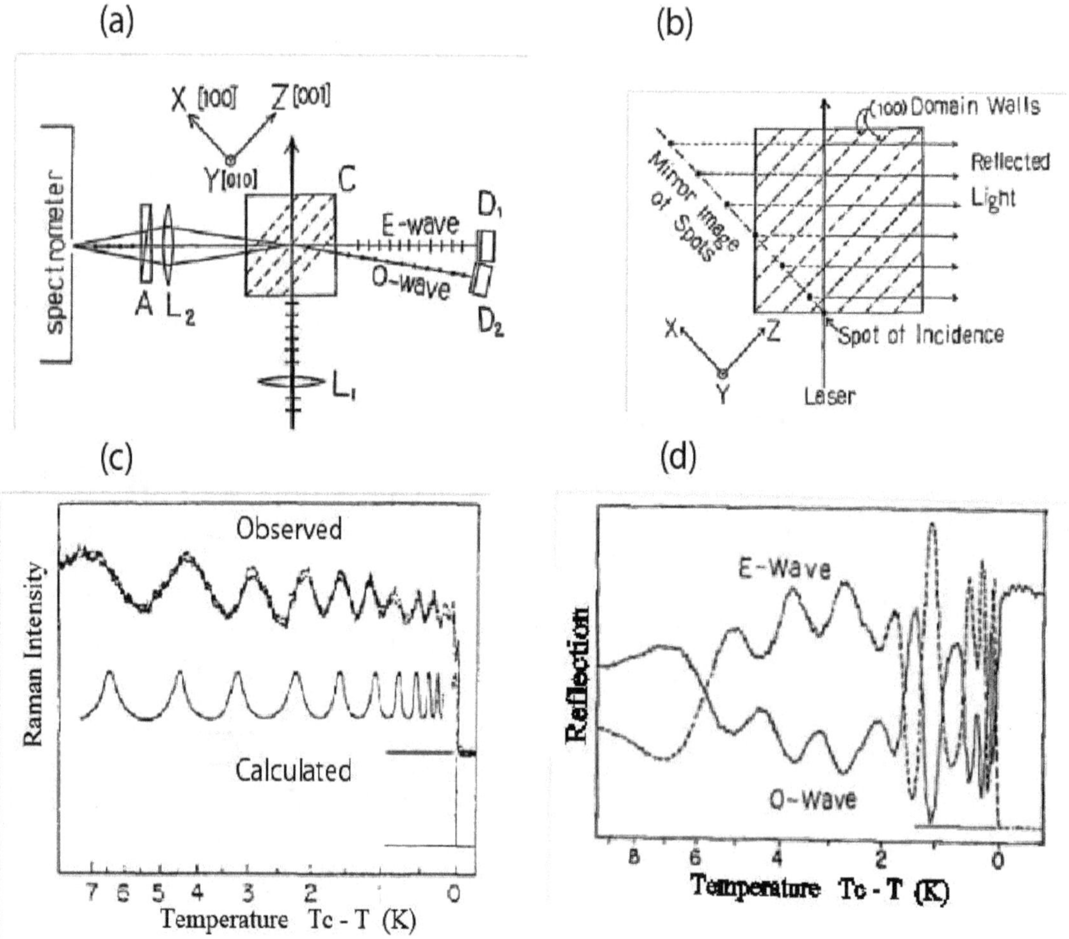


**Figure 7.** Influence of domains on Raman spectrum. (a) Scattering geometry. (b) Laser path and the reflection from domain walls. (c) Oscillatory change of Raman intensity with temperatture. (e) Intensity of the reflected light [23].

## 5. Ferroelectric $SrTiO_3$

$SrTiO_3$ is a typical crystal with perovskite structure (Fig. 8(a)) and it is easy to get a pure transparent single crystal. As a well-known incipient ferroelectric, it is one of the most widely studied dielectric crystals [24]. In 1999, Itoh et al. found the ferroelectricity in $SrTiO_3$ by the isotope substitution of $O^{16}$ by $O^{18}$ [25]. Since then, there has been considerable renewed interest in the low-temperature properties of $SrTiO_3$. Hereafter, we refer to $SrTiO_3^{16}$ and $SrTiO_3^{18}$ as STO16 and STO18, respectively. First, we will review the soft mode behavior in the normal STO16 and in the next subsection the soft mode in STO18 will be discussed.

### 5.1. Soft mode in STO16

$SrTiO_3$ undergoes a structural (anti-ferro-distortive) phase transition at $T_0$=105$K$ from cubic $O_h^1(Pm\bar{3}m)$ to tetragonal $D_{4h}^{18}(4/mcm)$ structure. This transition is induced by the freezing of a zone boundary nonpolar soft mode $R_{25}^{'}$ (structural soft mode), which is an alternative rotational vibration of $TO_6$ octahedra around one of the cubic axes (Z-axis $[001]_c$) [26]. As shown in Fig. 8(b), the crystal axes in the tetragonal phase are rotated by 45° from the cubic axes and the unit cell is a rectangular parallelepiped elongated along one of the cubic Z-axes. It should be noted, therefore, that the multi-domain effect must be carefully taken into account since a sample below $T_0$ is usually consisted from a number of tetragonal domains with different Z-directions. It is especially important for the case of STO18 as we shall show in the next section. Below $T_0$ in the tetragonal phase, the triply degenerate structural soft mode splits into two Raman-active $A_{1g}+E_g$ modes and their frequencies increase on further cooling reaching 44 and 11 $cm^{-1}$, respectively.

Besides the "structural soft mode" there is another soft mode, the so-called ferroelectric soft mode in $SrTiO_3$. It is the lowest transverse mode ($TO_1$) among the 4 zone-center $\Gamma_{15}$ modes in the cubic symmetry, which is known as the Slater mode (Fig. 8(a)). Frequency of the corresponding LO mode is much higher (170 $cm^{-1}$) than $TO_1$. The large LO/TO splitting indicates the very strong polar nature of this mode. Temperature dependence of the ferroelectric mode has been extensively studied. Above $T_0$ in the cubic phase, it is a doubly degenerate $E_u$ mode regardless of the propagation directions $\vec{K}_p$. Below $T_0$ in the tetragonal phase, however, the symmetry of this mode depends on the $\vec{K}_p$. If it propagates along Z-axis, the degeneracy remains but when it propagates in the X-Y plane, $TO_1$ splits into $A_{2u}+E_u$ [27] (also see Fig. 12(a) in the next section).

The softening of $A_{2u}$ and $E_u$ was first observed by IR spectroscopy [28] and later confirmed by neutron scattering [29] and hyper-Raman scattering [27, 30]. Its frequency (90$cm^{-1}$) at room temperature decreases to about 15 $cm^{-1}$ in low temperature. In spite of the significant softening, it rounds off at about 30 K and never freezes down to zero K. Thus, STO16 is known as quantum paraelectric crystal since quantum fluctuation at low temperatures hinders the freezing of the soft mode.

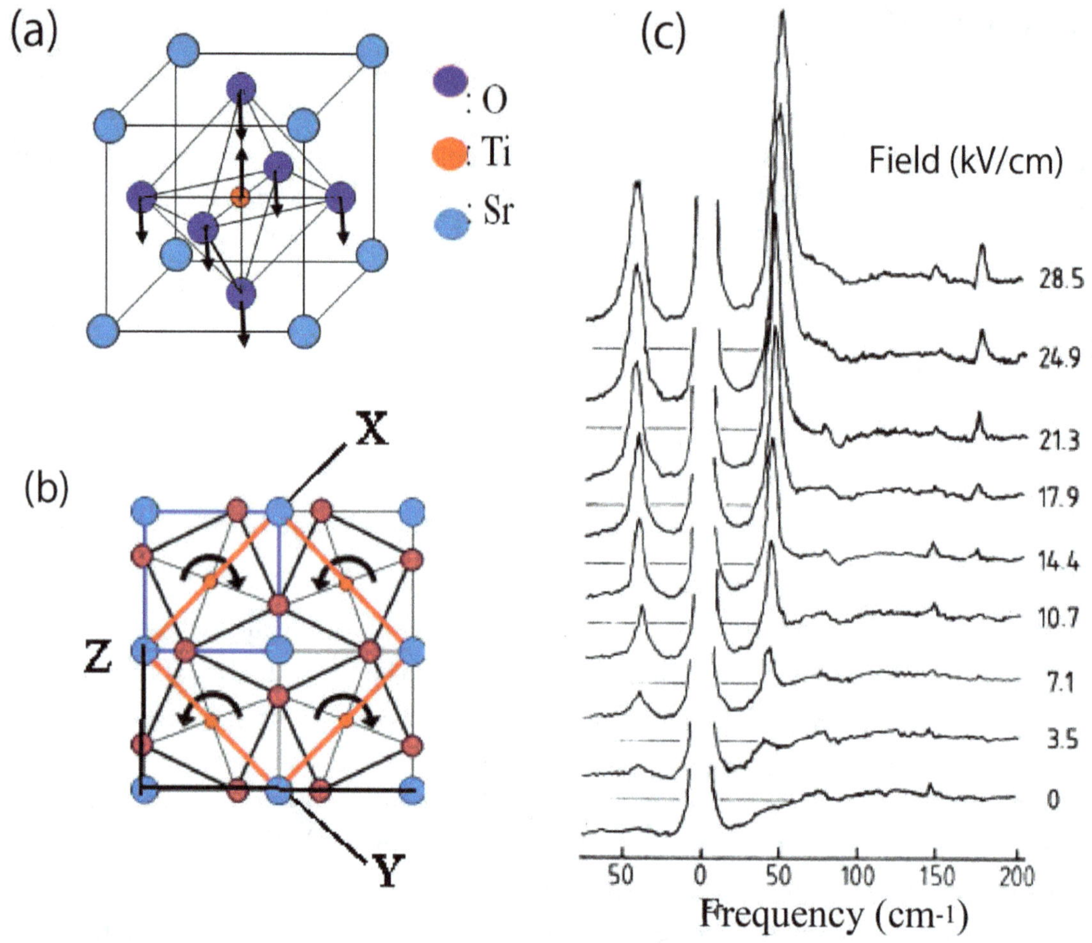


**Figure 8.** (a) Perovskite structure and the ferroelectric soft mode (Slater mode). (b) Blue and red squares are the unit cell in the cubic and the tetragonal phase, respectively. (c) Electric field effect on Raman spectra of ferroelectric soft mode in STO16 at $T$=101 K [32].

Although the ferroelectric soft mode is Raman-inactive, surprisingly it was found that if a DC electric field is applied its intensity and frequency drastically increase [31]. The field effect was observed even at higher temperature near $T_0$ as shown in Fig. 8(c) [32]. This is another evidence of the extremely strong polar nature of the ferroelectric soft mode in $SrTiO_3$. In other similar crystals, for example in $KNbO_3$, the effect is much weaker than in $SrTiO_3$ [33].

### 5.2. Soft mode in $SrTiO_3^{18}$ (STO18)

#### *5.2.1. Propagation dependence of the soft mode in STO18*

STO18 undergoes the structural phase transition from cubic to tetragonal at about $T_0$=108 K, which is slightly higher than $T_0$=105 K of STO16. On further cooling at $T_c$=24 K, however, STO18 transforms to the ferroelectric phase with orthorhombic structure ($C_{2v}$). [25] Ferroelectric properties of highly substituted STO18 (higher than 33%) were confirmed by the enormous increase of the dielectric constant and the appearance of the hysteresis curve.

All the samples used in many laboratories were provided from Itoh's laboratory. The samples are thin plates of size 0.3 x 2 (or 3) x 7 mm$^3$ with the widest plane $[110]_c$, and the longest and the shortest edges are parallel to $[001]_c$ and $[110]_c$, respectively. In the early stage, most samples used were not a single tetragonal domain but included many domains with different Z-directions. Thus, the reports measured with multi-domain samples were controversial and could not give consistent results.

Figure 9 is the summary of the temperature dependence of the soft mode frequency in STO18 measured by various experiments. The data are from samples with a single tetragonal domain except for the neutron diffraction.

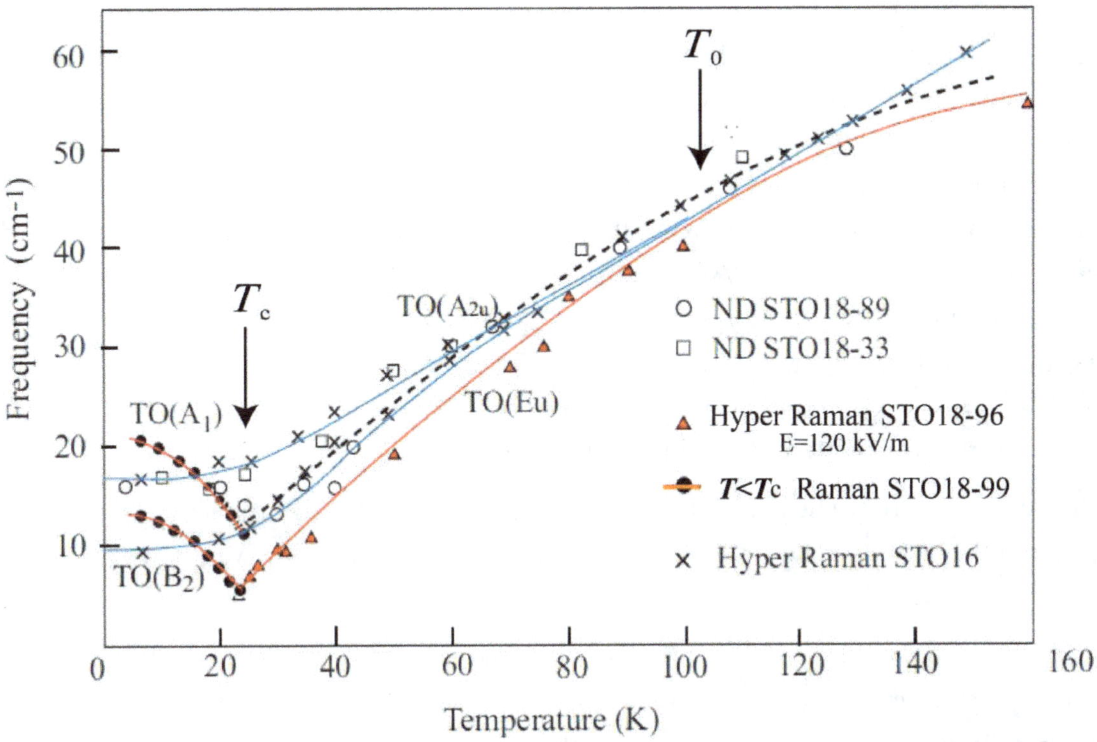


**Figure 9.** Temperature dependence of the ferroelectric soft mode of $SrTiO_3$. $T_c$ of STO18 is slightly different for various experiments but here it is assumed to be about 24 K. The solid lines are guide for eyes. ○ and □ are neutron diffraction of STO18 [34]. × are hyper Raman of STO16 [35] and ▲ are hyper Raman of STO18 above $T_c$ measured under the electric field [27]. ● are Raman scattering below $T_c$ in STO18-99 [36]. Below $T_c$, only the Raman data clearly show the soft mode behavior in STO18.

Hyper Raman data under DC-electric field from STO18-96 show the softening similar to STO16 above $T_c$, but they do not give reliable data below $T_c$.[27] Neutron diffraction data hardly detect the influence of the phase transition [34]. Thus, the space group of STO18 below $T_c$ has not yet been determined definitely. As we shall show later, it is probably because of the inhomogeneity of STO18 due to the existence of the paraelectric phase as a matrix of ferroelectric domains. Below $T_c$, only the Raman data succeeded in getting reliable soft mode frequencies [36]. Note that all these data show that the freezing of the soft mode (perfect softening) does *not* occur.

Later, highly substituted single-domain samples, STO18-99, became available. The results discussed below are the data obtained from such samples. For the correct assignment and interpretation, we found that the following features should be carefully taken into account [36]:

**a.** Temperature and polarization dependences of spectra are strongly dependent on the phonon propagation vector, $\vec{K}_p = \vec{K}_i - \vec{K}_s$ ($\vec{K}_i$ and $\vec{K}_s$ are the propagation vector of the incident and scattered photon, respectively).

**b.** Spectra measured in a single geometry or a single $\vec{K}_p$ is not enough for the correct assignment. Such a measurement would lead to incorrect conclusions.

**c.** The existence of the ferroelectric domains in the sample, particularly the direction of the spontaneous polarization $\vec{P}$ in the domains must be properly taken into account.

In the ferroelectric phase of STO18, five modes are observed in the low-frequency part below 50 cm$^{-1}$ as one can see in Fig.10 and 11. Three of them are from the structural soft modes, as $A_1$ (from $A_{1g}$) at 44 cm$^{-1}$ and $A_2$ and $B_1$ (from $E_g$) at 11 and 17.5 cm$^{-1}$, respectively. Since these modes are nonpolar, the line width is narrow and does not depend on $\vec{K}_p$ (Fig.11(b) and (c)). In contrast, the behavior of the ferroelectric soft modes, TO($A_1$) and TO($B_2$), very broad and sensitive to $\vec{K}_p$ (Fig.11). It is also sensitive to the scattering geometries. For example, the polarization dependence of the spectra with $\vec{K}_p$ // X at a temperature well below $T_c$ (at about 7K) is shown for three different geometries in Fig.10. Note that in the geometry (c) of Fig.10, the VV(Z, Z) spectrum is quite different from other geometries, since even for the same $\vec{K}_p$, the selection rules are different for different geometries.

Spectra for $\vec{K}_p$ // Y (not shown in Fig.11) were found to be exactly the same as $\vec{K}_p$ // X, suggesting that there are two kinds of small domains with $\vec{P}$ // X and $\vec{P}$ // Y equally distributed in the sample, as we will discuss later.

In Fig. 11, the spectra for $\vec{K}_p$ // Z and $\vec{K}_p$ // X look similar except that there are two broad peaks in $\vec{K}_p$ // Z spectra while in $\vec{K}_p$ // X there is only one. Moreover, their temperature dependences are essentially different as we shall show in the next section.

It should be emphasized that the $\vec{K}_p$ // X + Y spectra (Fig.11(c)) are qualitatively different from other directions in the following two points: (1) The strong and broad peak at 20 cm$^{-1}$ in the VV spectra of $\vec{K}_p$ // Z and $\vec{K}_p$ // X appears in $\vec{K}_p$ // X + Y at a lower frequency 17 cm$^{-1}$. This means that the depolarization field in $\vec{K}_p$ // X + Y is weaker than that in $\vec{K}_p$ // Z or $\vec{K}_p$ // X, so that the mode at 17 cm$^{-1}$ is not a pure TO($A_1$) mode. (2) In $\vec{K}_p$ // X + Y, there appear spurious modes (arrows in Fig. 11(c)) that are unable to assign to any mode in STO18. As we shall discuss later, these modes are originated from the matrix of the paraelectric phase.

Our final assignment of displacement and the symmetry of the ferroelectric soft modes are illustrated in Fig. 12(a) for temperature above $T_c$ and in Fig. 12(b) for well below $T_c$. All the features of the observed spectra, including the subtle differences in the polarized spectra and the expected selection rules, were consistently explained *only* if the spontaneous polarization $\vec{P}$ is either parallel to the tetragonal axes X or Y. $\vec{P}$ cannot be along one of the cubic directions

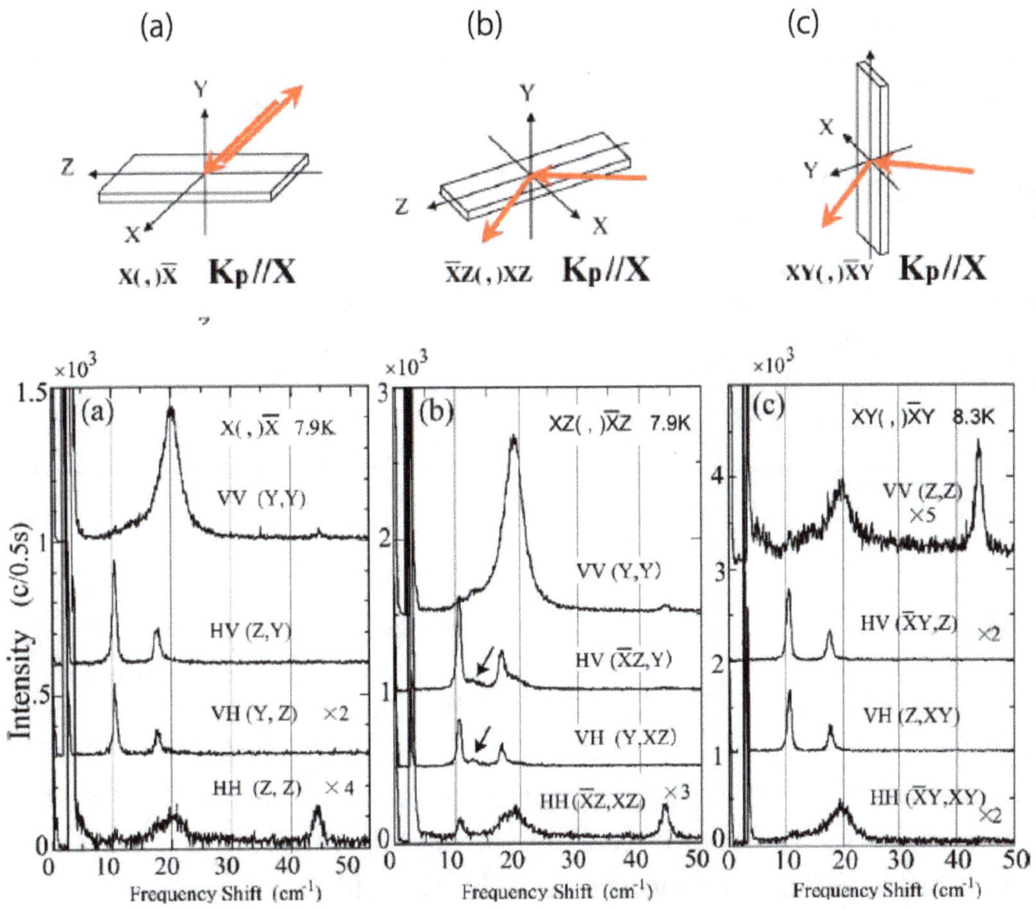


**Figure 10.** Three different scattering geometries for the same propagation direction, $\vec{K}_p$ // X. Broad mode at 20 cm$^{-1}$ is TO($A_1$) and the shoulder in HV and VH of (b) (shown by arrows) is TO($B_2$) [36] In the geometry (c), the VV(Z, Z) spectrum is quite different from other geometries.

as one might expect for the Slater mode [26], since it is neither parallel to $X \pm Y$ (=[100]c or [010]c ) nor Z (=[001]$_c$). If $\vec{P}$ were parallel to one of the cubic axes, observed spectra never satisfy any selection rules. Reason for this unexpected result will be given later.

As one can see in Fig. 12(b), the doubly degenerate TO(Eu) splits into TO($A_1$) + TO($B_2$) when it propagates along Z, but when it propagates in the X-Y plane, TO($A_1$) changes to TO($B_1$), of which displacement is perpendicular to $\vec{P}$ . Since the displacement of TO($A_1$) is parallel to $\vec{P}$ , it is reasonable that the frequency of TO($A_1$) (20 cm$^{-1}$) is higher than TO($B_1$) (17.5 cm$^{-1}$) and its intensity is stronger. Similar to the very strong external DC-field effect observed in STO16 (Fig. 8(c)), these differences are due to the effect of the depolarization field produced by $\vec{P}$ .

*5.2.2. Temperature dependence of soft mode in STO18*

Figure 13 is the temperature dependence of Raman spectra for different $\vec{K}_p$ directions.

In Fig. 13(a) only the broad and underdamped mode TO($A_1$) softens and becomes weaker on approaching $T_c$ as it should be. In contrast, Fig. 13(b) shows that two modes, TO($A_1$) and

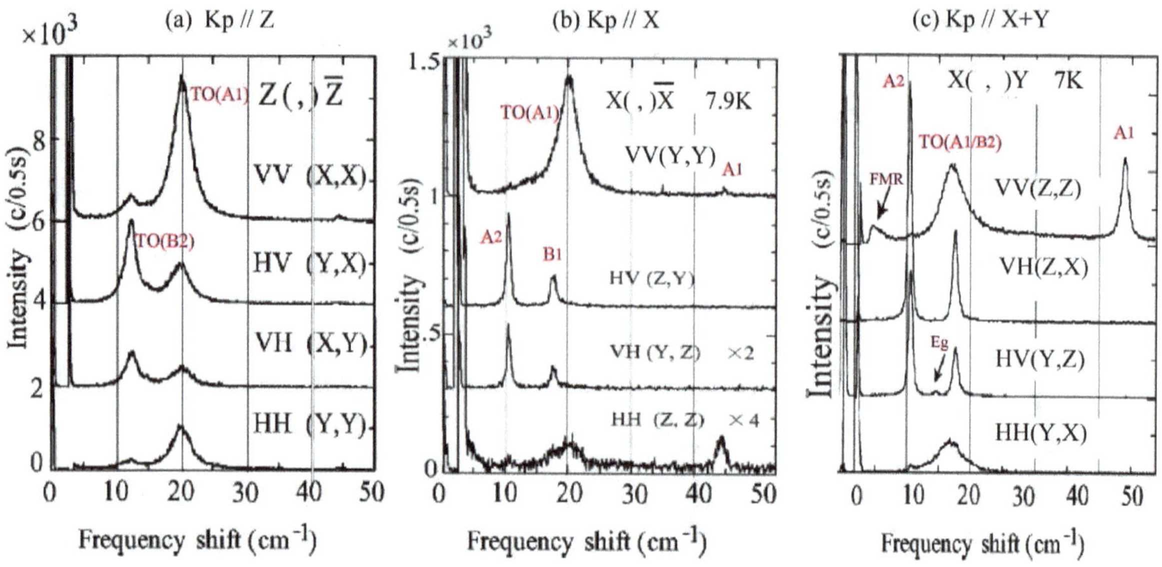


**Figure 11.** Polarized spectra at the lowest temperature for different propagation vector $\vec{K}_p$. (a) $\vec{K}_p$ // Z, (b) $\vec{K}_p$ // X, and (c) $\vec{K}_p$ // X + Y. Assignment of the peaks are given in red letters. Arrows in (c) are the spurious modes due to the paraelectric structure existing below $T_c$ as the matrix of ferroelectric domains.

TO($B_2$) clearly soften. It is clear that frequencies of the soft modes never go down to zero. Instead in the $\vec{K}_p$ // Z spectra, in addition to the soft modes, a quasi-elastic mode appears near $T_c$ (arrows in Fig. 13(b)). It is observed not only below $T_c$ but also above $T_c$. Moreover, it is observed only when $\vec{K}_p$ is out of the X-Y plane such as for $\vec{K}_p$ // Z and $\vec{K}_p$ // Z + Y [36, 37].

The absence of the quasi-elastic mode in $\vec{K}_p$ // X (Fig. 13(b)) implies that it is *not* an overdamped soft mode as in the KDP case (section 4.1) but is a relaxational mode originated from the large fluctuations of $\vec{P}$ near $T_c$ in the X-Y plane.

Another peculiar nature of the ferroelecticity of STO18 is the fact that even for a sample with 99% substitution, STO18 is *not* homogeneous contrary to the report by Taniguchi et al. [39], because the ferroelectric domains with $\vec{P}//X$ and $\vec{P}//Y$ coexist in the matrix of the paraelectric structure. The appearance of the spurious peaks in the $\vec{K}_p$ // X + Y spectra (Fig. 13(c)), which is FMR (Ferroelectric Micro-Domain) intrinsic to STO16, verifies the inhomogeneity of STO18.

The presence of the quasi-elastic mode in $\vec{K}_p$ // Z and its absence in $\vec{K}_p$ // X in STO18 were also reported by the independent two reports by Takesada et al. [17, 38] and Taniguchi et al. [39], respectively. As mentioned above, however, their conclusions are inconsistent with our results. Therefore, let us discuss briefly their results from our point of view.

**a.** Perfect softening of the Slater mode in STO18 was claimed in ref. [17] using a high-resolution spectrometer. Their results, however, are based on *only* the spectra with $\vec{K}_p$ // Z. Above $T_c$, they assign a *very weak bump* below 5 cm$^{-1}$ sitting on the strong quasi-elastic (our relaxational) mode as the underdamped soft mode TO($E_u$). But below $T_c$, no such a narrow mode is observed and the strong quasi-elastic (our relaxational) mode is

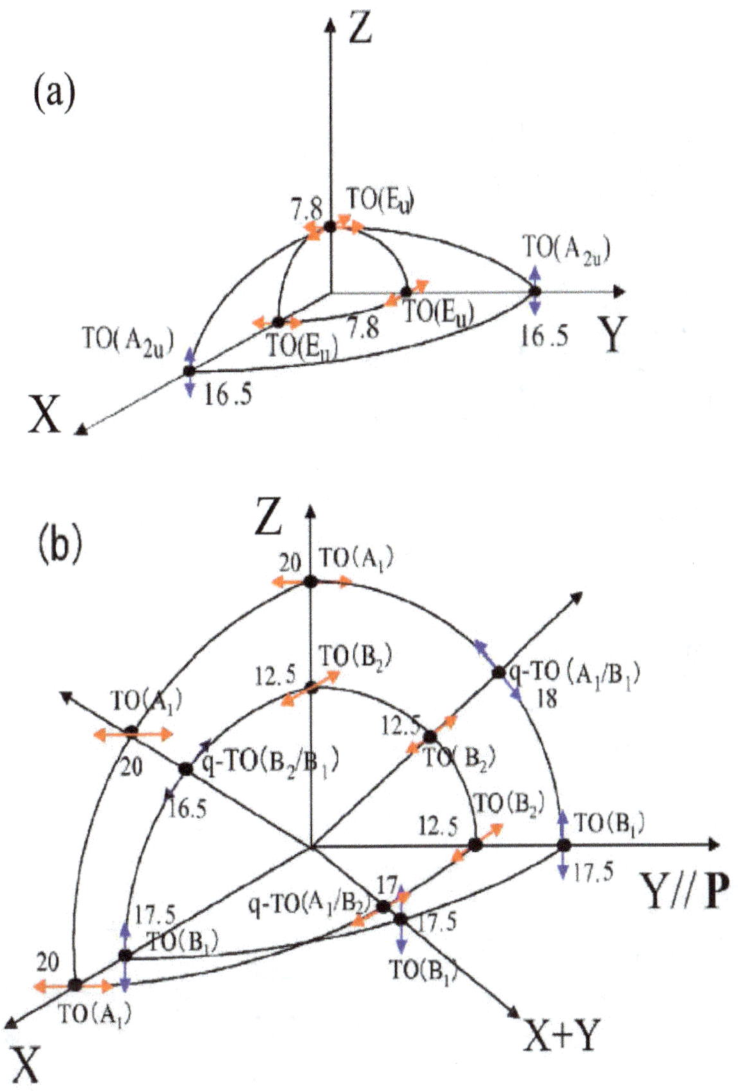


**Figure 12.** Ferroelectric soft-mode frequency surfaces for various directions of phonon propagation vector $\vec{K}_p$. $X = [110]_c$ and $Y = [1\bar{1}0]_c$ are the crystal axes of the tetragonal structure. (a) In the paraelectric phase near $T_c$. (b) In the ferroelectric phase well below $T_c$. Figure 12(b) is drawn for the case of $\vec{P}$ // Y. Since in a sample of STO18, domains with $\vec{P}$ // X and $\vec{P}$ // Y coexist (in the matrix of paraelectric phase), observed spectra are the sum of the similar figure with the X and Y axes exchanged. Red arrows are the displacements in the X-Y plane and blue arrows are those out of the X-Y plane.

suddenly assigned as the overdamped soft modes $TO(A_1)$ and $TO(B_2)$. (We note that the lowest mode is *not* $A_1$ *but* $B_2$ contrary to the assignment by Takesada et al. [17].) From the computer fitting of the overdamped spectra below $T_c$, they obtain the extremely steep dropping of the frequency of $TO(B_2)$ down to zero. However, as mentioned in the case of KDP, such analysis is very ambiguous and the sudden qualitative change of the soft mode at $T_c$ is unnatural. From these results they concluded that STO18 is an ideal displacive-type ferroelectrics induced by the Slater-type soft mode. If such a perfect softening took place at $T_c$ it would be much more clearly observed in the $\vec{K}_p$ // X spectra, since as shown

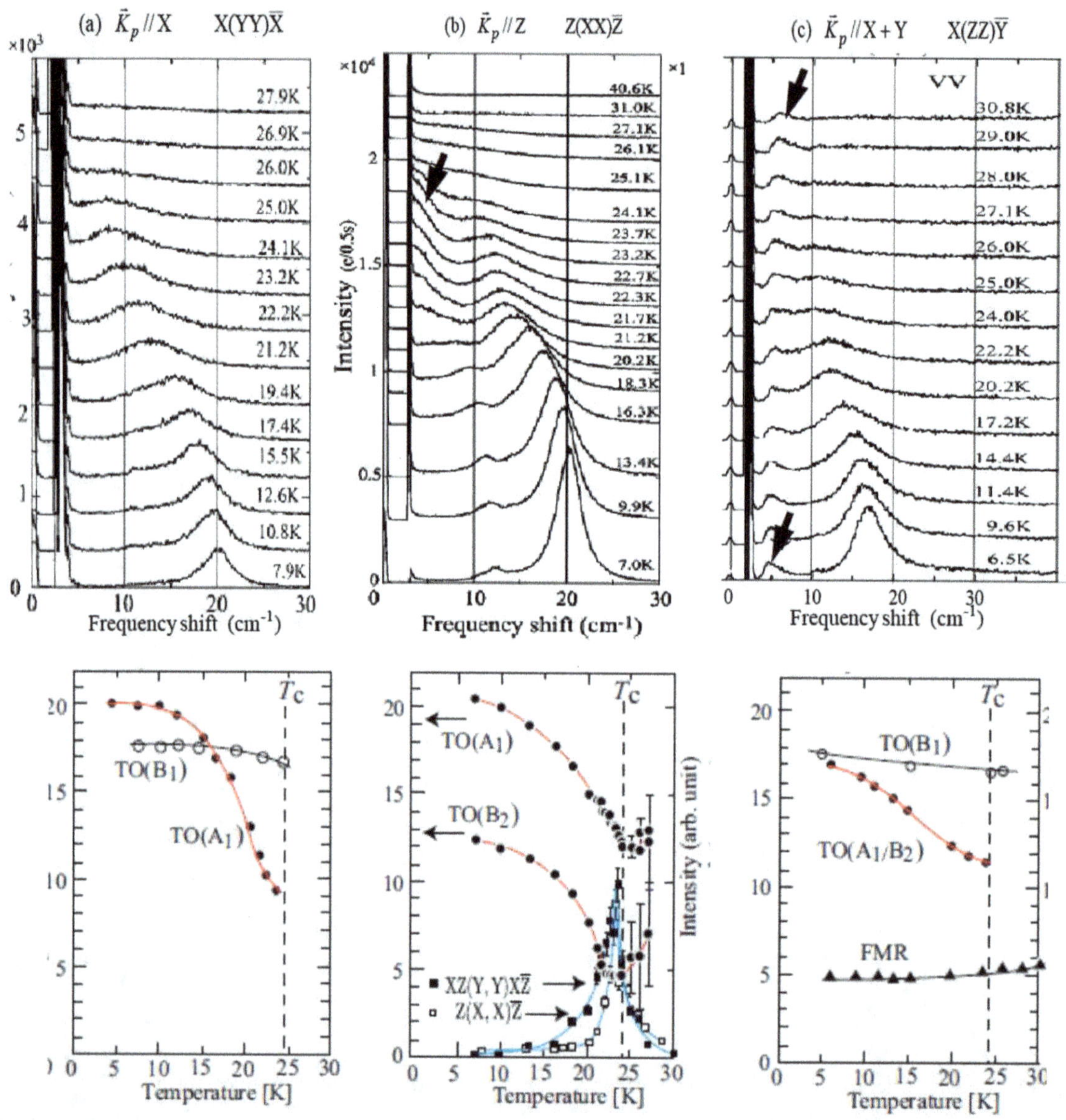

**Figure 13.** Temperature dependence of Raman spectra for different propagation directions: (a) $\vec{K}_p$ // X, (b) $\vec{K}_p$ //Z, (c) $\vec{K}_p$ // X + Y. The spectra for $\vec{K}_p$ // Y was identical to those for $\vec{K}_p$ // X. Note that in (b), a relaxational mode (blue lines) appears near $T_c$ and in (c), a spurious mode (the arrows) is seen at about 5 $cm^{-1}$ and no *pure* $TO(A_1)$ is observed.

in Fig. 13(a), in the $\vec{K}_p$ // X spectra no quasi-elastic mode appears near $T_c$. Furthermore, if the perfect softening were related to the Slater-type mode, the direction of $\vec{P}$ should be parallel to cubic axis, which contradicts our results. Therefore, the essential difference between their interpretation and ours cannot be attributed to the difference of the resolution of the spectra.

b. Another essentially incorrect result was reported by Taniguchi et al. related to the homogeneity of STO18 [39]. They measured $O^{18}$ concentration dependence of Raman spectra for $SrTi(O^{18}{}_x O^{16}{}_{1-x})_3$. In this paper, they measured the spectra again *only* with a single $\vec{K}_p$, not $\vec{K}_p$ // Z but $\vec{K}_p$ // X, in which the quasi-elastic mode is absent. Then it was concluded that the homogeneous ferroelectric phase changes into theferroelectric-paraelectric phase coexistence state as thesystem approaches quantum critical point $x=0.33$ and that highly substituted STO18 undergoes the ideal soft-mode-type quantum phase transition. As the evidence of the criticality, they claim that the mode at 15 $cm^{-1}$ which is the $E_g$ mode intrinsic to the paraelectric phase appears only in the low substitution samples below $x=0.33$. However, as we have shown in the $\vec{K}_p$ // X + Y spectra (Fig. 13(c)), in addition to the FMR at 5 $cm^{-1}$, the $E_g$ mode at 15 $cm^{-1}$ (in our case 14.5 $cm^{-1}$) is certainly observed even in a very highly substituted sample with $x=0.99$. This is a clear evidence of the existence of the paraelectric phase as the matrix of the ferroelectric domains. In other words, the *inhomogeneity is the intrinsic property* of STO18. Therefore, the transition of STO18 with high x is *not* the ideal (homogeneous) soft-mode-type quantum phase transition. In this case, they should have measured the $K_p$ dependence more carefully, especially the $\vec{K}_p$ // X + Y spectra, since they have noticed from our results [36] that $\vec{P}$ is not along the cubic axes which suggests that the soft mode cannot be the Slater mode.

### 5.2.3. *Why the soft mode is not the slater mode ?*

The reason why $\vec{P}$ is not along the cubic axes (as expected from the softening of the Slater mode) but along the tetragonal axes would be closely related to the existence of the rhombohedral (parallel to $[111]_c$) nano-scale clusters in STO18 ,which were found in the NMR experiment by Blinc et al. [40]. They revealed that below 70 K rhombohedral polar clusters are formed in the tetragonal matrix. These clusters subsequently grow in concentration, freeze out, and percolate, leading to an inhomogeneous ferroelectric state below $T_c$. This shows that the elusive ferroelectric transition in STO-18 is indeed connected with local symmetry lowering and implies the existence of an order–disorder component in addition to the displacive soft mode one.

Referring to this result we have shown in a recent paper [37] that the dipole interaction between the rhombohedral polar clusters is the very probable reason for the formation of $\vec{P}//[110]_c$. The calculation shows that a pair of dipoles located in the same X-Y plane gives the strongest attractive interaction than any other possible pair interaction. The unification of such a pair of rhombohedral polar clusters generates $\vec{P}$ parallel *not* to the cubic axes but to the tetragonal axes. Spontaneous polarization $\vec{P}$ generated by this process would percolate and grow to the macroscopic ferroelectric domains.

Thus, the dipole interaction model is consistent with various peculiarities observed in the Raman spectra [36], such as the inhomogeneity, the imperfect softening, and the appearance of the relaxational mode near $T_c$. The origin of the relaxational mode observed near $T_c$ would be the increase of the orientational fluctuations of $\vec{P}$ in the X-Y plane. Its coupling with the soft mode propagating along the Z axis would suppress the perfect softening.

Finally, we admit that the question "why the isotope substitution of $O^{16}$ by $O^{18}$ makes the ferroelectric transition possible in $SrTiO_3$" has not yet been fully solved.

# 6. LO/TO splitting in the ferroelectric Ice-XI

## 6.1. Structure of Ice-XI

The most common solid phase of water at normal pressure is the hexagonal ice (ice-Ih, $D^4_{6h}$, $P6_3/mmc$) which freezes at 273 K. In the Ih phase, however, due to the so-called ice-rules (Bernal–Fowler rules), protons on the hydrogen bonds are randomly distributed and below 100 K protons essentially freeze in place, leaving them disordered [41]. The residual entropy $S_{res} = R\ln(2/3)N$ [42] predicts the existence of the proton ordered state phase (known as ice-XI), but it would take a geological scale of time to transform into the ordered phase. In 1972, Kawada succeeded in obtaining the ice XI crystal at normal pressure by doping a small amount of KOH [43]. The calorimetric study [44] confirmed that ice-XI is stable for $H_2O$ below $T_c = 72$ K. As shown in Fig. 14, in both ice-Ih and ice-XI, four water molecules are in a unit cell. In ice-XI, the c-component of the dipole moment of each $H_2O$ molecule aligns ferroelectrically and the b-component aligns anti-ferroelectrically.

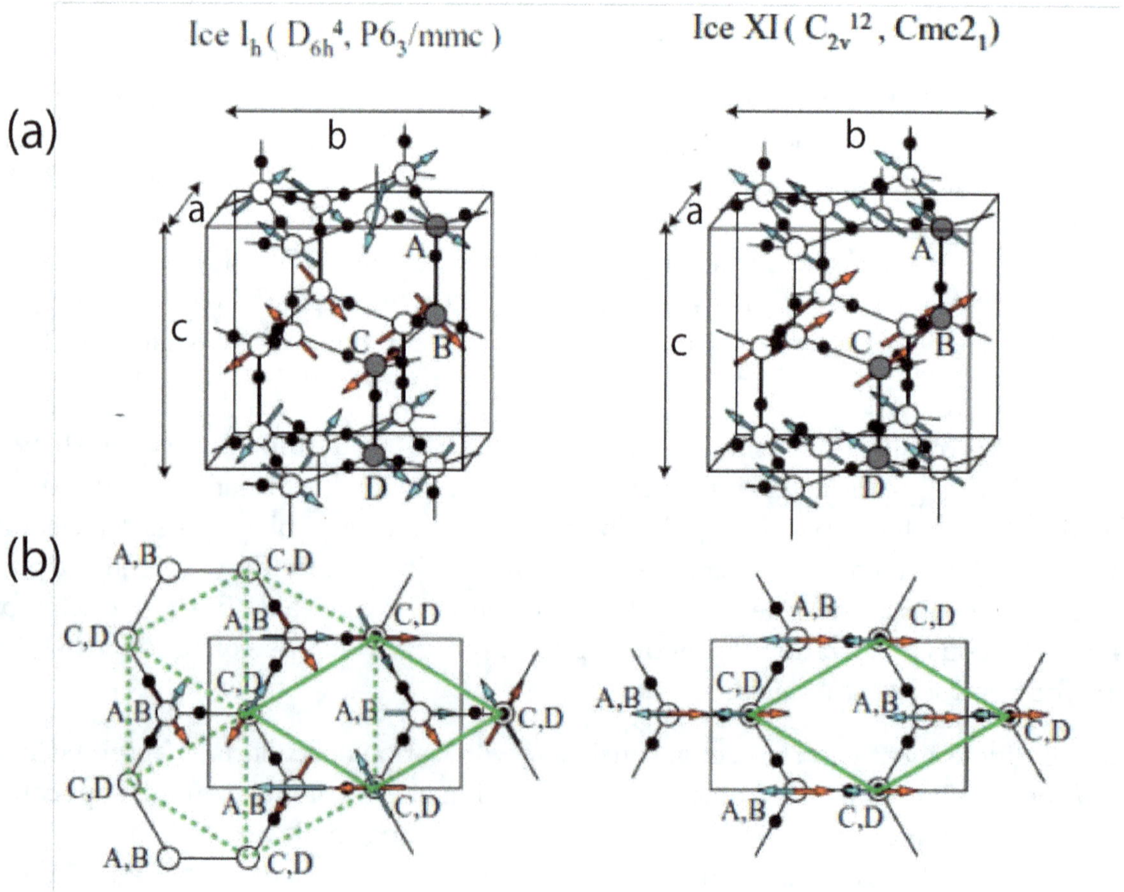


**Figure 14.** (a) Structures of ice-Ih and ice-XI. Four water molecules A, B, C, and D are in a unit cell [45]. (b) Projection to a–b plane. A unit cell is shown by the rhombus (thick green). The dotted hexagon shows the ice-Ih structure. Arrows represent the dipole moment of each molecule.

In contrast to neutron and theoretical studies, spectroscopic studies on ice-XI phase are very few. This is primarily because of the difficulty in growing a single crystal suitable for optical measurements. Recently, we have succeeded in growing a single crystal of ice-XI, which enabled us to measure the polarized Raman spectra [46]. Since the transition from Ih to XI is strongly of first order, no soft mode is expected for this transition. To observe whether the ferroelectric order was realized in our samples or not, the effect of the external electric field was tested on the sharp peak at 610 cm$^{-1}$ which appears only in ice-XI. Although the effect was not significant compared to the ferroelectric soft mode in $SrTiO_3$ (Fig. 8(c)), the electric field (max 4 kV/cm) applied along the c-axis certainly increased its Raman intensity.

Figure 15(a) is the comparison of the wide frequency range spectra between ice-Ih and ice-XI measured with low resolution. Changes in the spectra were clearly recognized in the translational (below 350 cm$^{-1}$), librational (350–1200 cm$^{-1}$) and the stretching (above 2800 cm$^{-1}$) mode range. The polarization dependences of these spectra were analyzed satisfactorily. The bending mode range (1300–2700cm$^{-1}$) was too complicated to assign them properly [46].

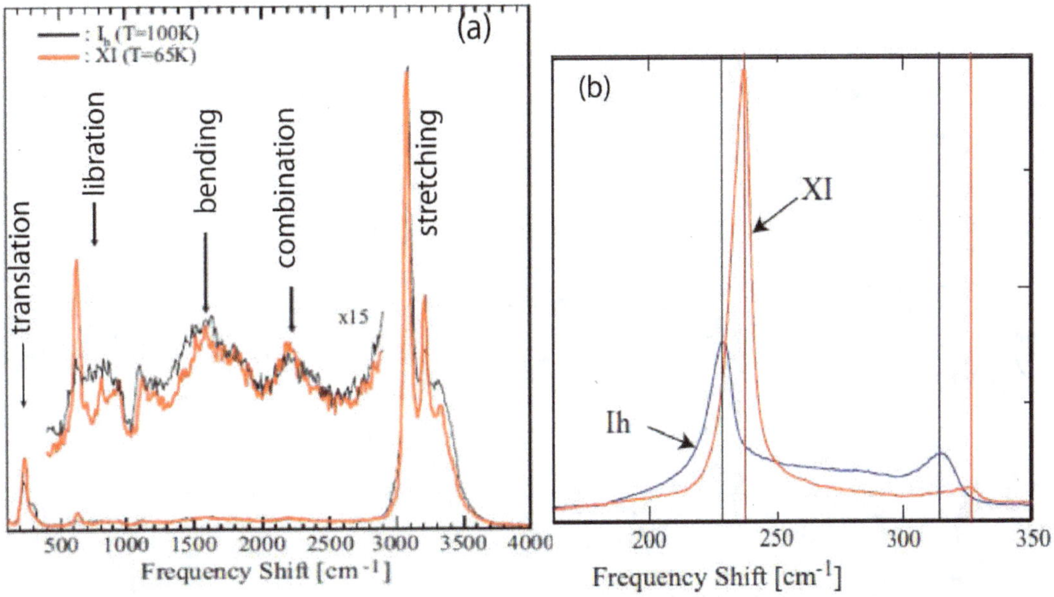


**Figure 15.** (a) Raman spectra in a wide frequency range. Scattering geometry is a(c, ∗)b with no polarization analyzer. Ice-XI is in red and ice-Ih (heated above $T_c$) is in black. (b) High-resolution Raman spectra in the translational mode range. Scattering geometry is a(cc)b Very strong peak at 237 cm$^{-1}$ which corresponds to the $A_1$ component of LO is observed [46].

## 6.2. LO/TO splitting of the translational mode

LO/TO splitting of polar modes in ice-Ih has been a topic of long-standing discussion and particularly the translational mode spectra below 350 cm$^{-1}$ were the controversial subjects. In 1991, Klug et al. [47] suggested from the analysis of IR reflection spectra of ice-Ih that a peak near 230 cm$^{-1}$ is TO mode and the corresponding LO mode would be about 4 cm$^{-1}$ higher than the TO. However, no direct evidence of the LO/TO splitting has yet been provided. Therefore, measurement of the Raman spectra of ice-XI and comparison with that of ice-Ih are important to solve the long-standing questions.

In the higher-resolution spectra in ice-XI, clear polarization dependences were observed and successfully assigned most of the translational modes by taking into account the depolarization effect based on the simplified point mass model for each water molecule [46].

Displacements of the water molecules of the related translational modes are shown in Fig. 16. Among the 9 translational modes in ice-Ih, the degenerate $E_{1g}$ and $E_{2g}$ are lifted as $E_{2g} \rightarrow A_1 + A_2$ and $E_{1g} \rightarrow B_1 + B_2$, respectively. Although the displacement patterns are similar for these modes, the frequencies of the modes parallel to $\pm b$ axes are expected to become higher than those parallel to the $a$ axes due to the depolarization field. Therefore, for the case of $\vec{K}_p // c$, LO/TO splitting is expected for the modes shown in the squares in Fig. 16.

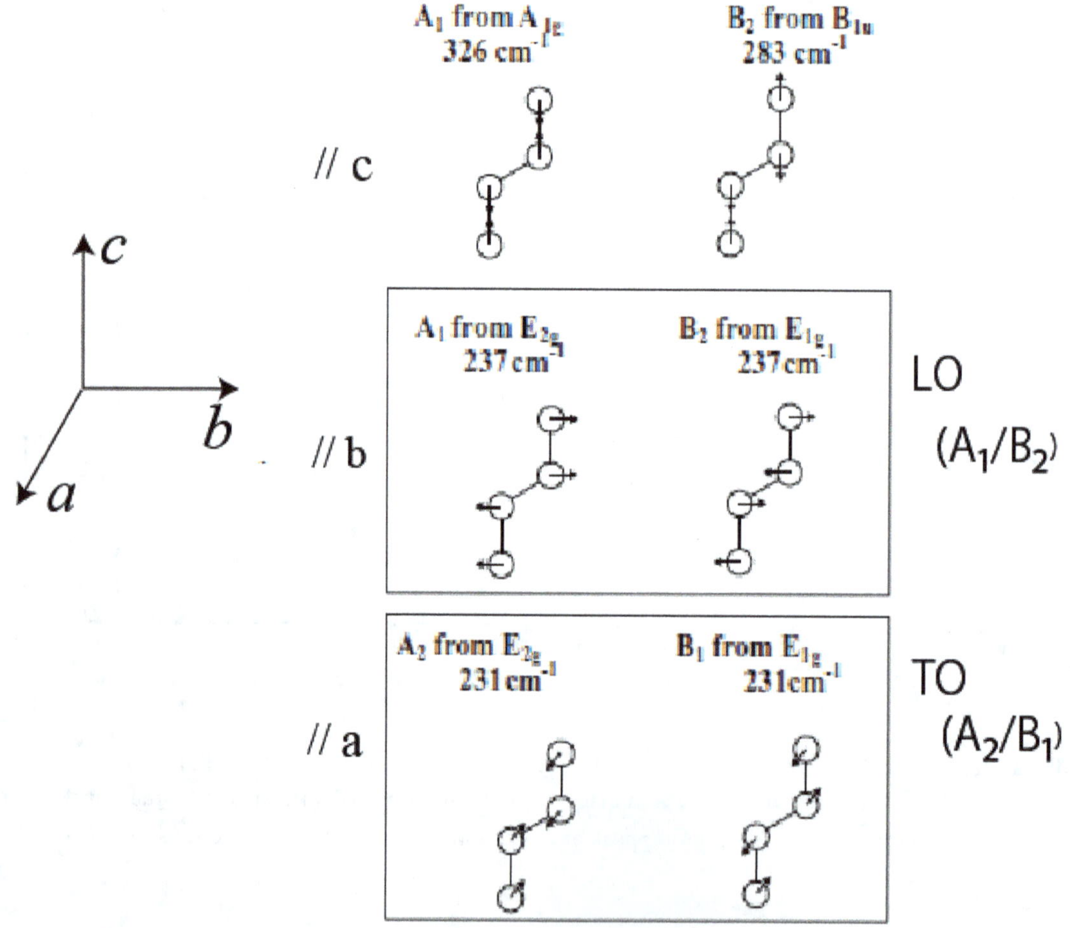


**Figure 16.** Dsplacement patterns and the symmetry of translational modes of ice-XI. The modes parallel to $\pm b$ axes are LO (237 $cm^{-1}$) and those parallel to $a$ axes are TO mode (231 $cm^{-1}$).

The assignment of LO/TO splitting was confirmed by the $\vec{K}_p$ dependence of Raman spectra shown in Fig. 17. The most important feature is the fact that the very strong mode at 237 $cm^{-1}$ is seen only in the geometry b(c, c)$\bar{b}$ (Fig. 17(a)) and in a(c, c)b in Fig. 15(b). In all other geometries and polarizations, no frequency shift is observed. This means that the shift of the peak from 231 cm$^{-1}$ to 237 cm$^{-1}$ takes place only when $\vec{K}_p$ has a component parallel to the *b*-axis. In other words, the modes at 237 cm$^{-1}$ should be assigned as the LO modes while the mode at 231 cm$^{-1}$, of which displacements are perpendicular to *b* is the corresponding TO mode. When the effect of the local depolarization electric field is larger than the crystal anisotropy, the LO/TO character dominates the symmetry of modes. Therefore, it is more appropriate to assign the peak at 237 cm$^{-1}$ as the LO mode with mixed symmetry of $A_1/B_2$ and the peak at 231 cm$^{-1}$ as the TO mode with mixed symmetry of $A_2/B_1$. Contribution of $B_2$ component to LO was confirmed in the a(b,c)c spectra shown in Fig.7 of ref.[46].

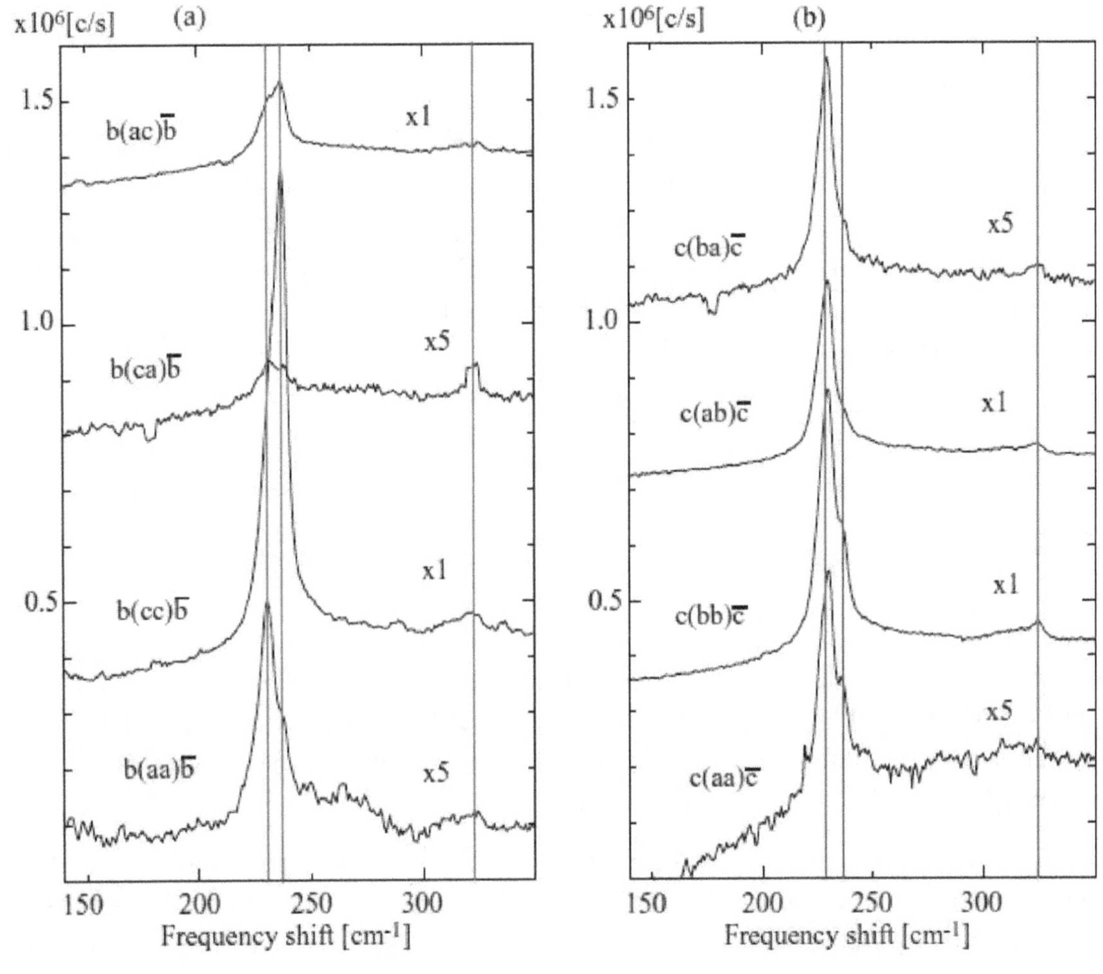


**Figure 17.** Dependence of the Raman spectra on the phonon propagation direction $\vec{K}_p$. (a) $\vec{K}_p // b$ and (b) $\vec{K}_p // c$. Very strong peak at 237 cm$^{-1}$ is observed only in (cc) spectrum in (a) corresponding to the $A_1$ component of LO.

The present result is the first experimental confirmation of the LO/TO splitting in ice. The 6±0.5 cm$^{-1}$ of the LO/TO splitting agrees with 4cm$^{-1}$ obtained in the IR reflection spectra of ice-Ih by Klug et al. [47] and also close to 10 cm$^{-1}$ of the calculation by Marchi et al. [48]. Another feature of the LO ($A_1/B_2$) mode at 237 cm$^{-1}$ is the fact that its intensity is strong only in the spectra with polarization (cc). From the selection rule, the $A_1$ mode is active also in (aa) and (bb). However, the intensity in (aa) and (bb) is much weaker than that in (cc). This indicates that the Raman polarizability tensor $R_{cc} = \partial\alpha_{cc}/\partial Q_{A1}$ is much larger than $R_{bb}$ or $R_{aa}$. Large $R_{cc}$ may be attributed to the depolarization field parallel to the c-axis caused by the partial ferroelectric order in ice-XI.

The behavior of the highest translational mode near 326 cm$^{-1}$ seems to be more complicated. It does not depend on $\vec{K}_p$ and polarization. Its intensity does not significantly increase by the transformation from ice-Ih to ice-XI. The differences from other translational modes cannot be explained by the simple point mass model. Maybe the modes with the displacement parallel to the c-axis (upper two modes in Fig. 16), are more complicated due to the long-range Coulomb force along the c-axis, which induces the interaction with other degrees of freedom such as the librational motion of water molecules.

### 6.3. Raman spectra of stretching modes in Ice-XI

Interpretation of Raman and infrared spectra of ice in the OH stretch region above 2800 cm$^{-1}$ has been also a topic of long-standing discussion. The lowest Raman peak and the central IR peak were assigned to in-phase symmetric $(\nu_1)$ and antisymmetric stretch $(\nu_3)$ vibrations of the two O-H bonds in a single water molecule, respectively (See Fig.18(c)). Whalley summarized the spectral features in ice-Ih and assigned the four stretching bands in terms of LO/TO splitting of $\nu_3$ vibrations as follows [49]:

band 1 3083 cm$^{-1}$ = $\nu_1$ (in-phase)

band 2 3209 cm$^{-1}$ = $\nu_3$ (TO)

band 3 3323 cm$^{-1}$ = $\nu_3$ (LO)

band 4 3420 cm$^{-1}$ = $\nu_1$ (out-of-phase)

Figure 18 shows the polarized Raman spectra in the range 2900–3600 cm$^{-1}$ of single crystal of ice-Ih and ice-XI at about 60 K. Spectra are composed of four bands and look similar for both phases except for the significant intensity increase in the depolarized spectra of the band 1 such as a(b, c)b [50].

In order to see whether Whalley's assignments are valid in ice-XI or not, spectra for different propagation vectors $\vec{K}_p$ were measured. Observed spectra for $\vec{K}_p//c$ and //b are, however, almost identical to the spectra for $\vec{K}_p \perp c$ shown in Fig. 18(b). The frequency of the band 3 (3327 cm$^{-1}$) does not vary in c(, )$\bar{c}$ or in b(, )$\bar{b}$. Furthermore, as shown in Fig.13 of ref. [50], the polarized Raman spectra (aa) and (bb) have the same intensity as that of (cc). This means that, in contrast to the translational mode case, Raman intensity of the stretching modes is not affected by the long-range electric field produced by the transformation to ice-XI. Thus, from the present work we could not find any direct evidence of band 3 being the $\nu_3$ (LO) mode. We assign the band 2 and 3 simply as the in-phase and out-of-phase of $\nu_3$ mode. Although the lowest frequency band 1 is no doubt the in-phase symmetric stretching mode $\nu_1$, the origin of the highest frequency weak band $\nu_4$ near 3400 cm$^{-1}$ is still not certain. Present results show that for the

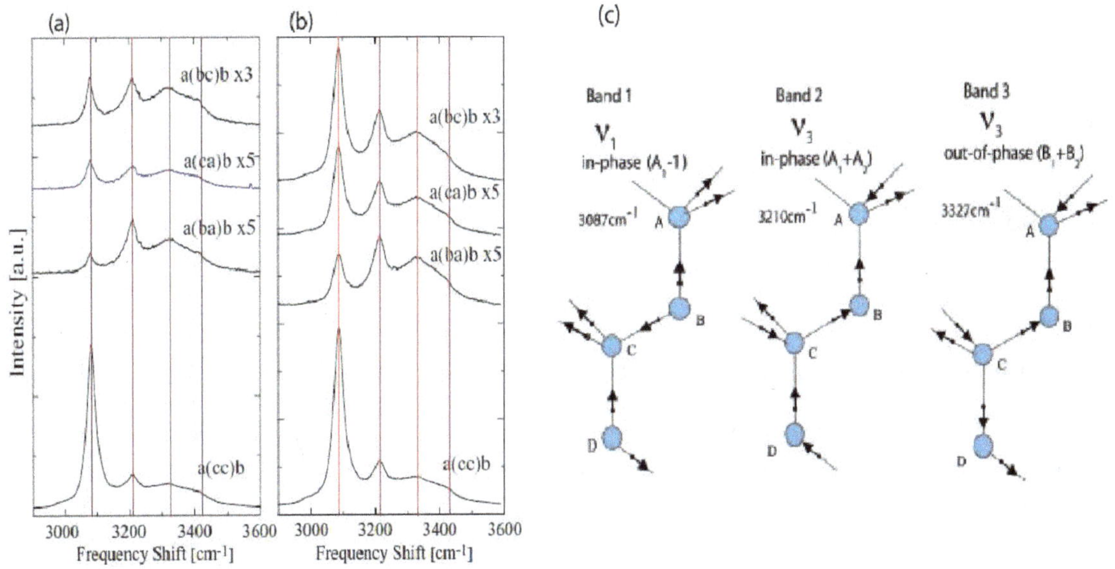


**Figure 18.** Polarization dependence of Raman spectra at about 60 K in the stretching mode range. (a) Ice-Ih and (b) ice-XI. Scattering geometries are a( , )b. (c) Displacement pattern of stretching mode corresponding to band 1,2 and 3. Only the depolarized spectra of band 1 significantly increase in ice-XI.

stretching vibrations in ice I (Ih and XI), the effect of the short-range intermolecular coupling is dominant than the long-range interaction produced by the proton ordering [50].

## 7. Summary and conclusions

Lattice vibrations that carry an electric dipole moment (polar phonons) have radically different long-wavelength properties from nonpolar vibrations. Soft modes in ferroelectric crystals are the typical polar modes. In this chapter, characteristics of soft modes and Raman spectra of polar modes were described mainly based on our experimental results.

In section 2, general properties of polar modes are given. In section 3, the physical meaning of the susceptibility $\chi_Q(\omega)$ such as the damped-harmonic-oscillator (DHO) and the Debye-type relaxation, which are often used in the analysis of the overdamped soft mode Raman spectra, was discussed in terms of the proposed generalized form of susceptibility (GVWF).

In section 4, the difficulty in discriminating between the displacive-type and the order–disorder-type phase transition is shown in the case of KDP. Particularly, the analysis of the overdamped soft mode spectra using an arbitrary susceptibility could be very ambiguous.

In section 5, the peculiar nature of the Raman spectra related to the ferroelectric transition of the isotope exchanged ferroelectric $SrTiO_3$ (STO18) was discussed. From the spectra with various propagation directions ($\overline{K}_p$), it was shown that the spontaneous polarization $\vec{P}$ is not a result of the freezing of the Slater mode. Therefore, the ferroelectric phase transition of STO18 is *not* an ideal soft mode type quantum phase transition.

In section 6, the first clear evidence of the LO/TO splitting in the Raman spectra of the translational mode in the proton-ordered (ferroelectric) ice-XI was given.

To conclude, the following points must be taken into account for the correct interpretation of Raman spectra in ferroelectric crystals:

1. The analysis of the overdamped mode by the computer fitting depends on the model and the used susceptibility function. The results would be often ambiguous as in the cases of KDP.

2. The effect of the depolarization field caused by the spontaneous polarization $\vec{P}$ below $T_c$ must be taken into account by the measurement of the propagation direction ($\overline{K}_p$) dependence, as in the cases of STO18 and ice-XI.

3. The influence of domains and the possible inhomogeneity in the sample must be taken into account. as in the cases of KDP and STO18.

4. The selection rules and the displacement patterns must be taken into account by the polarized Raman spectra using different scattering geometries, as in the case of STO18 and ice-XI.

5. The evidence of the LO/TO splitting should be obtained by the measurement of the propagation direction ($\overline{K}_p$) dependence, as shown in ice- XI.

## Acknowledgements

The author thanks Yasunari Takagi for the collaboration during the analysis of the Raman spectra of KDP and the generalization of the susceptibility. He also thanks Kohji Abe for his long-term collaboration and valuable discussions during the studies on STO18 and ice-XI. We wish to thank Prof. Mitsuru Ito of Tokyo Institute of Technology for providing the high-quality sample of STO18. The experiments were supported by the enthusiastic collaboration of a number of graduate students in UEC.

## Author details

Takeshi Shigenari*

Address all correspondence to: shigenar@pc.uec.ac.jp

Department of Applied Physics and Chemistry, The University of Electro-Communications, Chofu-shi, Tokyo, Japan

## References

[1] P.W.Anderson, Fizika dielektrikov (Akad. Nauk SSSR) 290 (1960) ; W.Cochran, Adv. in Phys. 9 387 (1960).

[2] J.M.Worlock, *Structural phase transitions and soft modes* (Universitetforlaget, Oslo). 329 (1971).

[3] J.F.Scott, Rev.Mod.Phys. 46 83 (1974).

[4] W.Hayes and R.Loudon, *Scattering of Light by Crystals*, Dover Publications, Inc. Mineola, New York 1978.

[5] C.H.Henry and J.J.Hopfeld, Phys.Rev.Lett. 15 964 (1965).

[6] Exception of the Worlock's conjecture is given by T.Shigenari, Phys.Lett. 46A 243 (1973).

[7] Y.Takagi and T.Shigenari, J.Phys.Soc.Jpn. 39 440 (1975).

[8] J.H.Van Vleck and V.F.Weisskopf, Rev.Mod.Phys. 17 227 (1945).

[9] Y.Takagi and T.Shigenari, J.Raman Spectroscopy 10 158 (1981) ; Y.Takagi, J.Phys.Soc.Jpn. 47 567 (1979).

[10] I.P.Kaminov and T.C.Damen, Phys.Rev.Lett. 20 1105 (1968).

[11] R.Blinc, J.Phys.Chem.Solids 13 204 (1960).

[12] W.Cochran, Phil.Mag. Suppl. 10 401 (1961).

[13] P.G.de Gennes, Solid St.Comm. 1 132 (1963).

[14] M.Tokunaga, Prog.Theor. Phys. (Kyoto) 36 357 (1966).

[15] Y.Takagi and T.Shigenari, J.Phys.Soc.Jpn. 47 576 (1979).

[16] Y.Tominaga and H.Urabe, Solid St.Comm. 41 561 (1982).

[17] M.Takesada, M.Itoh and T.Yagi, Phys.Rev.Lett. 96 227602 (2006).

[18] Y.Takagi, J.Phys.Soc.Jpn. 47 567 (1979).

[19] T.Shigenari and Y.Takagi, J.Phys.Soc.Jpn. 31 312 (1971). Temperatures were corrected in T.Shigenari, Butsuri 30 733 (1975). (in Japanese)

[20] Y.Tominaga, H.Urabe and M.Tokunaga, Solid State Comm. 48 265 (1983) ; Y.Tominaga and M.Tokunaga, KotaiButsuri(Solid state physics) 18 724 (1983). (in Japanese)

[21] T.Shigenari, Kotai Butsuri (Solid state physics) 23 600 (1988). (in Japanese)

[22] K.K.Kobayashi, J.Phys.Soc.Jpn. 24 497 (1968).

[23] T.Shigenari and Y.Takagi, Solid State Comm. 11 481 (1972) ; Y.Takagi and T.Shigenari, J.Opt.Soc.America 63 995 (1973).

[24] K.A.Müller and H.Burkard, Phys. Rev. B 19 3593 (1979): R.A.Cowley, Phys.Rev. 134 A981(1964).

[25] M.Itoh, R.Wang, Y.Inaguma, T.Yamaguchi, Y.-J.Shan and T.Nakamura, Phys. Rev. Lett. 82 3540 (1999).

[26] R.A.Cowley, Phys. Rev. 134, A981 (1964).

[27] A.Yamanaka, M.Kataoka, Y.Inaba, K.Inoue, B.Hehlen and E.Courtens, Europhys. Lett. 50 688 (2000).

[28] A.S.Barker, Jr. and R.Loudon, Rev.Mod. Phys. 44 18 (1972).

[29] G.Shirane and Y.Yamada, Phys. Rev. 177 858 (1969).

[30] K.Inoue, Ferroelectrics 52 253 (1983) ; H.Vogt, Phys. Rev. B51 8046 (1995).

[31] P.A.Fleury and J.M.Worlock, Phys. Rev. Lett. 18 655 (1968) ; P.A.Fleury and J.M.Worlock, Phys. Rev. 174 613 (1968).

[32] T.Shigenari and H.Ebashi, J.Phys.C: Solid State Phys. 14 969 (1981).

[33] T.Shigenari, Phys.Lett. 98A 63 (1983).

[34] Y.Noda, Private Communications. (2010).

[35] Y.Tsujimi, M.Kobayashi, Y.Minaki, M.Nakanishi, R.Wang, M.Itoh and T.Yagi, Ferroelectrics 304 71 (2004).

[36] T.Shigenari, K.Abe, T.Takemoto, O.Sanaka, T.Akaike, Y.Sakai, R.Wang and M.Itoh, Phys. Rev. B74 174121 (2006).

[37] T.Shigenari, T.Nakano and K.Abe, Europhys.Lett. 94 57001 (2011).

[38] M.Takesada, M.Itoh and T.Yagi, Phys.Rev.Lett. 96 227602 (2006) ; M.Takesada, M.Itoh, A.Onodera and T.Yagi, Ferroelectrics. 346 20 (2007).

[39] H.Taniguchi, M.Itoh and T. Yagi, Phys.Rev.Lett. 99 017602 (2007).

[40] R.Blinc, B.Zalar, V.V.Laguta and M.Itoh, Phys.Rev.Lett. 94 147601 (2005).

[41] J.D.Bernal and R.H.Fowler, J. Chem. Phys. 1 515 (1933).

[42] L.Pauling, J. Am. Chem. Soc. 57 2680 (1935).

[43] S.Kawada, J. Phys. Soc. Jpn. 32 1442 (1972).

[44] T.Matsuo, Y.Tajima and H.Suga, J. Phys. Chem. Solids. 47 165 (1986).

[45] R.Howe and R.W. Whitworth, J. Chem. Phys. 90 4450 (1989).

[46] K.Abe and T.Shigenari, J. Chem. Phys. 134 104506 (2011).

[47] D.D.Klug, J.S.Tse and E.Whalley, J. Chem. Phys. 95 7011 (1991).

[48] M.Marchi, J.S.Tse and M.L.Klein, J. Chem. Phys. 85 2414 (1986).

[49] E.Whalley, Can. J. Chem. 55 3429 (1977).

[50] T.Shigenari and K.Abe, J. Chem. Phys. 136 174504 (2012).

# 7

# Origin of Piezoelectricity in Piezoelectric Ceramics from the Viewpoints of Elastic Constants Measured by Acoustic Wave Velocities

Toshio Ogawa

**Abstract**

This chapter consists of two parts; the first part describes *"how can high piezoelectricity be realized from measuring acoustic wave velocities?"* That is, the measurement of sound velocities resulted in an effective tool for researching and developing piezoelectric materials, and, furthermore, it was possible to design the material compositions of lead-free piezoelectric ceramics as well as lead-containing ceramics. The second part describes the *"effects of firing and DC poling treatments on elastic constants measured from acoustic wave velocities in barium titanate piezoelectric ceramics"*. Namely, it could be applied to manufacturing processes such as firing and DC poling treatments in piezoelectric ceramics, which is accompanied with densification of ceramics and ferroelectric domain aligment, respectively. It is clarified through the two parts that the evaluation of elastic constants leads to the origin of piezoelectricity in piezoelectric ceramics.

**Keywords:** origin of piezoelectricity, piezoelectric ceramics, acoustic wave velocities, elastic constants, domain alignment

# 1. Introduction

## 1.1. How can high piezoelectricity be realized from measuring acoustic wave velocities?

Lead-free piezoelectric ceramics have been studied by many researchers [1-4], because of replacing Pb(Zr, Ti)$O_3$ (PZT) ceramics. There are three major chemical compositions: alkali niobate [5], alkali bismuth titanate [6], and barium titanate [7]. While relatively high piezoelectricity is realized in alkali niobate (the piezoelectric strain $d_{33}$ constant is 307 pC/N in 0.95(Na, K, Li, Ba)($Nb_{0.9}Ta_{0.1}$)$O_3$-0.05Sr$ZrO_3$ with a small amount of MnO [5, 8]) and barium titanate, low piezoelectricity with low dielectric constant and high mechanical quality factor is obtained in alkali bismuth titanate.

Improving the piezoelectricity in lead-free ceramics, a study on Young's modulus (Y) vs. piezoelectricity is important how to realize higher piezoelectricity in piezoelectric materials. We have already reported Y in PZT [9-13], Pb$TiO_3$ (PT) [14], Ba$TiO_3$ (BT) [7], alkali niobate ceramics composed of (Na, K, Li, Ba)($Nb_{0.9}Ta_{0.1}$)$O_3$-Sr$ZrO_3$ (SZ) [5, 8] and in a relaxor single crystal of Pb[($Zn_{1/3}Nb_{2/3}$)$_{0.91}$$Ti_{0.09}$]$O_3$ (PZNT) [15-17] by measuring the impedance responses in various kinds of piezoelectric resonators. Figure 1 shows the relationships between Y and electromechanical coupling factors of transverse mode ($k_{31}$) and longitudinal mode ($k_{33}$) in piezoelectric materials. From these figures, it is clarified that the decrease in Y increased the piezoelectricity such as $k_{31}$ and $k_{33}$, because materials with lower Y were easy to deform by DC poling field. Therefore, it is said that the measurement of Y was important to obtain high piezoelectricity.

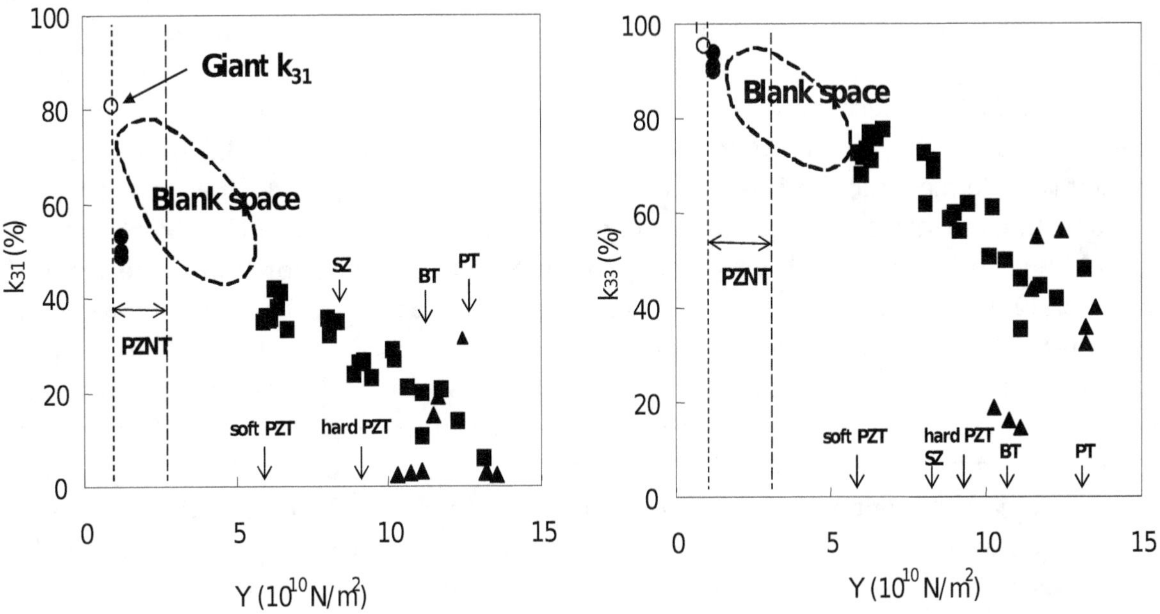


**Figure 1.** Relationships between Young's modulus (Y) and coupling factors of transverse mode ($k_{31}$) and longitudinal mode ($k_{33}$) in piezoelectric ceramics and relaxor single crystals; giant $k_{31}$ was realized in Pb[($Zn_{1/3}Nb_{2/3}$)$_{0.91}$$Ti_{0.09}$]$O_3$ (PZNT) single-crystal plate and there is blank space for coupling factors in the range of Y = 1 – 5 × $10^{10}$ N/m$^2$; PZT, PT, BT, and SZ mean Pb(Zr, Ti)$O_3$, Pb$TiO_3$, Ba$TiO_3$, and alkali niobate, respectively.

Recently, we developed a novel method to easily measure acoustic wave velocities suitable for conventional disk samples with ordinary dimensions (10-20 mm diameter and 0.5-2.0 mm thickness) by an ultrasonic thickness gauge with high-frequency pulse oscillation [18-20]. Therefore, this method was applied to hard and soft PZT [9-13, 21] and lead-free ceramics composed of alkali niobate [5, 8] and alkali bismuth titanate [6]. In this pursuit, we report the acoustic wave velocities in piezoelectric ceramics measured by our developed method and the calculation results of Young's modulus, Poisson's ratio, modulus of rigidity and bulk modulus, especially to obtain high piezoelectricity in lead-free ceramics. Furthermore, we propose the design for R&D on piezoelectric materials from a viewpoint of measuring acoustic wave velocities.

### 1.2. Experimental procedure

The piezoelectric ceramic compositions measured were as follows: $0.05Pb(Sn_{0.5}Sb_{0.5})O_3$-(0.95-$x$)$PbTiO_3$-$x$$PbZrO_3$ ($x$ = 0.33, 0.45, 0.48, 0.66, 0.75) with (hard PZT) and without 0.4 wt% $MnO_2$ (soft PZT) [9-13, 21]; $0.90PbTiO_3$-$0.10La_{2/3}TiO_3$ (PLT) and $0.975PbTiO_3$-$0.025La_{2/3}TiO_3$ (PT) [14]; (1-$x$)(Na, K, Li, Ba)$(Nb_{0.9}Ta_{0.1})O_3$-$x$$SrZrO_3$ (SZ) ($x$ = 0.00, 0.02, 0.04, 0.05, 0.06, 0.07) [5, 8]; (1-$x$)$(Na_{0.5}Bi_{0.5})TiO_3$ (NBT)-$x$$(K_{0.5}Bi_{0.5})TiO_3$ (KBT) ($x$ = 0.08, 0.18) and 0.79NBT-0.20KBT-$0.01Bi(Fe_{0.5}Ti_{0.5})O_3$ (BFT) ($x$ = 0.20) [6]; and (1-$x$)NBT-$x$$BaTiO_3$ (BT) ($x$ = 0.03, 0.07, 0.11) [6].

DC poling was conducted for 30 minutes at the most suitable poling temperature ($T_P$) depending on the Curie points of the ceramic materials. The DC poling field (E) depended on the coercive fields and the insulation resistance of the piezoelectric ceramics. The DC poling conditions are as follows: E = 3, 000 V/mm and $T_P$ = 150 ℃ for SZ; E = 2, 500-3, 000 V/mm and $T_P$ = 70 ℃ for KBT and BT; E = 3, 000 V/mm and $T_P$ = 80 ℃ for hard and soft PZT; E = 4, 000 V/mm and $T_P$ = 80 ℃ for PLT; E = 4, 000 V/mm and $T_P$ = 200 ℃ for PT, respectively. Before and after DC poling, the dielectric and piezoelectric properties were measured at room temperature using an LCR meter (HP4263A), a precision impedance analyzer (Agilent 4294A), and a piezo-$d_{33}$ meter (Academia Sinica ZJ-3D). Furthermore, the acoustic wave velocities were measured using an ultrasonic precision thickness gauge (Olympus 35DL), which has PZT transducers with 30 MHz for longitudinal wave ($V_L$) generation and 20 MHz for transverse wave ($V_S$) generation [22]. The acoustic wave velocities were evaluated on the basis of the propagation time between the second-pulse echoes in the thickness of ceramic disks parallel to the poling field with dimensions of 14 mm diameter and 0.5-1.5 mm thickness [18-20]. The sample thickness was measured using a precision micrometer (Mitutoyo MDE-25PJ). The number ($n$) of disk samples measured was $n$ = 5-8, and the data in the figures indicate the average of individual measured values. In addition, Young's modulus (Y), Poisson's ratio (σ), modulus of rigidity (G), and bulk modulus (K) in the thickness direction of ceramic disks were calculated on the basis of the $V_L$ and $V_S$, as shown in the following equations [23, 24]:

$$Y = 3\rho V_S^2 \frac{V_L^2 - \frac{4}{3}V_S^2}{V_L^2 - V_S^2} \tag{1}$$

$$\sigma = \frac{1}{2}\left\{1 - \frac{1}{\left(\frac{V_L}{V_S}\right)^2 - 1}\right\} \quad (2)$$

$$G = \rho V_S^2 \quad (3)$$

$$K = \rho\left(V_L^2 - \frac{4}{3}V_S^2\right) \quad (4)$$

where ρ is the bulk density of the ceramic disks. Figure 2 shows the relationships between the ratio of sound velocities ($V_S/V_L$), the elastic constants, and dielectric and piezoelectric constants.

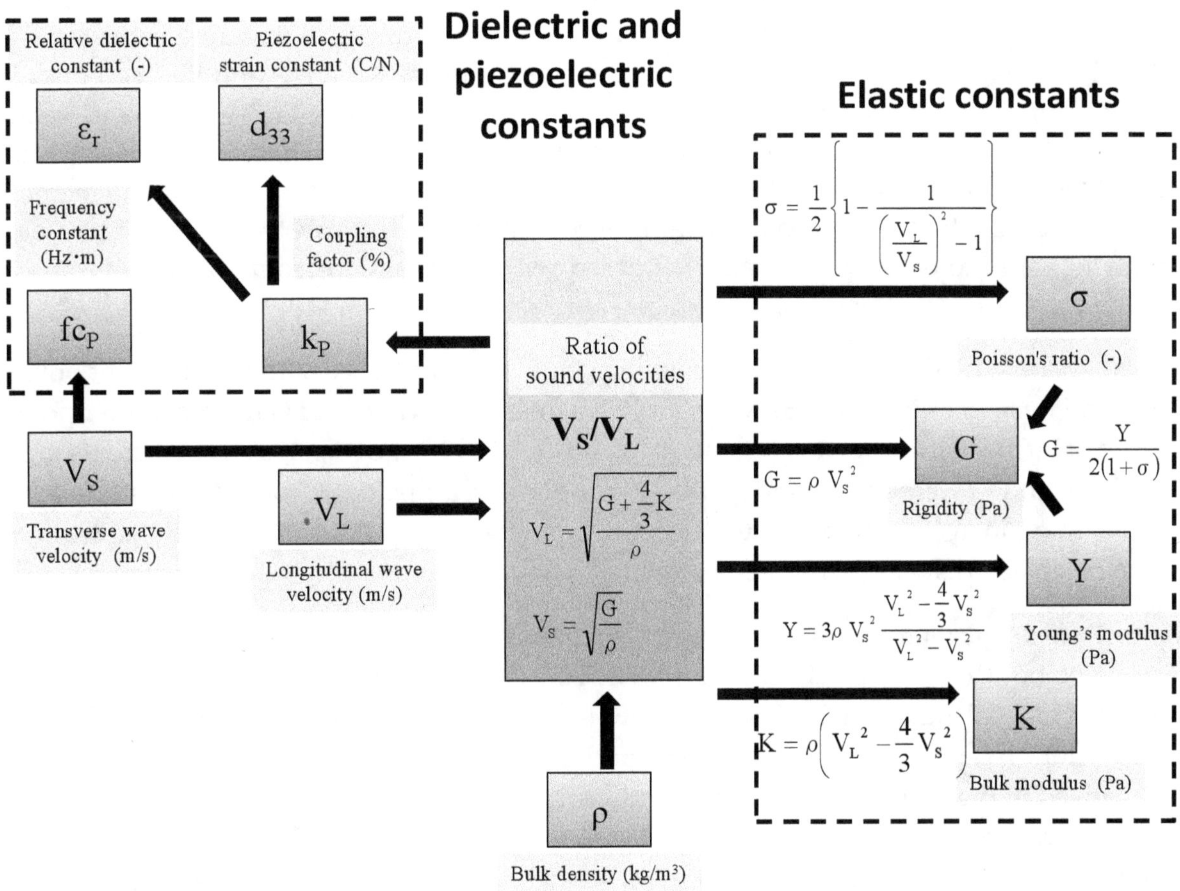


**Figure 2.** Relationships between the ratio of sound velocities ($V_S/V_L$), elastic constant and dielectric and piezoelectric constants.

## 1.3. Results and discussion

### *1.3.1. Dependence of planar coupling factor on elastic constant*

Figure 3 shows the relationships between longitudinal ($V_L$) and transverse ($V_S$) wave velocities, Young's modulus (Y), Poisson's ratio ($\sigma$), modulus of rigidity (G), and bulk modulus (K) vs. planar coupling factors ($k_p$) of disk in (1- *x*)(Na, K, Li, Ba)($Nb_{0.9}Ta_{0.1}$)$O_3$-*x*SZ (abbreviated to "SZ"), (1-*x*)NBT-*x*KBT ("KBT"), 0.79NBT-0.20KBT-0.01BFT ("KBT"), and (1-*x*)NBT-*x*BT ("BT") lead-free ceramics compared with 0.05Pb($Sn_{0.5}Sb_{0.5}$)$O_3$-(0.95-*x*)$PbTiO_3$-*x*$PbZrO_3$ ceramics with ("hard PZT") and without 0.4 wt% $MnO_2$ ("soft PZT"), and with 0.90$PbTiO_3$-0.10$La_{2/3}TiO_3$ ("PLT") and 0.975$PbTiO_3$-0.025$La_{2/3}TiO_3$ ("PT") ceramics before and after fully DC poling. In the case of after poling (marks and solid lines in Figure 3), although the $V_L$ values of the PZT ceramics were almost constant at approximately 4, 600-4, 800 m/s independently of the composition *x*, their $V_S$ values linearly decreased from 2, 500 to 1, 600 m/s with increasing $k_p$ from 20% to 65% (solid line). In addition, the $V_L$ and $V_S$ values of the PZT ceramics were smaller than those of the lead-free ceramics ($V_L$ = 5, 000-5, 800 m/s and $V_S$ = 2, 600-3, 000 m/s; solid lines). Although the $V_L$ values of the PT ceramics were almost the same (4, 800 m/s) as those of the PZT ceramics, the $V_S$ values of the PT ceramics were approximately 2, 700 m/s. On the other hand, the $V_L$ values of the SZ ceramics were relatively high (5, 500-5, 800 m/s); furthermore, the $V_S$ values of the SZ ceramics also increased (2, 600-2, 700 m/s) and linearly decreased with increasing $k_p$ from 25% to 50% (solid line), the behavior of which was almost the same as that of the $V_S$ values of the PZT ceramics. The $V_L$ values of the KBT, BT, and PLT ceramics (5, 000-5, 400 m/s) were between those of the PZT, PT, and SZ ceramics. However, the $V_S$ values of the KBT, BT, and PLT ceramics (2, 800-3, 000 m/s) were the highest. Therefore, it was possible to divide $V_L$ and $V_S$ into three material groups, namely, PZT and PT/ KBT, BT (alkali bismuth titanate), and PLT/ SZ (alkali niobate). In addition, $k_p$ increased from 4% to 65% with decreasing Y from 15 × $10^{10}$ to 6 × $10^{10}$ N/$m^2$ and increased with increasing $\sigma$ from 0.25 to 0.43. It was clarified that higher $k_p$ values can be realized at lower Y and G, and higher $\sigma$ and K.

In comparison with the values of before poling (the $k_p$ was made use of the values after poling; marks and dash lines in Figure 3), $V_L$, $\sigma$, and K increase and $V_S$, Y, and G decrease after poling because of ferroelectric domain alignment. In addition, while the correlation coefficients in the $k_p$ vs. Y, $\sigma$, and G were almost independent of poling treatment, the coefficients in the $k_p$ vs. K after poling increases from 0.49 to 0.74. It is thought that the increase in K is significant to realize piezoelectricity as mentioned below.

Figure 4 shows the relationships between $k_p$ vs. changes ($\Delta$) in longitudinal ($V_L$) and transverse wave velocities ($V_S$) [$\Delta V_L/\Delta V_S$], and changes in Young's modulus (Y), Poisson's ratio ($\sigma$), bulk modulus (K), and rigidity (G) [$\Delta Y/\Delta\sigma/\Delta K/\Delta G$] before and after DC poling in soft and hard PZT, $PbTiO_3$ (PT/PLT), alkali niobate (SZ), and alkali bismuth titanate (KBT/ BT). Higher $k_p$ was realized in the regions of large +$\Delta V_L$ and +$\Delta K$, and larger -$\Delta V_S$, -$\Delta Y$, and -$\Delta G$. There were thresholds regarding $k_p$ vs. $\Delta V_S$, $\Delta Y$, and $\Delta G$ around -5%, -7%, and -10%, respectively. On the other hand, there were no thresholds in the cases of $k_p$ vs. $\Delta V_L$ and $\Delta K$, especially $\Delta\sigma$. As there were $k_p$ peaks regarding $\Delta\sigma$ in hard and soft PZT ceramics and $k_p$ maximum in alkali niobate (SZ) ceramics, the compositions in $k_p$ peaks and $k_p$ maximum correspond to a morphotropic phase boundary (MPB) in PZT [25] and to take lowest value of $V_S/V_L$ in SZ (see the following

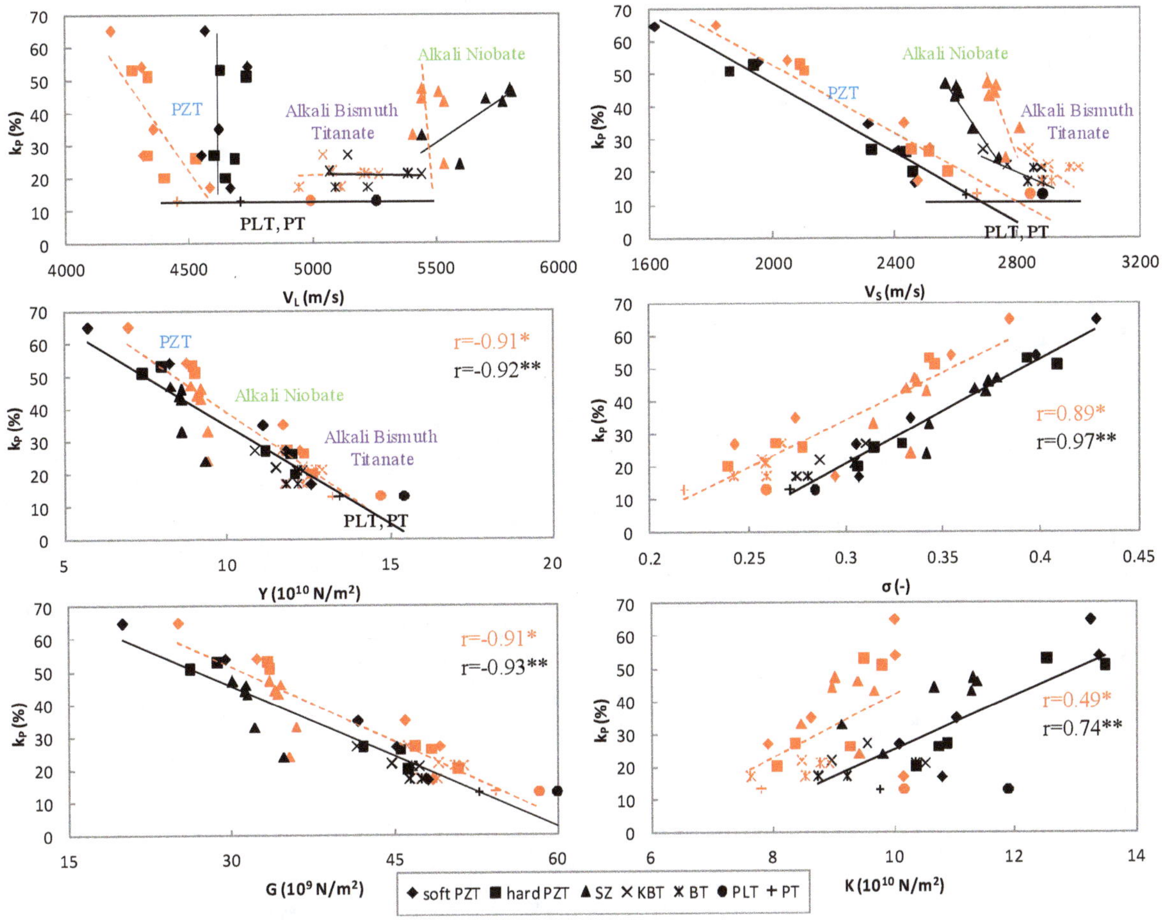


**Figure 3.** $V_L$, $V_S$, Y, σ, G, and K vs. $k_p$ before* (red marks, dash lines) and after poling** (black marks, solid lines) in lead-containing and lead-free ceramics (r: correlation coefficient).

Figure 10). We believe that the origin of piezoelectricity in piezoelectric ceramics was due to large change in $V_S$ ($-\Delta V_S$) while applying DC poling field parallel to the thickness of disks. Therefore, the larger changes in $V_S$ ($-\Delta V_S$) correspond to larger changes in Y ($-\Delta Y$) and G ($-\Delta G$). These phenomena mean that the origin of high piezoelectricity was due to the mechanical softness of the materials under compress stress (large $+\Delta V_L$ and $+\Delta K$). In addition, the realization of high piezoelectricity is easy deformation by DC poling field in diameter (large $-\Delta G$) as well in thickness (large $-\Delta Y$).

Figure 5 shows schematic charts between domain alignment and the changes in Y, σ, G, and K after DC poling. Comparing domain alignment before poling to the alignment after poling, same charges (+ or -) are generated and gathered in the regions of each ends by opposite charge due to DC poling field, namely orientation polarization which occurs by domain alignment. The orientation polarization acts by reducing of Y and G by repulsion to each other because there are same charges in domain alignment. The enhancing σ and K can be explained by the same phenomena. Therefore, it can be said that higher domain alignment leads to large changes in Y, G, and K. However, large change in σ ($+\Delta\sigma$ in Figure 4) does not lead to higher domain alignment since σ value is decided by the combinations of Y (G) and K after poling.

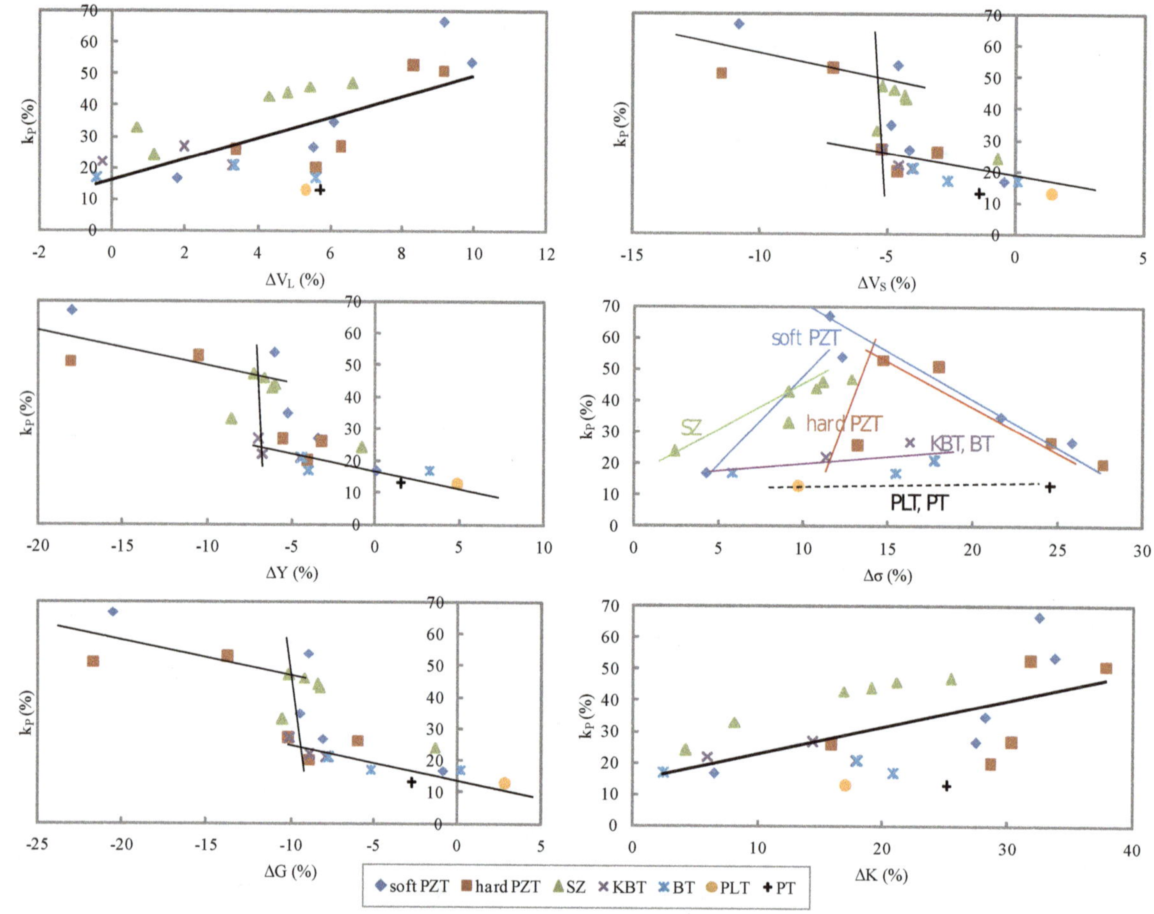


**Figure 4.** Changes (Δ) in $V_L$, $V_S$, Y, σ, G, and K vs. $k_p$ before and after poling in lead-containing and lead-free ceramics.

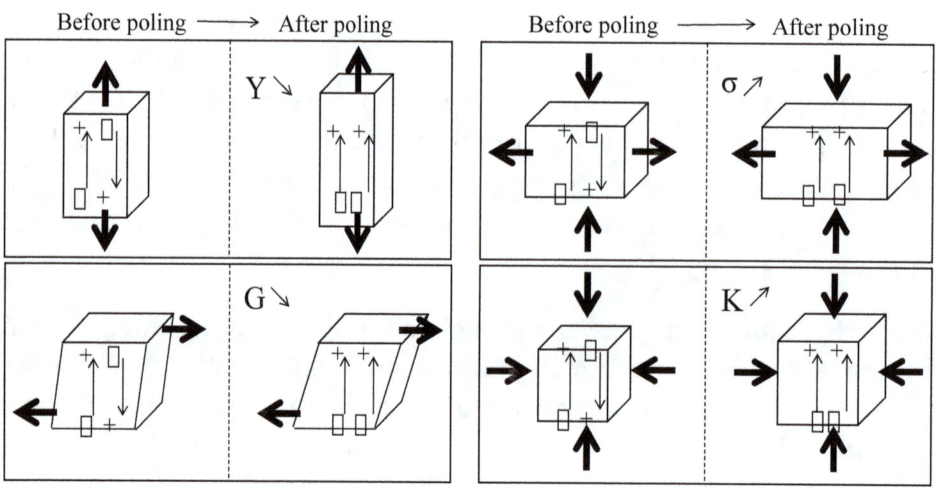


**Figure 5.** Effect of ferroelectric domain alignment by DC poling on Y, σ, G, and K; generated charges (- → +) accompanied with domain alignments, displacement (→) and changes in Y, σ, G, and K (↘: decrease, ↗: increase) after poling are also shown in this figure.

### 1.3.2. Design for research and development on lead-free piezoelectric ceramics

Figure 6 shows the relationship between $V_S/V_L$ vs. $k_p$. The $k_p$ linearly increased with decreasing $V_S/V_L$ in lead-free ceramics as well as lead-containing ceramics such as PZT, PLT, and PT. Furthermore, it was confirmed that $V_S/V_L$ was an effective figure to evaluate both the elastic constants and the piezoelectric constants.

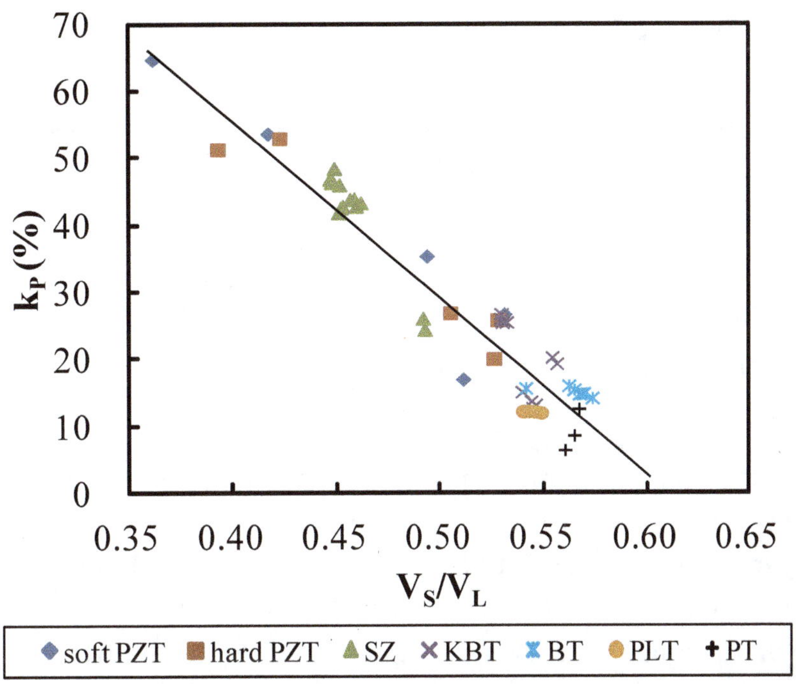


**Figure 6.** Relationship between ratio of sound velocities ($V_S/V_L$) and planar coupling factor ($k_p$) in lead-free and lead-containing ceramics.

When we research and develop new piezoelectric ceramics with high piezoelectricity in lead-free ceramics, we must need a new concept different from the conventional research looking for chemical compositions such as MPB [25]. From the equations (1)-(4) and the change in $V_L$ and $V_S$ before and after poling as mentioned previously (Figure 4), we focused on the ceramic bulk density (ϱ). Figure 7 shows the relationship $V_S/V_L$ vs. ϱ: ϱ of lead-containing ceramics (PZT, PLT, and PT) was independent of $V_S/V_L$. However, in lead-free ceramics (SZ, KBT, and BT) ϱ decreased with decreasing $V_S/V_L$. In the case of ϱ vs. $k_p$ in Figure 8, although $k_p$ in lead-containing ceramics (PZT, PLT, and PT) is independent of $V_S/V_L$, $k_p$ in lead-free ceramics (SZ, KBT, and BT) increased with decreasing ϱ. From the above results, we came to an important concept to obtain lead-free ceramics with high piezoelectricity, namely the R&D on lead-free ceramics with lower bulk density. As a result, it confirmed the importance of measuring $V_L$ and $V_S$ to evaluate the piezoelectricity.

It was said that the direction of the R&D on lead-free piezoelectric ceramics with high piezoelectricity was looking for ceramics with lower bulk density. For example, in perovskite structure, small cations at A and B sites in perovskite structure of $ABO_3$ were selected. In

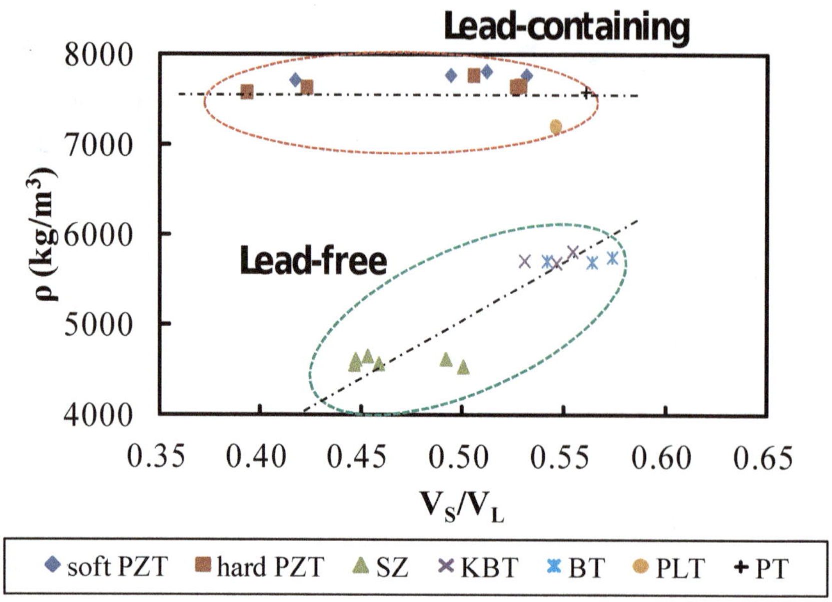


**Figure 7.** Relationship between ratio of sound velocities ($V_S/V_L$) and ceramic bulk density (ρ) in lead-free and lead-containing ceramics.

addition, for practical use, the Curie point (Tc) or depolarization temperature must be over 250 ℃. From the two items we will expect the candidates for new lead-free ceramics such as $SrTeO_3$ (ρ = 4.82 g/cm³, Tc = 485 ℃) and $YMnO_3$ (Tc = 640 ℃), in addition to $KNbO_3$ (ρ = 4.62 g/cm³, Tc = 418 ℃) and $LiNbO_3$ (ρ = 4.46 g/cm³, Tc = 1, 210 ℃), respectively [26, 27]. In fact, the bulk density (ρ) of SZ, which possessed the highest $k_p$ [5, 8] in lead-free ceramics we investigated, was around 4.6 g/cm³.

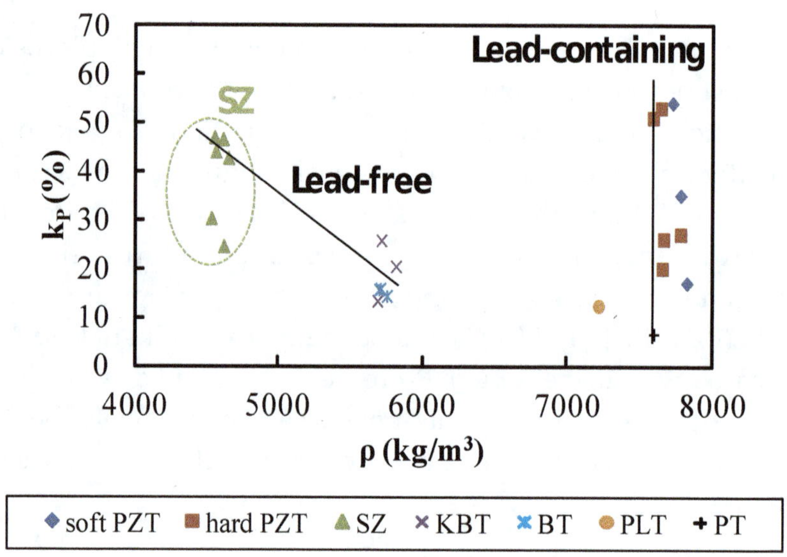


**Figure 8.** Relationship between ceramic bulk density (ρ) and planar coupling factor ($k_p$) in lead-free and lead-containing ceramics.

### *1.3.3. Ferroelectricity in almost same ceramic bulk density*

In addition to the relationships between $V_S/V_L$ vs. $k_p$ in Table 1, we evaluated the chemical composition dependence of $V_S/V_L$ in piezoelectric ceramics. Figure 9 shows the composition $x$ dependence of $k_p$ and $d_{33}$ in (1-$x$)(Na, K, Li, Ba)($Nb_{0.9}Ta_{0.1}$)$O_3$-$x$$SrZrO_3$ (SZ) ($x$ = 0.00, 0.02, 0.04, 0.05, 0.06, 0.07). While the $k_p$ in SZ compositions of $x$ = 0.00 - 0.07 was independent of ϱ in Figure 8, the maximum $k_p$ and $d_{33}$ were obtained at $x$ = 0.04 and 0.05, respectively (Figure 9). Figure 10 shows the $x$ dependence of $V_S/V_L$ in SZ before and after poling. The minimum $V_S/V_L$ after poling were obtained at $x$ = 0.04 - 0.05. From both figures, it was concluded that the highest $k_p$ appeared in the case of the minimum $V_S/V_L$. However, there is a composition without MPB at $x$ = 0.04 - 0.05 in SZ [5, 8]. Therefore, we could introduce the novel method to evaluate the piezoelectricity by measuring acoustic wave velocities of $V_L$ and $V_S$ in spite of the existence of MPB.

| | Lead-free | Lead-containing |
|---|---|---|
| Longitudinal wave velocity ($V_L$) | $V_L$↗ $k_P$↗ | $V_L$; independent of $k_P$ |
| Transverse wave velocity ($V_S$) | $V_S$↘ $k_P$↗ | $V_S$↘ $k_P$↗ |
| $V_S/V_L$ | $V_S/V_L$↘ $k_P$↗ | $V_S/V_L$↘ $k_P$↗ |
| Ceramic bulk density (ρ) | $V_S/V_L$↘ ρ↘<br>ρ↘ $k_P$↗ | ρ; independent of $V_S/V_L$ and $k_p$ |

**Table 1.** Relationships between longitudinal wave velocity ($V_L$), transverse wave velocity ($V_S$), ratio of sand velocities ($V_S/V_L$), ceramic bulk density (ϱ), and planar coupling factor ($k_p$) in lead-free and lead-containing ceramics.

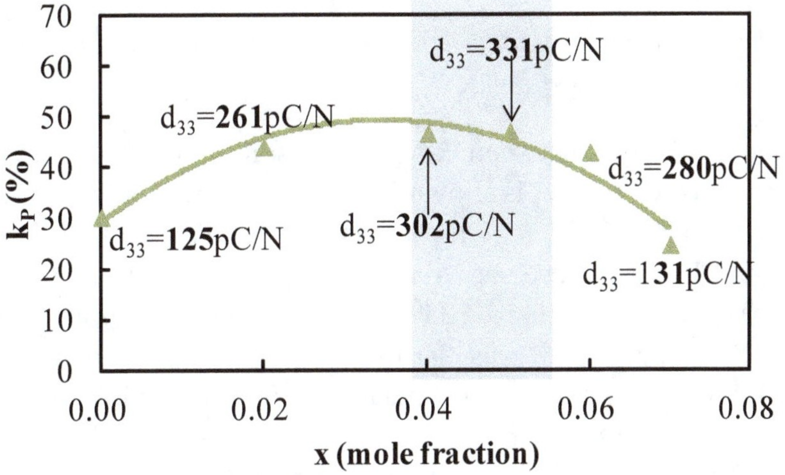


**Figure 9.** Composition $x$ dependence of $k_p$ and $d_{33}$ in (1–$x$)(Na, K, Li, Ba)($Nb_{0.9}Ta_{0.1}$)$O_3$-$x$$SrZrO_3$ (SZ) ($x$ = 0.00, 0.02, 0.04, 0.05, 0.06, 0.07).

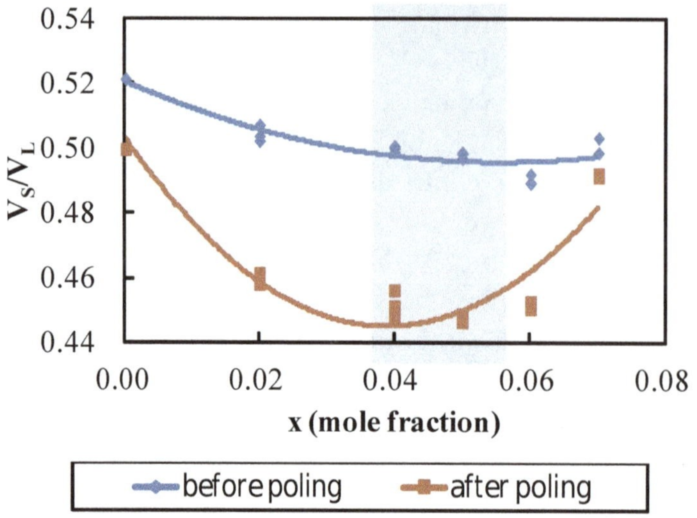


**Figure 10.** Composition $x$ dependence of ratio of sound velocities ($V_S/V_L$) in (1–$x$)(Na, K, Li, Ba)($Nb_{0.9}Ta_{0.1}$)$O_3$-$x$$SrZrO_3$ (SZ) ($x$ = 0.00, 0.02, 0.04, 0.05, 0.06, 0.07).

We applied this novel method to 0.05Pb($Sn_{0.5}Sb_{0.5}$)$O_3$-(0.95-$x$)$PbTiO_3$-$x$$PbZrO_3$ ($x$ = 0.33, 0.45, 0.48, 0.66, 0.75) with (hard PZT) and without 0.4 wt% $MnO_2$ (soft PZT) to confirm the effectiveness of our developed $V_S/V_L$ evaluation. While $k_p$ in PZT compositions of $x$ = 0.33 - 0.75 was independent of ϱ in Figure 8, the maximum $k_p$ was obtained at MPB around $x$ = 0.45 - 0.48 [11, 21]. Figure 11 shows the $x$ dependence of $V_S/V_L$ in PZT before and after poling. The minimum $V_S/V_L$ before and after poling appeared around $x$ = 0.45 - 0.48. From the relationships between $x$ vs. $k_p$ and $V_S/V_L$, it was concluded that the highest $k_p$ was realized in the case of minimum $V_S/V_L$. These compositions around $x$ = 0.45 - 0.48 correspond to MPB in hard and soft PZT. Therefore, it was said that we could introduce the novel method to evaluate the piezoelectricity by measuring acoustic wave velocities of $V_L$ and $V_S$ in the compositions with MPB.

### *1.3.4. Relationship between the ratio of transverse wave velocity to longitudinal wave velocity and Poisson's ratio*

Figure 12 shows the relationship between $V_S/V_L$ vs. σ in solids including piezoelectric ceramics. The equation between σ, $V_L$, and $V_S$ is shown in this figure. All of the σ in solids was plotted on a line of the equation. The $V_S/V_L$ regions of Poisson's ratio in soft PZT, hard PZT, lead-free and $PbTiO_3$ (PLT and PT) were also shown in this figure. The regions of σ regarding $V_S/V_L$ increased from $PbTiO_3$, lead-free, hard PZT to soft PZT with increasing piezoelectricity. In addition to higher Poisson's ratio, it was clarified that higher $k_p$ can be realized in lager K as shown in Figure 3. We believe the physical meaning of this behavior toward σ is as follows: increasing mechanical softness (lower Y) in piezoelectric materials, it becomes easy to deform by DC poling field. The σ becomes larger while the materials become softer, and furthermore, the K must become larger in order to transmit effectively from longitudinal deformation (the directions of poling and applying electric field are thickness direction of disk) to transverse deformation (the radial direction of disk). In this study, it was described a road map in

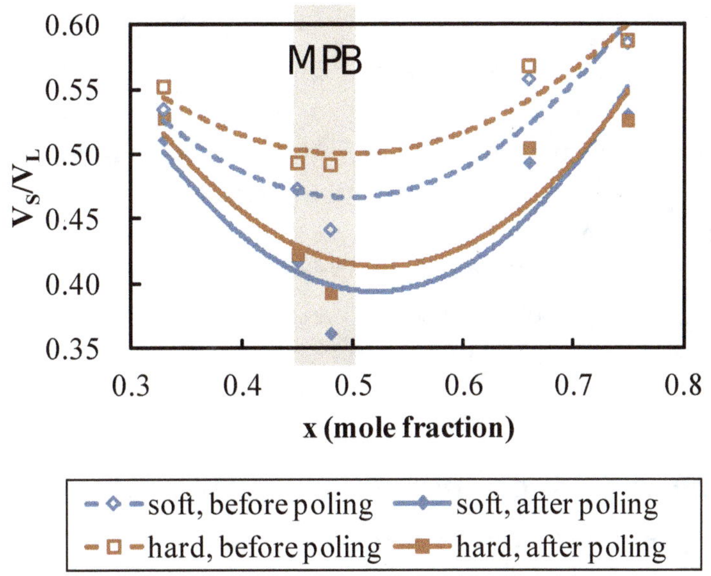


**Figure 11.** Composition $x$ dependence of ratio of sound velocities ($V_S/V_L$) in $0.05Pb(Sn_{0.5}Sb_{0.5})O_3$-$(0.95–x)PbTiO_3$-$xPbZrO_3$ ($x$ = 0.33, 0.45, 0.48, 0.66, 0.75) with (hard PZT) and without 0.4 wt% $MnO_2$ (soft PZT) ceramics before and after poling.

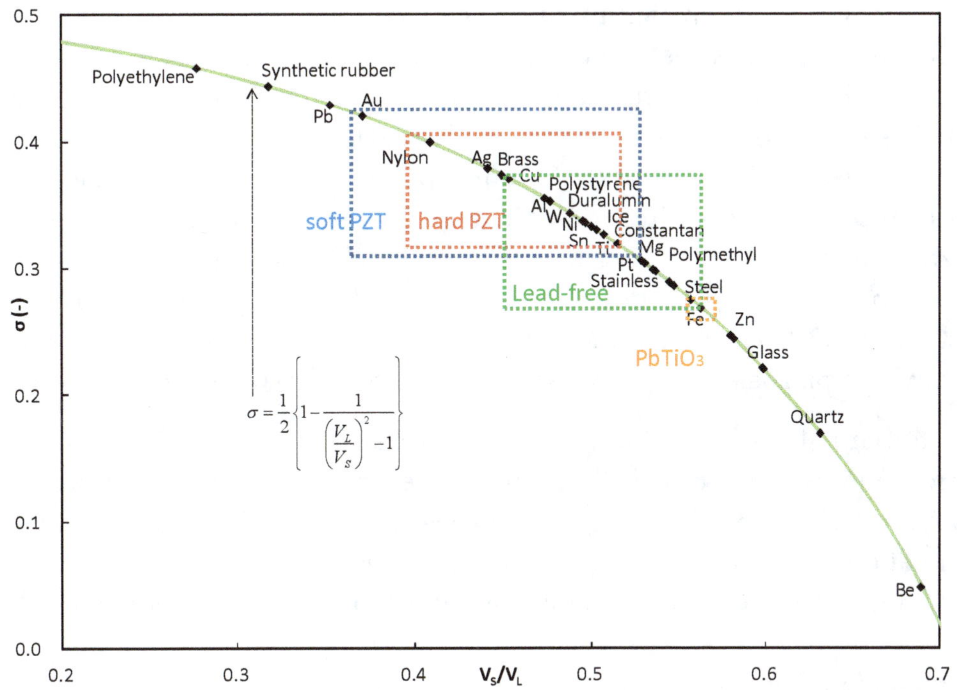


**Figure 12.** Relationship between ratio of sound velocities ($V_S/V_L$) and Poisson's ratio ($\sigma$) in solids including piezoelectric ceramics.

piezoelectric ceramics regarding the relationships between longitudinal and transverse wave velocities, Young's modulus, Poisson's ratio, modulus of rigidity and bulk modulus to research and develop new piezoelectric ceramic materials, especially lead-free ceramics with high piezoelectricity.

### 1.4. Conclusions in this part

Longitudinal and transverse wave velocities in PZT, lead titanate, and lead-free ceramics were measured by an ultrasonic precision thickness gauge with high-frequency pulse generation to calculate elastic constants such as Young's modulus, Poisson's ratio, and so forth. Since the ceramic bulk density was focused on to improve piezoelectricity in lead-free ceramics, the candidates of lead-free ceramic compositions with high piezoelectricity were proposed. It was confirmed that our evaluation method was an effective tool for R&D on piezoelectric material. Furthermore, the origin of piezoelectricity in piezoelectric ceramics could be explained by the elastic constants before and after DC poling.

## 2. Effects of firing and DC poling treatments on elastic constants measured from acoustic wave velocities in barium titanate piezoelectric ceramics

### 2.1. Introduction

Recently, we have developed a method to evaluate elastic constants, such as Young's modulus and Poisson's ratio, by measuring longitudinal and transverse wave velocities using an ultrasonic thickness gauge with high-frequency generation in comparison with a conventional method [28]. This method has been confirmed to be useful for measuring ceramic disks with diameters of 10-20 mm and thicknesses of 0.5-2.0 mm. In addition, it has been clarified that this method is suitable for evaluating cases involving (1) firing process analysis, such as the analysis of the effect of lead oxide (PbO) atmosphere during firing on the dielectric and piezoelectric properties of lead zirconate titanate (PZT) ceramics, and the oxygen atmosphere firing of PZT ceramics to realize pore-free ceramics, (2) DC poling process analysis, such as the analysis of the DC poling field dependence in as-fired (before poling) ceramics and relaxor single crystals, and (3) piezoelectric materials research and development (R&D) for lead-free ceramics with high piezoelectricity from the viewpoints of elastic constants [18-20].

In this part, to clarify the effects of firing temperature and DC poling on barium titanate (BT) piezoelectric ceramics, we studied the firing temperature and DC poling dependences on acoustic wave velocities and dielectric and piezoelectric properties of BT ceramics. Here, we report the relationships between firing temperature and DC poling effect vs. acoustic wave velocities, Young's modulus, Poisson's ratio, modulus of rigidity, and bulk modulus.

### 2.2. Experimental procedure

The BT raw materials in this study were utilized for two types of powder particle with high purities above 99.95% and average particle sizes of 0.2 μm (abbreviated to BT02) and 0.5 μm (BT05) (Sakai Chemical Industry). After firing at 1, 300-1, 350 ℃ for BT02 and at 1, 300-1, 360 ℃ for BT05 for 2 h, the bulk density ($\varrho$) and microstructure of the obtained ceramic disks were evaluated. DC poling was conducted at a temperature of 60 ℃ and a field of 2.0 kV/mm for 30 min. After DC poling, dielectric and piezoelectric properties were measured at room temperature using an LCR meter (HP4263A), a precision impedance analyzer (Agilent 4294A), and a piezo-$d_{33}$ meter (Academia Sinica ZJ-3D). Furthermore, the acoustic wave velocities of the BT ceramics before and after poling were measured using an ultrasonic precision thickness gauge (Olympus 35DL), which has PZT transducers with a frequency of 30 MHz for longitudinal wave generation and a frequency of 20 MHz for transverse wave generation. The acoustic wave velocities were evaluated on the basis of the propagation time between the second-pulse and the third-pulse echoes in the thickness direction parallel to the DC poling field for the ceramic disks with 14 mm diameter and 0.9-1.2 mm thickness [18-20]. The sample thickness was measured using a precision micrometer (Mitutoyo MDE-25PJ). The number ($n$) of disk samples measured was $n = 5$-8, and the data in the figures indicate the average of individual measured values. Furthermore, Young's modulus (Y), Poisson's ratio ($\sigma$), modulus of rigidity (G), and bulk modulus (K) in the thickness direction of ceramic disks were calculated on the basis of the longitudinal ($V_L$) and transverse ($V_S$) wave velocities using the equations (1)-(4) in Section 1.2. We investigated the relationships between firing temperature and DC poling effect vs. $V_L$, $V_S$, Y, $\sigma$, G, and K; furthermore, we clarified the relationships between $\varrho$, the microstructure, and the elastic constants.

### 2.3. Results and discussion

#### 2.3.1. *Firing temperature dependence of dielectric and piezoelectric properties*

Figures 13(a)-(d) show the relationships between firing temperature ($F_T$) and (a) relative dielectric constant ($\varepsilon_r$), (b) planar coupling factor ($k_p$) for the radial mode on the disks, (c) frequency constant ($fc_p$), and (d) piezoelectric strain $d_{33}$ constant in the cases of BT02 and BT05 ceramics before and after DC poling. Although $\varepsilon_r$ decreases with increasing $F_T$, $k_p$, $fc_p$, and $d_{33}$ with increasing $F_T$; furthermore, there is an optimum $F_T$ of 1, 340 ℃ in BT02 for obtaining the highest $k_p$ and $d_{33}$. The differences in the dielectric and piezoelectric properties vs. $F_T$ between BT02 and BT05 were due to the ceramic bulk density, as mentioned later (see the following Figure 17), and the powder particle activity during firing because the specific surface areas of BT02 and BT05 powder particles measured by the Brunauer, Emmett, and Teller (BET) method were 9.4 and 2.3 $m^2/g$, respectively.

#### 2.3.2. *Effects of firing temperature and DC poling on acoustic wave velocities and elastic constants*

Figures 14(a)-(d) show the relationships between $F_T$ and (a) longitudinal wave velocity ($V_L$), (b) transverse wave velocity ($V_S$), (c) Young's modulus (Y), and (d) Poisson's ratio ($\sigma$) in the cases of BT02 and BT05 ceramics. Although $V_L$, $V_S$, and Y before and after DC poling increase with increasing $F_T$, $\sigma$ after poling is almost independent of $F_T$; furthermore, there is an optimum $F_T$ of 1, 340 ℃ in BT02 from the plots of $F_T$ vs. $V_L$ and Y. The increase in Y with increasing $F_T$ indicates the increase in the mechanical hardness of the ceramic disks. By comparing $F_T$ vs. $V_L$ in Figure 14(a) with $F_T$ vs. $2fc_p$ ($fc_p$ is shown in Figure 13(c), the dependences of $V_L$ and $2fc_p$ on $F_T$ were almost the same, because both of them correspond to longitudinal wave velocities, as shown in Figure 15. In addition, we confirmed that $V_L$ precisely corresponded to $2fc_t$, which is twice the frequency constant ($fc_t$) of the coupling factor ($k_t$) for the thickness mode on the disks measured using the typical impedance vs. frequency response (Figure 15) [29].

Therefore, it is considered that our measurement method using the ultrasonic precision thickness gauge is suitable for evaluating acoustic wave velocities, especially in piezoelectric ceramics.

Figures 16(a)-(b) show the changes in Y (ΔY) and σ (Δσ) during poling, respectively. ΔY decreases with increasing $F_T$, regardless of the types of BT powder particle. This phenomenon indicates that the ceramics become mechanically softer after DC poling. It is considered that such a change is due to the ferroelectric domain (↑) alignment induced by the poling field, such as a change in the configuration from a random orientation (↑↓) before poling to a directional orientation (↑↑) after poling [5-7, 9-15]. Δσ abruptly increases from $F_T$ of 1, 320 ℃ and shows almost the same tendency as in the case of $F_T$ vs. $k_p$ in Figure 13(b). By comparing $F_T$ vs. $k_p$ (Figure 13(b)) with $F_T$ vs. Δσ (Fig. 16(b)), a higher $k_p$ was obtained in the case of a larger Δσ. The physical meaning of the phenomenon regarding σ was deduced, that is, a larger Δσ was needed to realize a higher $k_p$ because of the easy deformation of the bulk perpendicular to the poling field (radial direction in the disks) as well as parallel to the poling field (thickness direction in the disks).

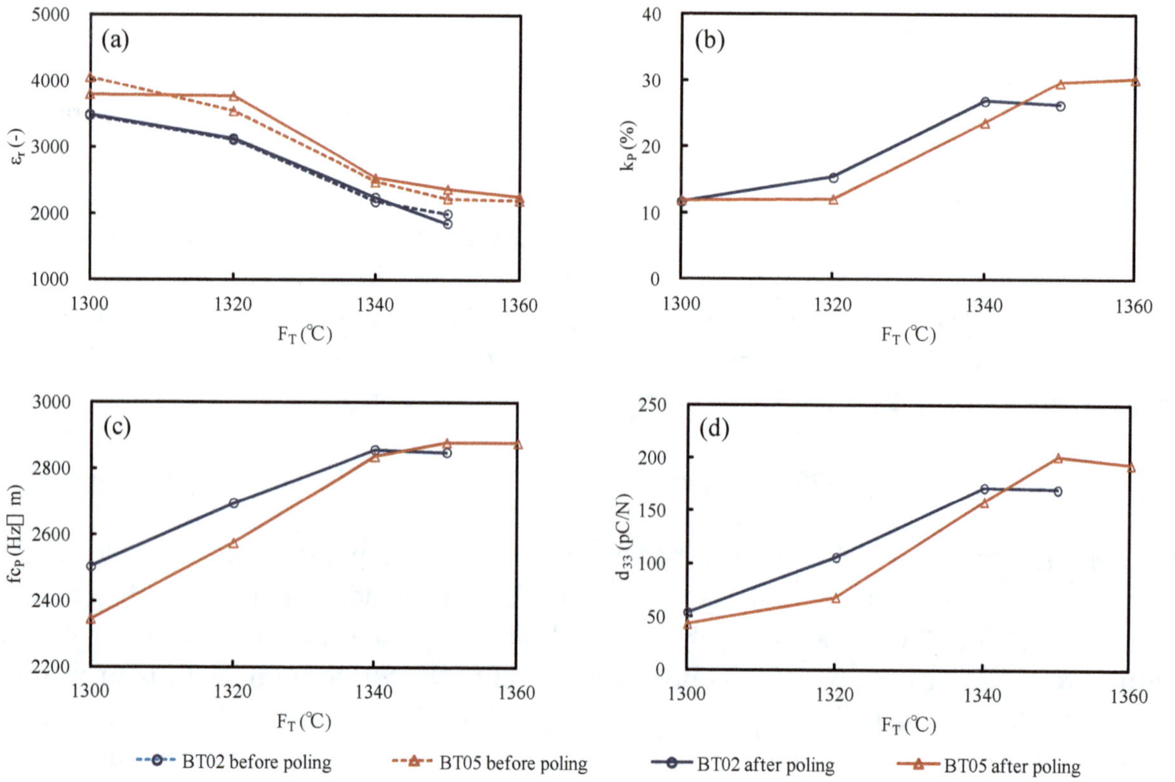


**Figure 13.** Firing temperature ($F_T$) vs. (a) relative dielectric constant ($\varepsilon_r$), (b) planar coupling factor ($k_p$) for radial mode on disks, (c) frequency constant ($fc_p$), and (d) piezoelectric strain $d_{33}$ constant in BT02 and BT05 ceramics before and after DC poling.

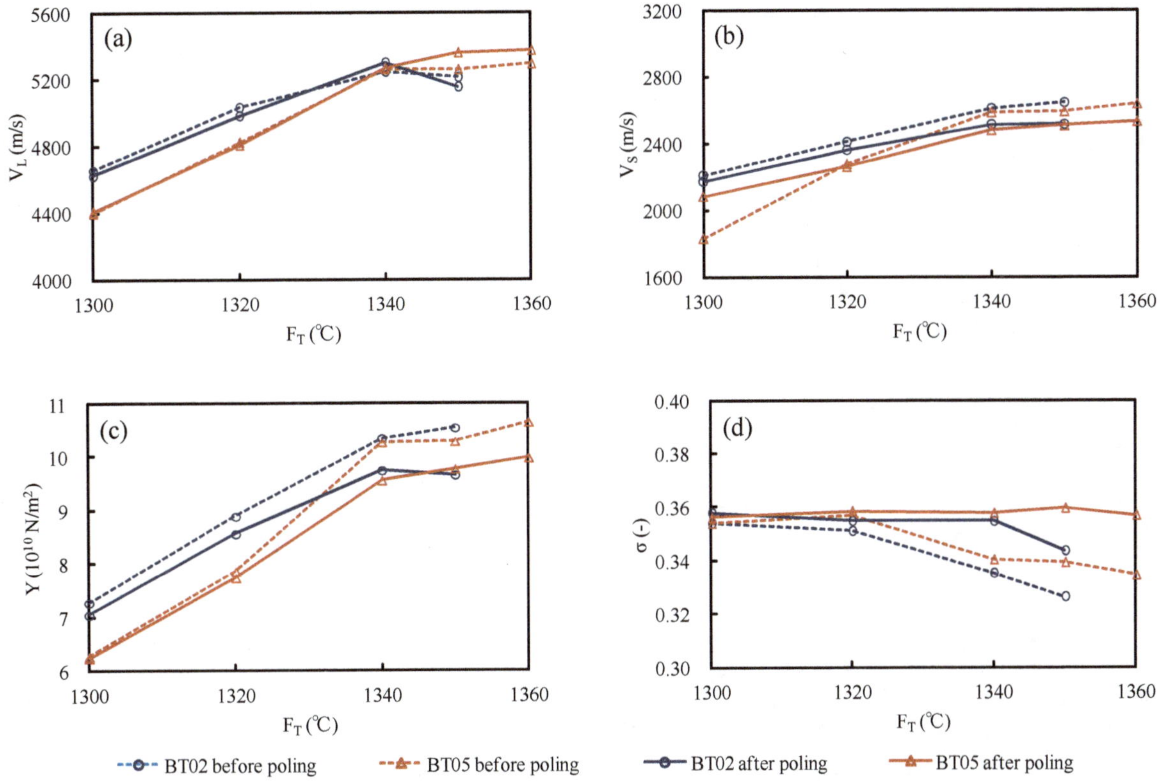


**Figure 14.** Firing temperature ($F_T$) vs. (a) longitudinal wave velocity ($V_L$), (b) transverse wave velocity ($V_S$), (c) Young's modulus (Y), and (d) Poisson's ratio (σ) in BT02 and BT05 ceramics before and after DC poling.

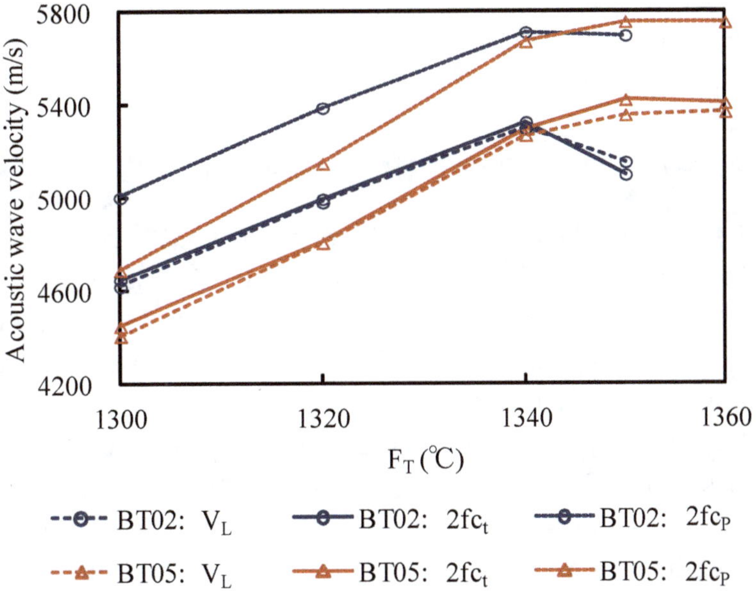


**Figure 15.** Comparison between longitudinal wave velocity ($V_L$), $2fc_t$, and $2fc_p$, which are twice the frequency constants of $fc_t$ for the thickness mode and $fc_p$ for the radial mode on disks in BT02 and BT05 ceramics after DC poling; $2fc_t$ and $2fc_p$ correspond to acoustic wave velocities.

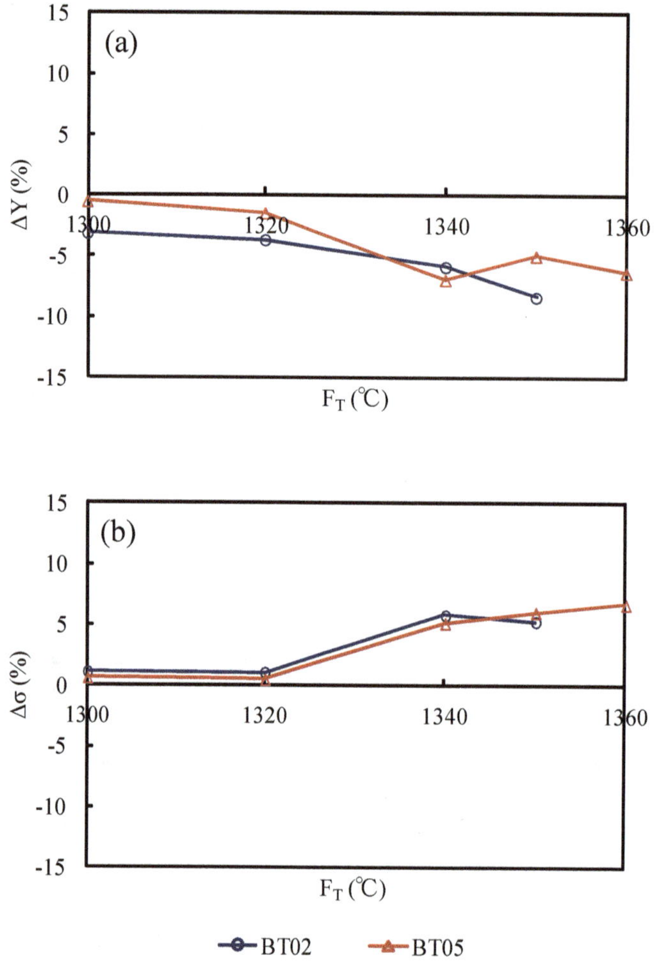


**Figure 16.** Changes in (a) Young's modulus (ΔY) and (b) Poisson's ratio (Δσ) in BT02 and BT05 ceramics during DC poling.

*2.3.3. Relationship between ceramic microstructure and elastic constants*

Figure 17 shows the $F_T$ dependence of the bulk density (ϱ) of the ceramic disks. The relative bulk density ($\varrho/\varrho_0$) became over 94% for both BT02 and BT05 during firing at 1, 340-1, 360 ℃, where $\varrho_0$ (6.02 g/cm$^3$) is the theoretical density of BT ceramics [25]. From our previous study on the relationships between $k_p$ vs. Y and σ in piezoelectric ceramics, a higher $k_p$ was realized at a lower Y and a higher σ [18-20]. However, there is no correspondence between the lower Y (Fig. 14(c)), the higher σ (Fig. 14(d)), the lower $k_p$ (Fig. 13(b)), and the lower $d_{33}$ (Fig. 13(d)) during firing at 1, 300-1, 320 ℃ because of the low relative bulk density of $\varrho/\varrho_0$ (< 0.94). On the other hand, while Y became almost constant during firing at 1, 340-1, 360 ℃ (Fig. 14(c)), σ before poling decreased with increasing $F_T$ (Fig. 14(d)). It was considered that the decrease in σ with increasing $F_T$ was due to the increase in the density of coarse grains with average diameters of 50 μm as mentioned later. As $F_T$ vs. σ before poling (Fig. 14(d)) shows the same tendency as $F_T$ vs. $\varepsilon_r$ (Fig. 13(a)), the $\varepsilon_r$ of which is directly related to ferroelectric domain structures, the σ dependence of $F_T$ was considered to be due to the domain structures and anisotropy of coarse grains being different from those of fine grains with average diameters of 1.2 μm. Since the ferroelectric domain size increases with increasing grain size of BT ceramics, the domain density decreases in the case of coarse grains [30-33]. Therefore, the ceramics with coarse grains exhibit a lower $\varepsilon_r$ and a larger crystal anisotropy than the ceramics with fine grains [34-38]. Furthermore, $\sigma$ decreases with increasing $F_T$ because it approaches a behavior similar to that of a BT single crystal, $\sigma$ of which is 0.29 while $\sigma$ of the ceramics is 0.31 [25]. In order to evaluate the physical characteristics of ceramic grains, the modulus of rigidity (G) and bulk modulus (K) were calculated by the equations (3) and (4) in Section 1.2. [23, 24]: G and K before poling increase with increasing $F_T$ during firing at 1,300-1,360 ℃, as shown in Figures 18(a)-(b), respectively, because the ceramic bulk density is improved in BT ceramics as shown in Figure 17; as a result, G and K are obtained from the equations (3) and (4), as well as Y in Figure 14(c) from equation (1). On the other hand, $\sigma$ is independent of the ceramic bulk density, as shown in equation (2). Figures 19(a)-(d) show the relationships between $\varrho/\varrho_0$ vs. Y, $\sigma$, G, and K, respectively. Y, G, and K depend on $\varrho/\varrho_0$ because they are affected by increasing the mechanical strength, especially the hardness of the ceramics, with $F_T$. As mentioned previously regarding Figure 14(d), $\sigma$ is independent of $\varrho/\varrho_0$ except when $\varrho/\varrho_0$ is above 0.94, as shown in Figure 19(b), which corresponds to ceramics with a larger domain size and a lower domain density with coarse grains. G decreased and K increased after poling owing to the domain alignment as shown in Figure 18. In addition, the change in G ($\Delta$G) in BT ceramics during poling linearly decreased with increasing $F_T$, and peaks of the changes in K ($\Delta$K), which correspond to the peaks of $\Delta\sigma$ (Figure 16(b)), were obtained at $F_T$ of 1340 ℃ in the BT02 ceramics and at $F_T$ of 1, 350-1, 360 ℃ in the BT05 ceramics, as shown in Figures 20(a)-(b). Therefore, a higher $k_p$ could be realized at $F_T$ at the peaks of $\Delta$K and $\Delta\sigma$. Since ceramic grains with a higher K due to domain alignment during DC poling, indicating a high ceramic bulk density for obtaining a higher $k_p$, are difficult to change in terms of their volume while applying external stress, the deformation of grains is practically transferred from the parallel direction to the perpendicular direction toward the direction of the applied stress; as a result, a higher $\sigma$ is achieved. We believe that the increase in $\sigma$ ($\Delta\sigma$) as a result of the increase in K ($\Delta$K) indicates fundamental issues regarding the poling of ceramic grains to obtain a higher $k_p$; therefore, the R&D of piezoelectric ceramics with high piezoelectricity must be focused on to realize a lower G and a higher K during DC poling from the viewpoints of elastic constants. From the above-mentioned results, it was considered that the values of $F_T$ vs. G and K in BT02 and BT05 ceramics after poling (Figures 18(a)-(b)) correspond to the values of $k_p$ vs. $F_T$ (Figure 13(b)).

Figure 21(a) shows the $F_T$ dependence of microstructures in the BT02 ceramics. With the increase in $k_p$ with $F_T$, ceramic grains grew from 1.2 μm (fine grains shown as white parts in the figures) to 50 μm (coarse grains shown as black parts in the figures) in diameter; moreover, the ratio of black parts to whole parts (black parts plus white parts) increased. The border between fine and coarse grains is shown in Figures 21(b)-(c). Furthermore, it was found that the black parts consist of several coarse grains, as shown in Figure 21(b). This phenomenon was almost the same as in the case of the microstructure in the BT05 ceramics. Figure 22 shows the relationship between $k_p$ and the area ratio of coarse grains (black parts) measured by image-analyzing software (WinROOF [39]). In this figure, when the area ratio increases, $k_p$ increases because of the increase in the density of coarse grains. Therefore, the coarse grains in the dense ceramic bulk contribute to a higher $k_p$ and $V_L$, $V_S$, Y, and $\sigma$ in the coarse grains correspond to those at $F_T$ of 1, 360 ℃ in Figures 14(a)-(d), which almost agree with the values previously reported [25].

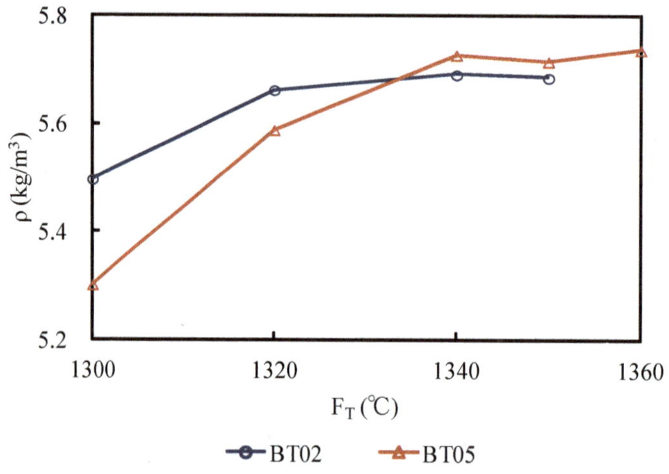


**Figure 17.** Firing temperature ($F_T$) dependence of bulk density (ϱ) in BT02 and BT05 ceramic disks.

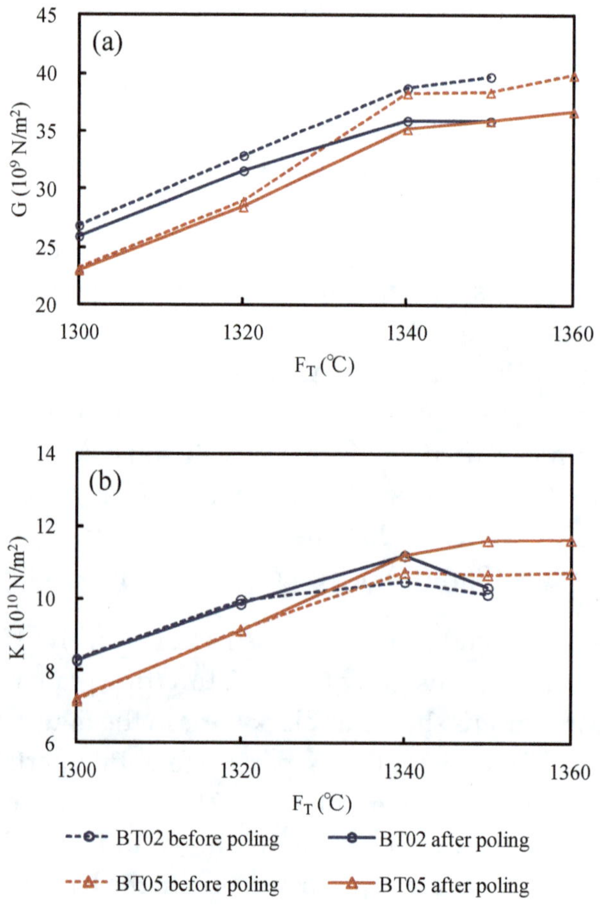


**Figure 18.** Firing temperature ($F_T$) vs. (a) modulus of rigidity (G) and (b) bulk modulus (K) in BT02 and BT05 ceramics before and after DC poling.

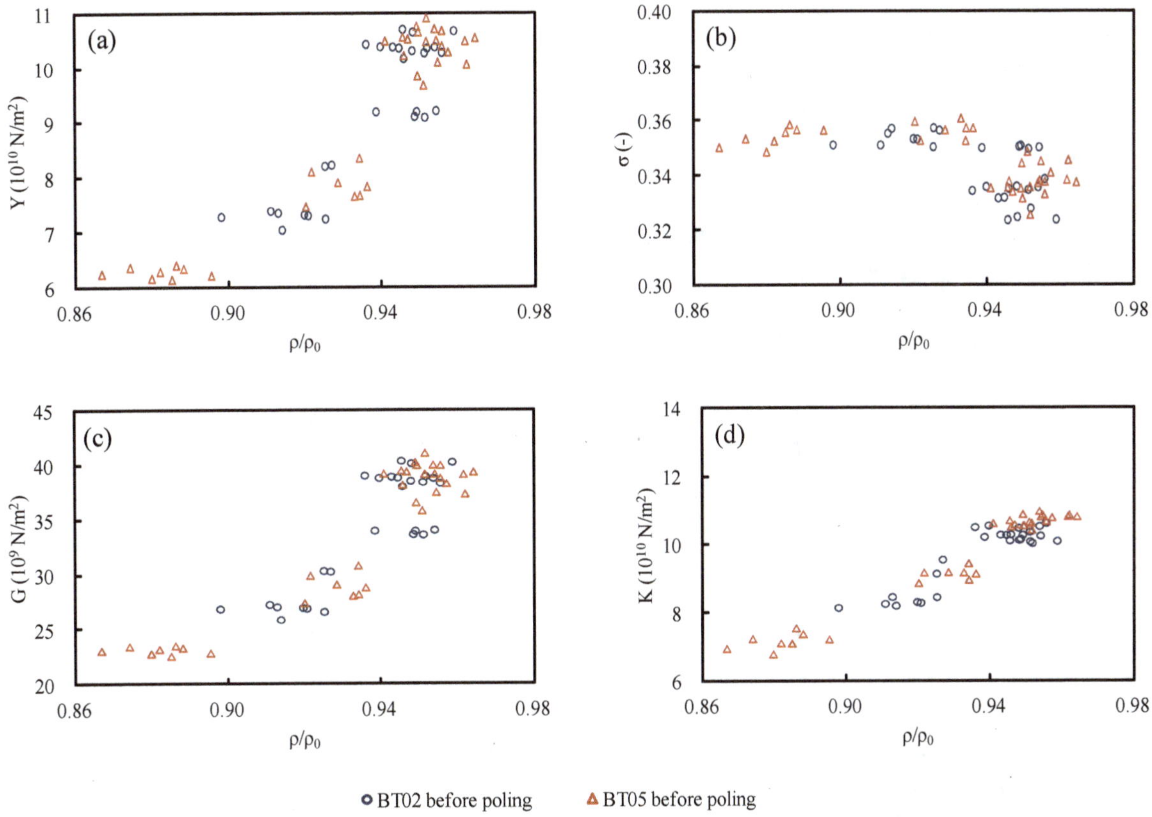


**Figure 19.** Relative bulk density ($\varrho/\varrho_0$) dependences of (a) Young's modulus (Y), (b) Poisson's ratio ($\sigma$), (c) modulus of rigidity, and (d) bulk modulus (K) in BT02 and BT05 as-fired (before poling) ceramics sintered during firing at 1, 300-, 360 ℃, where $\varrho$ and $\varrho_0$ (6.02 g/cm³) are the measured bulk density and theoretical density [25] of BT ceramics, respectively.

### 2.3.4. *Firing temperature dependence of elastic constants in barium titanate ceramics*

Figure 23 shows that the relationships between $k_p$ vs. Y, $\sigma$, G, and K in BT02 and BT05 ceramics fired at different temperatures were inserted into the relationships between $k_p$ vs. Y, $\sigma$, G, and K in lead-containing and lead-free piezoelectric ceramics (Figure 3), which were fired at the optimal temperatures for each composition to realize the maximum $k_p$. Higher $k_p$ is obtained in the cases of lower Y and G, and furthermore, higher $\sigma$ and K, which are indicated by yellow arrows in Figure 23. This figure also indicates the firing temperature dependence of the bulk density vs. Y, $\sigma$, G, and K in BT02 and BT05 ceramics in Figure 19. $k_p$ increases with the increase of Y, G, and K because of the increase in bulk density with increasing firing temperature. On the other hand, $\sigma$ was independent of $k_p$ because $\sigma$ is an intrinsic material constant. These phenomena are indicated by green arrows in Figure 23. It is predicted the same phenomena regarding the firing temperature and bulk density dependences on $k_p$ vs. Y, $\sigma$, G, and K in cases of lead-containing and lead-free ceramics.

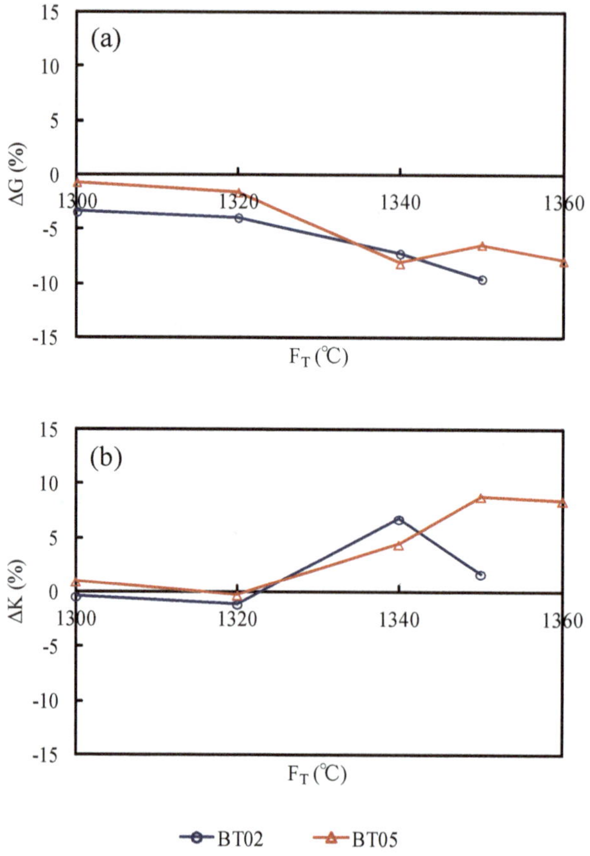


**Figure 20.** Changes in modulus of rigidity (ΔG) and bulk modulus (ΔK) in BT02 and BT05 ceramics during DC poling.

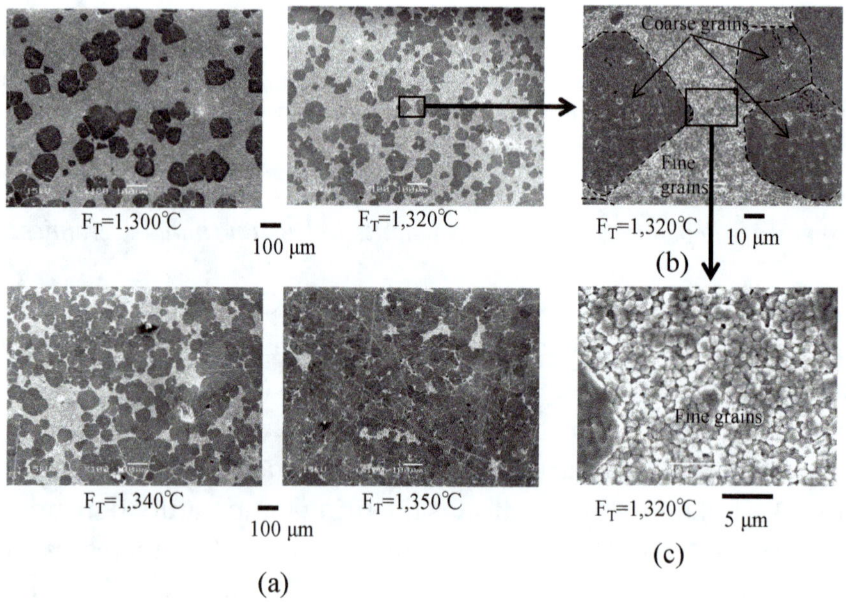


**Figure 21.** (a) Firing temperature ($F_T$) dependence of microstructures in BT02 ceramics, (b) border between fine grain [see (c), average grain size of 1.2 μm] and coarse grain [see (b), average grain size of 50 μm], and relationship between black parts and coarse grains.

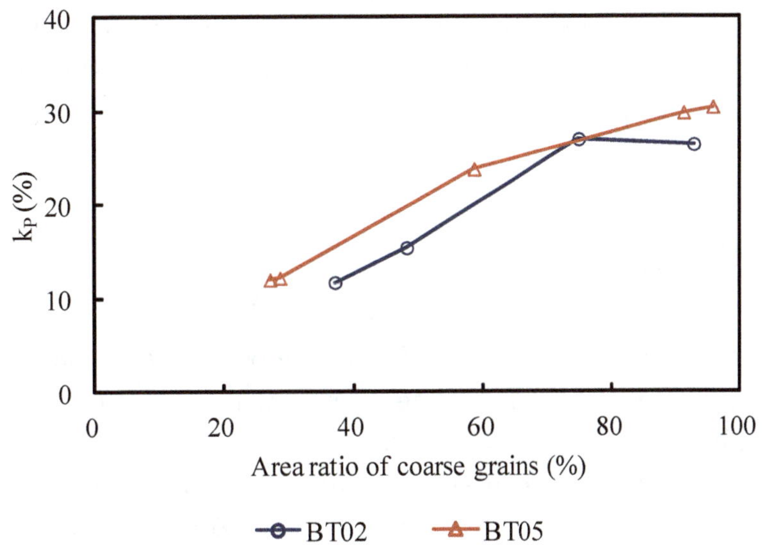


**Figure 22.** Planar coupling factor ($k_p$) vs. area ratio of coarse grains of black parts measured by an image-analyzing software program (WinRoof [39]).

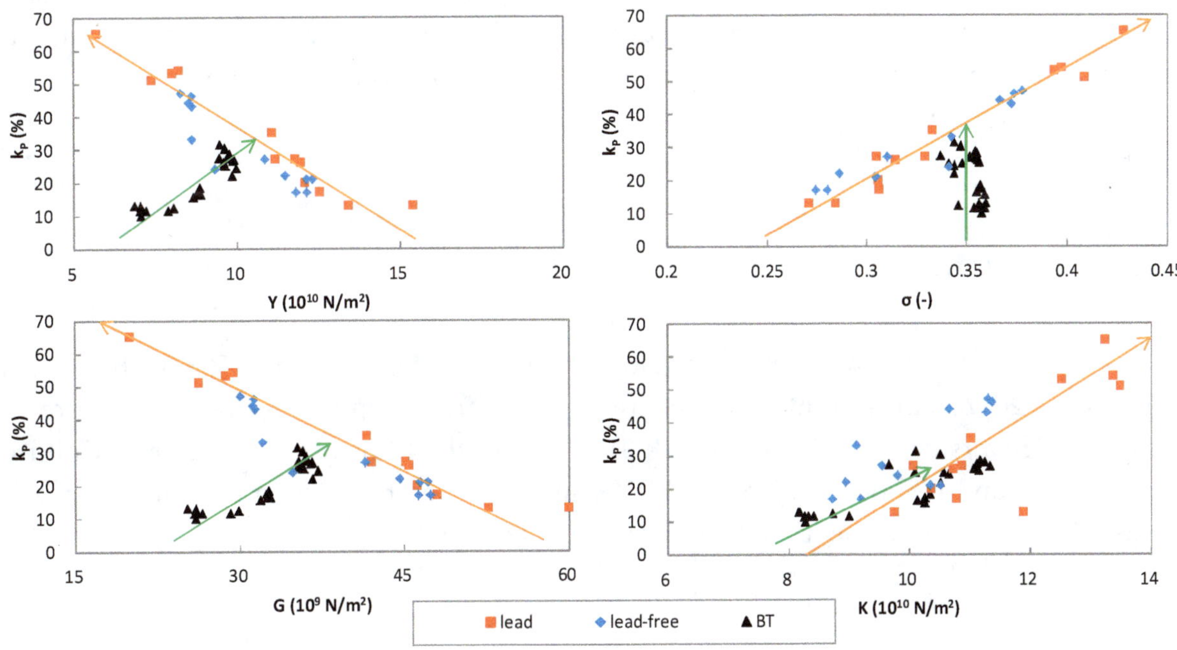


**Figure 23.** Planar coupling factor ($k_p$) vs. Y, σ, G, and K in BT02 and BT05 ceramics fired at different temperatures. These data were inserted into the relationships between $k_p$ vs. Y, σ, G, and K in lead-containing and lead-free piezoelectric ceramics (Figure 3), which were fired at the optimal temperatures for each composition to realize maximum $k_p$.

## 2.4. Conclusions in this part

The effects of firing temperature and DC poling on the longitudinal and transverse wave velocities in barium titanate ceramics were investigated using an ultrasonic precision thickness gauge with high-frequency pulse generation. The results could explain the relationships between acoustic wave velocities, Young's modulus, Poisson's ratio, the modulus of rigidity, and the bulk modulus vs. firing temperature, and the changes in elastic constants during DC poling.

## 3. Summary of this chapter

Sound velocities were evaluated in ceramic disks composed of lead-containing and lead-free ceramics using an ultrasonic precision thickness gauge with high-frequency pulse generation, and furthermore, dielectric and piezoelectric constants were simultaneously measured utilizing the same disk samples. Calculating elastic constants by using the sound velocities, higher piezoelectricity in ceramics were obtained in lower Young's modulus and rigidity, and furthermore, higher Poisson's ratio and bulk modulus. Piezoelectric ceramics with lower Young's modulus and rigidity caused by ferroelectric domain alignment while DC poling were easy to deform by electric field and external force. In addition to these phenomena, higher bulk modulus needs to realize higher Poisson's ratio. Lower Young's modulus means mechanical soft in ceramics; however, higher bulk modulus, which means mechanical hard in ceramics, need to obtain higher piezoelectricity. It was thought that these characteristics run counter to the mechanical characteristics of piezoelectric ceramics; however, it was the origin of piezoelectricity in ceramics.

## Acknowledgements

This work was partially supported by two Grants-in-Aid for Scientific Research C (Nos. 21560340 and 26420282), a Grant from the Strategic Research Foundation Grant-Aided Project for Private Universities 2010-2014 (No. S1001032) from the Ministry of Education, Culture, Sports, Science and Technology, Japan (MEXT), and a Cooperation Research Foundation 2014 between Academy and Industry of Fukuroi City, Shizuoka, Japan.

## Author details

Toshio Ogawa

Address all correspondence to: ogawa@ee.sist.ac.jp

Department of Electrical and Electronic Engineering, Shizuoka Institute of Science and Technology, Toyosawa, Fukuroi, Shizuoka, Japan

## References


[1] Saito Y, Takao H, Tani T, Nonoyama T, Takatori K, Homma T, Nagaya T, Nakamura M. Lead-free piezoceramics. Nature 2004;432:84-7.

[2] Yang ZP, Liu B, Wei LL, Hou YT. Structure and electrical properties of (1-x)$Bi_{0.5}Na_{0.5}TiO_3$-x$Bi_{0.5}K_{0.5}TiO_3$ ceramics near morphotropic phase boundary. Mater Res Bull 2008;43:81-9.

[3] Zhao W, Zhou HP, Yan YK, Liu D. Morphotropic phase boundary study of the BNT-BKT lead-free piezoelectric ceramics. Key Eng Mater 2008;368-372:1908-10.

[4] Dai YJ, Pan JS, Zhang XW. Composition range of morphotropic phase boundary and electrical properties of NBT-BT system. Key Eng Mater 2007;336-338:206-9.

[5] Ogawa T, Furukawa M, Tsukada T. Poling field dependence of piezoelectric properties and hysteresis loops of polarization versus electric field in alkali niobate ceramics. Jpn J Appl Phys 2009;48:709KD07-1-5.

[6] Ogawa T, Nishina T, Furukawa M, Tsukada T. Poling field dependence of ferroelectric properties in alkali bismuth titanate lead-free ceramics. Jpn J Appl Phys 2010;49:09MD07-1-4.

[7] Ogawa T. Poling field dependence of ferroelectric properties in barium titanate ceramics. Jpn J Appl Phys 2001;40:5630-3.

[8] Furukawa M, Tsukada T, Tanaka D, Sakamoto N. Alkaline niobate-based lead-free piezoelectric ceramics. Proc 24th Int Japan-Korea Semin Ceramics 2007;339-42.

[9] Ogawa T, Yamada A, Chung YK, Chun DI. Effect of domain structures on electrical properties in tetragonal PZT ceramics. J Korean Phys Soc 1998;32:S724-S726.

[10] Ogawa T, Nakamura K. Poling field dependence of ferroelectric properties and crystal orientation in rhombohedral lead zirconate titanate ceramics. Jpn J Appl Phys 1998;37:5241-5.

[11] Ogawa T, Nakamura K. Effect of domain switching and rotation on dielectric and piezoelectric properties in lead zirconate titanate ceramics. Jpn J Appl Phys 1999;38:5465-9.

[12] Ogawa T. Domain switching and rotation in lead zirconate titanate ceramics by poling fields. Ferroelectrics 2000;240:75-82.

[13] Ogawa T. Domain structure of ferroelectric ceramics. Ceram Int 2000;25:383-90.

[14] Ogawa T. Poling field dependence of crystal orientation and ferroelectric properties in lead titanate ceramics. Jpn J Appl Phys 2000;39:5538-41.

[15] Ogawa T. Poling field dependence of ferroelectric properties in piezoelectric ceramics and single crystal. Ferroelectrics 2002;273:371-6.

[16] Kato R, Ogawa T. Chemical composition dependence of giant piezoelectricity on $k_{31}$ mode in $Pb(Mg_{1/3}Nb_{2/3})O_3$-$PbTiO_3$ single crystals. Jpn. J. Appl. Phys. 2006;45:7418-21.

[17] Ogawa T. Giant $k_{31}$ relaxor single-crystal plate and their applications. In Lallart M. (ed.) Ferroelectrics –Applications–. Rijeka: Intech; 2011, pp.3-34.

[18] Ogawa T, Nishina T. Acoustic wave velocities measurement on piezoelectric ceramics to evaluate Young's modulus and Poisson's ratio for realization of high piezoelectricity. In Nair KM, Priya S. (eds.) Ceramic Transactions, Advances and Applications in Electroceramics II. Hoboken: Wiley; 2012;235:pp.105-12.

[19] Ogawa T, Ishii K, Matsumoto T, Nishina T. Poling field dependence of longitudinal and transverse wave velocities, Young's modulus, and Poisson's ratio in piezoelectric ceramics. J Appl Phys 2012;51:09LD03-1-5.

[20] Ogawa T. Acoustic wave velocity measurement on piezoelectric ceramics. In Ebrahimi F. (ed.) Piezoelectric Materials and Devices –Practice and Applications–. Rijeka: Intech; 2013, pp.35-52.

[21] Ogawa T. Highly functional and high-performance piezoelectric ceramics. Ceramic Bull 1991;70:1042-9.

[22] Olympus Co. Ultrasonic Precision Thickness Gages User's Manual. Part No. 910-261-EN, Revision H. Waltham: Olympus NDT Inc; 2008.

[23] Mason WP. Piezoelectric Crystals and Their Application to Ultrasonics. Princeton: Nostrand; 1950, p.390.

[24] Philippoff W, Brodnyan J. Preliminary results in measuring dynamic compressibilities. J Appl Phys 1955;26:846.

[25] Jaffe B, Cook WR, Jaffe H. Piezoelectric Ceramics. London: Academic Press; 1971.

[26] Landolt-Bornstein Group III. Condensed Matter, Ferroelectrics and Related Substances. Berlin: Springer; 2001, 36A1.

[27] Landolt-Bornstein Group III. Condensed Matter, Ferroelectrics and Related Substances. Berlin: Springer; 2002, 36A2.

[28] Mason WP. Physical Acoustics. New York: Academic Press; 1964, Part A, p.182.

[29] IEEE Standard. 1988;pp.176-1987.

[30] Salje E. Domain mobilities and elastic instabilities in ferroelastic and co-elastic materials. In Phase Transitions in Ferroelastic and Coelastic Crystals. Cambridge: Cambridge University Press; 1990, p.76.

[31] Arlt G. The influence of microstructure on the properties of ferroelectric ceramics. Ferroelectrics 1990;104:217.

[32] Randall CA, Kim N, Kucera JP, Cao W, Shrout TR. Intrinsic and extrinsic size effects in fine-grained morphotropic-phase-boundary lead zirconate titanate ceramics. J Am Ceram Soc 1988;81:677.

[33] Cao W, Randall CA. Grain size and domain size relations in bulk ceramic ferroelectric materials. J Phys Chem Solids 1996;10:1499.

[34] Arlt G, Hennings D, With GD. Dielectric properties of fine-grained barium titanate ceramics. J Appl Phys 1985;58:1619.

[35] Buessem WR, Cross LE, Goswami AK. Phenomenological theory of high permittivity in fine-grained barium titanate. J Am Ceram Soc 1966;49:33.

[36] Kinoshita K, Yamaji A. Grain-size effects on dielectric properties in barium titanate ceramics. J Appl Phys 1976;47:371.

[37] Arlt G, Peusens H. The dielectric constant of coarse grained $BaTiO_3$ ceramics. Ferroelectrics 1983;48:213.

[38] Demartin M, Damjanovic D. Dependence of the direct piezoelectric effect in coarse and fine grain barium titanate ceramics on dynamic and static pressure. Appl Phys Lett 1996;68:3046.

[39] http://mitani-visual.jp/winroof01.html.

# 8

# Role of Ca off-Centering in Tuning Ferroelectric Phase Transitions in Ba(Zr,Ti)$O_3$ System

Desheng Fu and Mitsuru Itoh

**Abstract**

We here report the substitution effects of the smaller Ca for the bulky Ba in the $(Ba_{1-x}Ca_x)(Ti_{1-y}Zr_y)O_3$ perovskite oxides for two systems $(Ba_{1-x}Ca_x)TiO_3$ with $y$=0 and $(Ba_{1-x}Ca_x)(Ti_{0.9}Zr_{0.1})O_3$ with $y$=0.1. Ca off-centering was found to play a critical role in stabilizing the ferroelectric phase and tuning the polarization states in both systems. It was demonstrated that the atomic displacement due to Ca off-centering in the bulky Ba-sites in the perovskite structure provides an effective approach to compensate for the reduction of ferroelectricity due to chemical pressure, which allows to keep the Curie point nearly constant in the $(Ba_{1-x}Ca_x)TiO_3$ system and increase the Curie point in the $(Ba_{1-x}Ca_x)(Ti_{0.9}Zr_{0.1})O_3$ system. It was commonly observed that the Ca off-centering effects lead to the shift of the rhombohedral–orthorhombic and orthorhombic–tetragonal phase transitions toward lower temperatures and the ferroelectric stability of the tetragonal phase, resulting in the occurrence of quantum phase transitions with interesting physical phenomena at low temperatures in the $(Ba_{1-x}Ca_x)TiO_3$ system and remarkable enhancement of electromechanical coupling effects around room temperature in the $(Ba_{1-x}Ca_x)(Ti_{0.9}Zr_{0.1})O_3$ systems over a wide range of Ca-concentrations. These findings may be of great interest for the design of green piezoelectric materials.

**Keywords:** Ca off–centering, $BaTiO_3$, Ba(Zr,Ti)$O_3$, perovskite oxides, ferroelectric, piezoelectric, phase transition, quantum effects, electromechanical coupling effects

## 1. Introduction

There is increased interest in developing green piezoelectric materials in the field of electronics due to environmental concerns regarding the Pb-toxicity in commercially used lead-based $Pb(Zr,Ti)O_3$ (PZT) piezoelectric ceramics. As listed in Table 1 (Ref. 1-4), $BaTiO_3$ single crystals have the highest piezoelectric coefficients among single crystals of lead-free piezoelectrics. Although the reported values of its piezoelectric coefficient vary somewhat, recent investigations on the mono-domain of a single crystal by high energy synchrotron x-ray radiation show that $BaTiO_3$ has a $d_{33}$ value of 149±54 pm/V at least at the level of lattice distortion (Fig. 1).[4] The large piezoelectric response makes $BaTiO_3$ a promising material for novel green piezoelectric ceramics.[5-9]

| Crystal | Point group | $d_{33}$ (pC/N) | $\varepsilon_{33}$ | k (%) |
|---|---|---|---|---|
| quartz | 32 | 2.3($d_{11}$) | 4.6 | 10 (X-cut) |
| ZnO | 6mm | 10.6 | 11 | 41 ($k_{33}$) |
| $LiNbO_3$ | 3m | 6 | 30 | 17 (Z-cut) |
| $PbTiO_3$ | 4mm | 19.3 | 121 | 64($k_{33}$) |
| $BaTiO_3$ | 4mm | 149 | 168 | 65($k_{33}$) |
| PZN-PT8% | 3m[111] | 84 | 1000 | 0.39($k_{33}$) |
| | [001] | 2500 | 5000 | 0.94($k_{33}$) |

**Table 1.** Typical piezoelectric crystals and their piezoelectric ($d_{33}$ or $d_{11}$) & dielectric ($\varepsilon_{33}$) constants, and electromechanical coupling factor *k*.[1-4]

Piezoelectricity is the ability of a single crystal with non-centrosymmetry (with the exception of point group 432) to develop an electric charge proportional to a mechanical stress or to produce a deformation proportional to an electric field. The piezoelectricity in $BaTiO_3$ is a direct result of its ferroelectricity, originating from the Ti atomic displacement in the oxygen octahedron of the $ABO_3$ perovskite structure[10, 11] (Fig. 2). As can be inferred from the depiction of the variation of the dielectric permittivity with temperature in Fig. 2, there are two challenging issues that remain to be solved for $BaTiO_3$: (1) the temperature instability of physical properties around room temperature due to the tetragonal (*T*)-orthorhombic (*O*) phase transition; and (2) its relatively lower Curie point of ~400 K in comparison with lead-based piezoelectrics. A-site substitution of Pb for Ba is able to increase the Curie point; however, such an approach is undesirable for green piezoelectrics. Principally, A-site and/or B-site substitution can be used to modify the ferroelectricity of $BaTiO_3$. Here, we show that A-site substitution of Ca for Ba in the Ba-based perovskite oxides can lead to a variety of interesting phenomena: (1) the dramatic improvement of temperature stability of its physical properties, (2) the occurrence of quantum fluctuation at low temperatures, and (3) remarkable enhancement of electromechanical responses.[6-9]

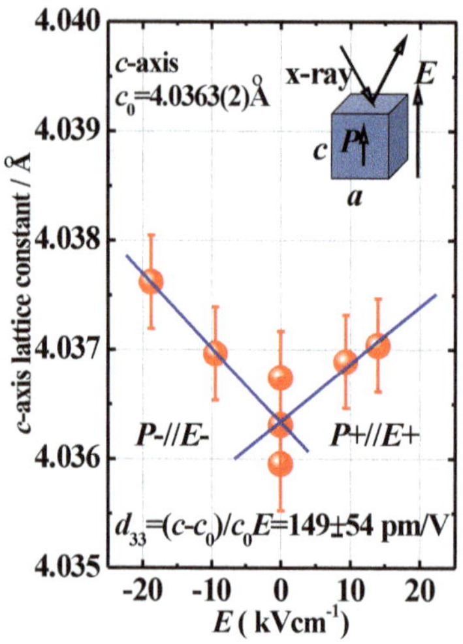


**Figure 1.** Lattice distortion of mono-domain of a $BaTiO_3$ crystal under the application of an electric field.[4]

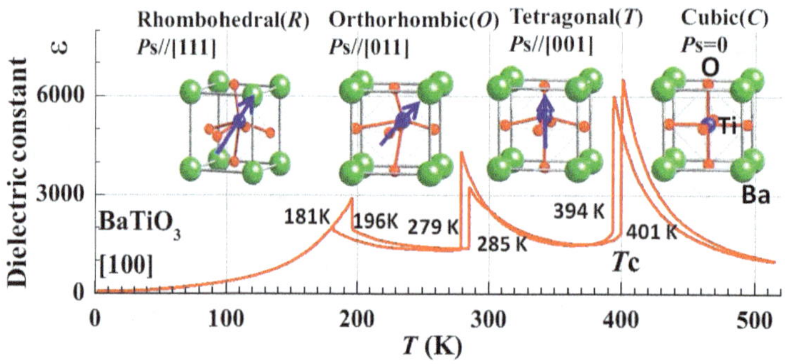


**Figure 2.** Change with temperature of the dielectric permittivity of a $BaTiO_3$ single crystal. The schematics of Ti displacement in the oxygen octahedron of the perovskite structure are also shown.

## 2. Effects of Ca substitution in $BaTiO_3$

In the early years of 1960, Mitsui and Westphal investigated the influence of Ca substitution on the phase transitions in $BaTiO_3$.[12] Using the ceramic samples, they established a phase diagram of $(Ba_{1-x}Ca_x)TiO_3$ in the composition range of x<0.25 mole for temperatures higher than ~100 K. One interesting finding is that the Curie point remained nearly unchanged within the studied composition range. Such behavior is unexpected when considering that $CaTiO_3$ is paraelectric and that the ionic radius of Ca (~1.34Å) is smaller than that of Ba (~1.60Å),[13] which would lead to the shrinkage of the unit cell and a reduction in ferroelectricity of systems with substitution of Ca for Ba. To gain new insight into the role of Ca substitution in Ba-based perovskite oxides, we re-examined the $(Ba_{1-x}Ca_x)TiO_3$ system using single crystal samples, which allowed us to observe the intrinsic phenomena of the system.[6-8, 14]

## 2.1. Crystal growth

To obtain a single crystal of $(Ba_{1-x}Ca_x)TiO_3$, we used the floating zone (FZ) technique (Fig. 3(a)) that allowed us to grow a single crystal with high purity. According to the reported phase equilibria in the system $(1-x)BaTiO_3$-$xCaTiO_3$ (Fig. 4), only the crystal with a congruent melting composition ($x$=0.27 report by DeVries and Roy[15] or $x$=0.227 report by Kuper et al.[16]) can be directly grown from the melt. Surprisingly, using the FZ technique, we could grow a single crystal of $(Ba_{1-x}Ca_x)TiO_3$ with a perovskite structure for a wide composition range of $0.02 \leq x \leq 0.34$ with a high growth rate of 20 mm/h.[6] Fig. 3(b) shows a rod of crystal obtained by this method. It was found that the crystal can be stably grown under an atmosphere of Ar or $N_2$ gas. The crystal was yellowish but transparent. The Laue X-ray diffraction patterns clearly indicated that the obtained $(Ba_{1-x}Ca_x)TiO_3$ crystal had a perovskite structure.

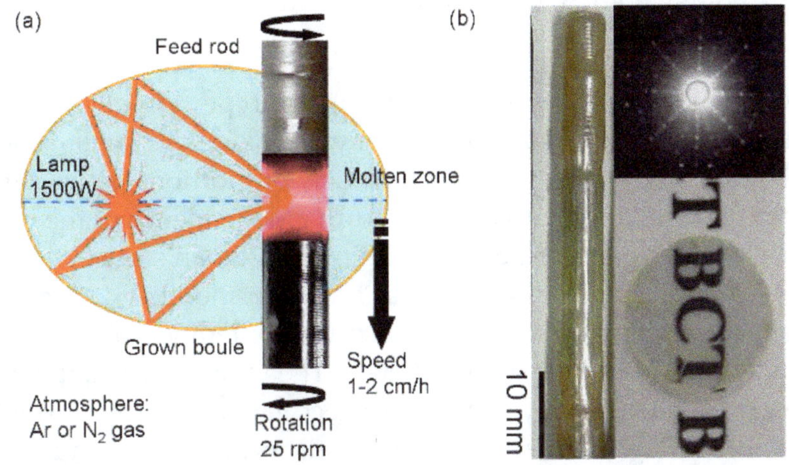


**Figure 3.** (a)A schematic drawing of the floating zone (FZ) technique used to grow the $(Ba_{1-x}Ca_x)TiO_3$ single crystal. (b) Photograph of the $(Ba_{1-x}Ca_x)TiO_3$ single crystal grown by the FZ technique and its Laue X-ray back diffraction patterns along the $[001]_c$ direction of the perovskite structure.

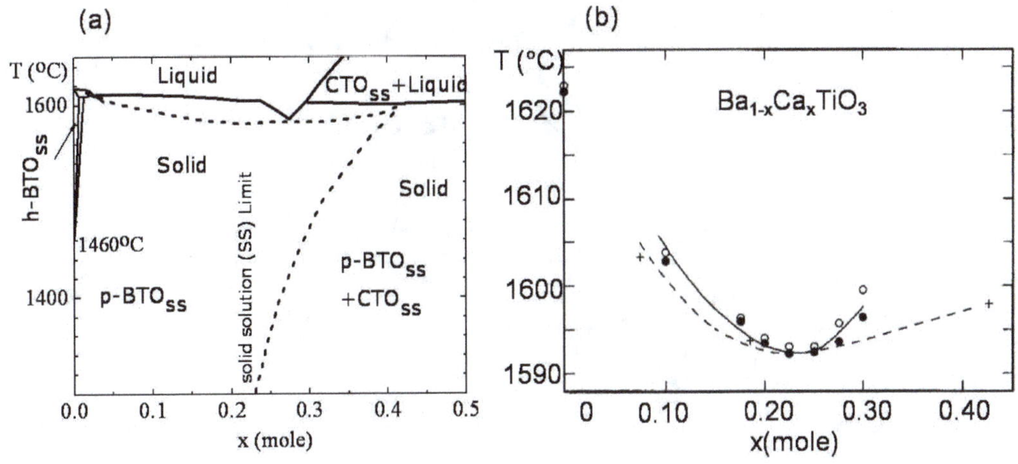


**Figure 4.** Phase equilibria in the system $(1-x)BaTiO_3$-$xCaTiO_3$ reported by (a) DeVries and Roy[15] and (b) Kuper et al.[16]

## 2.2. Dielectric behaviors

Figure 5 shows the temperature dependence of the dielectric permittivity of a $(Ba_{1-x}Ca_x)TiO_3$ single crystals in the temperature range from 2K to 400K. Compared with the polycrystalline ceramics, the $(Ba_{1-x}Ca_x)TiO_3$ single crystal showed a very sharp change of dielectric response at the phase transition, allowing easy determination of the transition temperatures. Similar to the polycrystalline ceramics[12], the Curie point was nearly independent of the Ca concentration. However, the *T–O* and *O*-rhombohedral (*R*) ferroelectric transitions shifted to lower temperatures as the Ca concentration increased. For compositions of $x>0.23$, these two transitions completely disappeared, and the *T*-phase was the only stable ferroelectric phase in the crystal. This situation is very similar to that of $PbTiO_3$, in which the *T*-phase is the only stable ferroelectric structure. Another interesting finding was that the dielectric permittivity was nearly unchanged for temperatures lower than the Curie point for $x>0.23$. This unique behavior provides the possibility of designing electronic devices that operate stably in a wide temperature range from the boiling point of water all the way down to absolute zero Kelvin using $(Ba_{1-x}Ca_x)TiO_3$.

As is well known, the dielectric response in displacive-type ferroelectrics is dominated by phonon dynamics, particularly the soft-mode behavior. Lyddane–Sachs–Teller (LST) relationship predicts that the dielectric permittivity is inversely proportional to the soft-mode frequency. As a step toward understanding the temperature independence of the dielectric response of $(Ba_{1-x}Ca_x)TiO_3$ ($x$ >0.23), we performed confocal micro-Raman scattering measurements for the $(Ba_{1-x}Ca_x)TiO_3$ ($x$ =0.233) single crystals to clarify its soft-mode dynamics.[14] In contrast to $BaTiO_3$, a well-defined soft phonon mode was observed for temperatures lower than the Curie point in the $(Ba_{1-x}Ca_x)TiO_3$ ($x$ =0.233) single crystals. The temperature dependence of the soft-mode frequency agreed qualitatively with the dielectric permittivity through the Lyddane–Sachs–Teller relationship (Fig. 6). This result clearly indicates that the unique dielectric response of $(Ba_{1-x}Ca_x)TiO_3$ ($x$=0.233) is directly derived from its soft-mode dynamics.

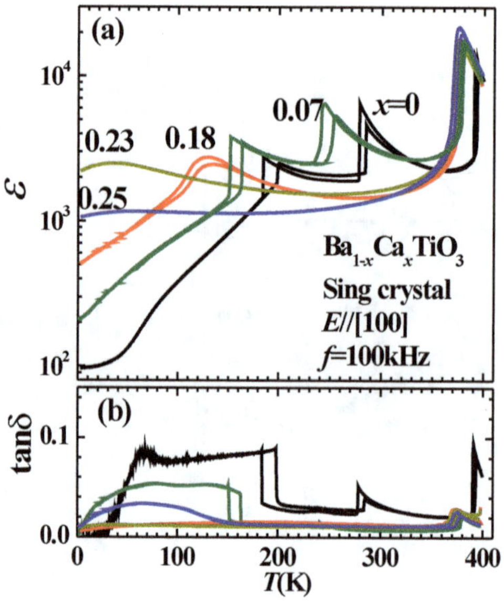


**Figure 5.** Dielectric behaviors of $(Ba_{1-x}Ca_x)TiO_3$ single crystals.

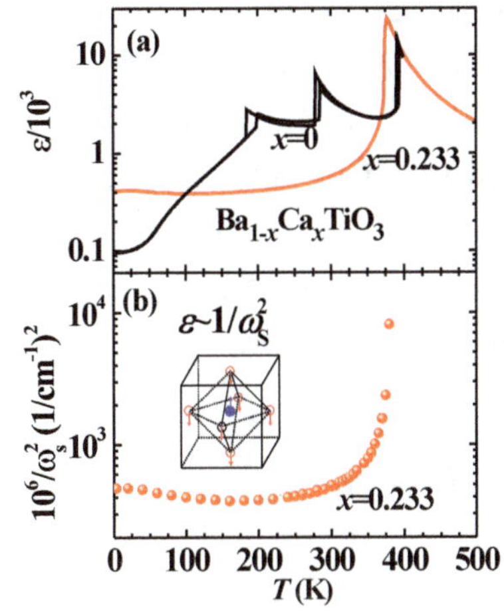


**Figure 6.** Comparison of (a) dielectric permittivity of $(Ba_{1-x}Ca_x)TiO_3$ single crystals and (b) the phonon frequency of the Slater soft-mode.

## 2.3. Phase diagram and quantum phase transitions

From the temperature change of dielectric permittivity of a $(Ba_{1-x}Ca_x)TiO_3$ crystal, we have established its phase diagram in a composition range of $x \leq 0.34$ for temperature down to 2 K (Fig. 7). Compared with the phase diagram proposed by Mitsui and Westphal for ceramics,[12] our phase diagram has been expanded to a composition up to $x$=0.34 mole and to temperatures as low as 2 K. These expansions of composition and temperature allow us to reveal some unexpected phenomena in this system: (1) ferroelectric *R*- and *O*-phases become unstable as the Ca concentration increased, and they are predicted to disappear at $x>x_c^{O-R}=0.18$ and $x>x_c^{T-O}=0.233$, respectively; and (2) the ferroelectric *T*-phase is a ground state for $x> x_c^{T-O}$.

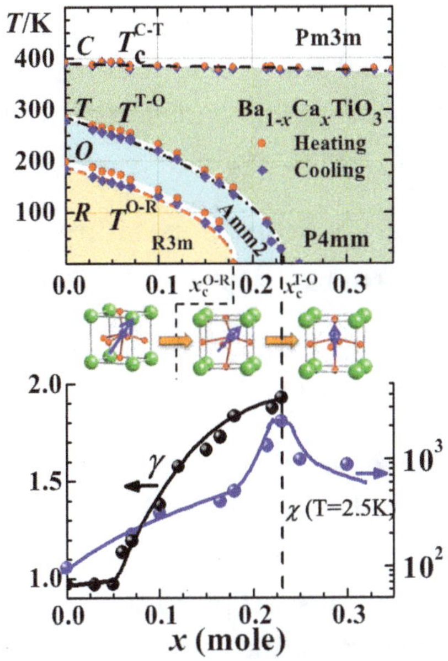


**Figure 7.** Top panel: phase diagram of $(Ba_{1-x}Ca_x)TiO_3$ crystals. Left of bottom panel: Change of critical exponent $\gamma$ for the dielectric susceptibility in the *T-O* phase transition with the composition. Right of bottom panel: Variation of the dielectric susceptibility measured at 2.5 K with the composition.

One important finding is that ferroelectric–ferroelectric quantum phase transitions occur in $(Ba_{1-x}Ca_x)TiO_3$ crystals. The occurrence of ferroelectric–ferroelectric quantum phase transitions is supported by two experimental facts: the compositional dependence of $T^{O\text{-}R}$ and $T^{T\text{-}O}$ transition temperatures and the temperature dependence of the dielectric susceptibility in the crystals at compositions close to $x_c^{T\text{-}O}$. The theoretical and experimental studies[17-19] on quantum phase transitions indicate that (a) for a quantum ferroelectric, the transition temperature depends on the substitution concentration (i.e., on an effective order parameter) as

$$T_c \propto (x - x_c)^{1/2}, \tag{1}$$

as opposed to the classical relationship,

$$T_c \propto (x - x_c). \tag{2}$$

(b) The inverse dielectric susceptibility varies with temperature as

$$\chi^{-1} \propto (T - T_c)^2 \tag{3}$$

for the quantum mechanical limit instead of the classical Curie law

$$\chi^{-1} \propto (T - T_c). \tag{4}$$

Our phase diagram clearly shows that the *T-O* and *O-R* phase transitions deviate from the classical relationship (equation (2)) for $x \leq 0.06$, and exactly follow equation (1) for $x > 0.06$ with $x_c$ equal to $x_c^{T\text{-}O}=0.233$ and $x_c^{O\text{-}R}=0.18$, respectively.

To examine point (b), we have analyzed the temperature variation of the inverse dielectric susceptibility of *T-O* phase transition in the $(Ba_{1-x}Ca_x)TiO_3$ crystals with the following equation,

$$\chi^{-1} \propto (T - T_c)^{\gamma}. \tag{5}$$

Figure 8 shows two typical examples: one for $x=0$ and another for $x=0.23$ close to $x_c^{T\text{-}O}$. For pure $BaTiO_3$ with $x=0$, the classical Curie law with $\gamma=1$ was observed to be operative. In contrast, for $x=0.23$, the critical exponent $\gamma$ for the susceptibility was found to have a value of 2, which is predicted for the quantum phase transition (equation (3)). The left of the bottom panel in Fig. 7 shows the variation of $\gamma$ with $x$. It is clear that the $\gamma$ value changes from the value of the classical limit to that of the quantum limit as $x$ increases from 0 to $x_c^{T\text{-}O}$. This fact again indicates that a quantum phase transition indeed occurs at zero Kelvin in the system when the Ca concentration increases. Interestingly, a dielectric anomaly was observed for the *T-O* quantum phase transition at $x = x_c^{T\text{-}O}$ as shown at the right of the bottom panel of Fig. 7. At a temperature of 2.5 K, close to zero Kelvin, the crystal with $x$=0.23 close to $x_c^{T\text{-}O}$ shows a maximum value of dielectric susceptibility in the system.

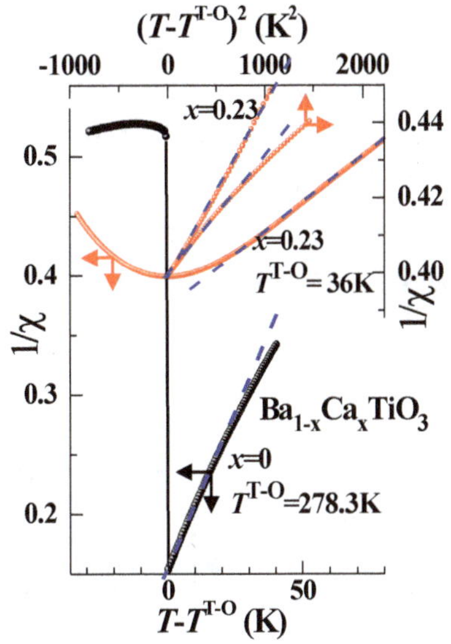


**Figure 8.** Change of the inverse dielectric susceptibility ($\chi$=$\varepsilon$-1) near the *T-O* phase transition in $(Ba_{1-x}Ca_x)TiO_3$ crystals.

## 2.4. Ca off-centering predicted from first principles calculations

As mentioned above, the ionic radius of Ca is approximately 16% smaller than that of Ba. The substitution of Ca for Ba will absolutely result in the shrinkage of the perovskite unit cell. As shown in Fig. 9(a), both *a*- and *c*-axes of the tetragonal structure shrink with increasing Ca concentration, resulting in the reduction of the unit cell. The volume of the unit cell of $x$=0.34 is approximately 3.6% smaller than that of pure $BaTiO_3$. This chemical-pressure-induced reduction of the unit cell does not have significant influence on the Curie point ($T_c^{C\text{-}T}$) of the system (Fig. 9(c)).

In contrast, for the case of hydrostatic pressure, at the same level of unit-cell reduction, the Curie point is reduced to ~180 K, which is greatly lower than the ~400 K of pure $BaTiO_3$ (Fig. 9(c)). The hydrostatic pressure gradually reduces the Curie point, leading to the complete disappearance of ferroelectricity in $BaTiO_3$ at a level of 5% reduction of the unit cell. Although the chemical substitution of the smaller Ca for the bulky Ba also leads to the reduction of the unit cell, the effects of chemical pressure on the ferroelectricity in the $(Ba_{1-x}Ca_x)TiO_3$ system were very different from what we would expect with the application of hydrostatic pressure.

Apparently, the reduction of the unit cell by the chemical pressure shrinks the oxygen octahedron in the perovskite structure, resulting in the reduction of available space for a Ti off-centering shift in the oxygen octahedron. Since the ferroelectricity is derived from the Ti-shift in the oxygen octahedron in the perovskite structure of $BaTiO_3$, it is naturally expected that chemical-pressure-induced reduction of the unit cell in the $(Ba_{1-x}Ca_x)TiO_3$ system should weaken the ferroelectricity of the system and result in a decrease in its Curie point. However, as shown in the phase diagram in Fig. 7 and Fig. 9, the Curie point remains nearly unchanged with the Ca substitution. Also, as shown in Fig. 9(b), the tetragonality (*c*/*a*), which is generally used to estimate the atomic displacement and thus the ferroelectricity and polarization in the tetragonal ferroelectric,[20, 21] maintains a constant value in the whole composition range in the $(Ba_{1-x}Ca_x)TiO_3$ system. In accordance with such unchanged tetragonality, we observed that the saturation polarization is also nearly independent on the Ca substitution within the limits of a solid solution (see Fig. 11).[8] These facts suggest that in addition to the Ti displacement in the oxygen octahedron, there should be an additional atomic displacement to contribute to the ferroelectricity of the $(Ba_{1-x}Ca_x)TiO_3$ system. The ionic radii of Ca and Ba are 1.34 Å and 1.61 Å, respectively. There is a difference of 0.27 Å between them. This suggests that the smaller Ca ion may have an off-centering displacement in the bulky Ba sites (Fig. 10(b)).

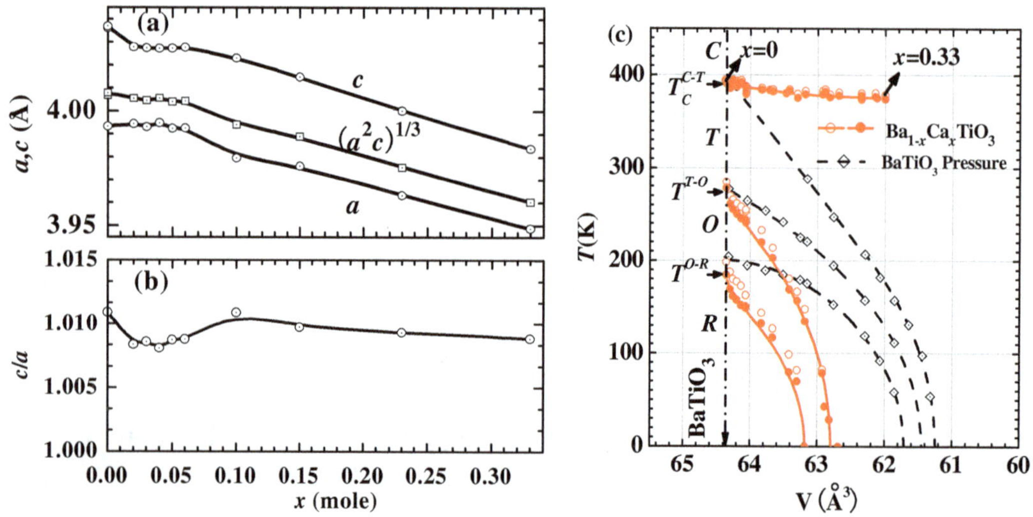


**Figure 9.** Change of (a) lattice constants and (b) tetragonality (*c*/*a*) of $(Ba_{1-x}Ca_x)TiO_3$ crystal. (c) Change of phase transition temperature as a function of unit cell volume (determined at room temperature);the hydrostatic pressure effect for pure $BaTiO_3$ is also shown for comparison.[7, 22, 23]

To examine the idea of Ca off-centering in the bulky Ba sites, we performed first principles calculations for this system.[7, 24] Since Ca substitution tends to stabilize the tetragonal structure, we focused on the calculations in this structure to get information about Ca displacement. The results are summarized in Fig. 10. As shown in Fig. 10(a), a Ti shift along the [001] direction leads to the stability of the tetragonal phase in $BaTiO_3$ ($x$=0). In our calculations for the substitution of Ca for Ba ($x$=1/8), we calculated the relative change in potential energy for various locations of the Ca ion, as shown schematically in Fig. 10(c). As shown in Fig. 11(d), a Ca off-centering shift results in the lowering of the potential energy of the system. This result clearly indicates that the Ca off-centering stabilizes the structure of the $(Ba_{1-x}Ca_x)TiO_3$ system. After tracing the variation of potential energy for moving Ca along various paths, we found that the most likely direction for a Ca shift is [113] since the potential energy is the lowest when Ca is shifted along this direction (Fig. 10(e)). The Ca-shift along the [113] direction seems to be incompatible with the tetragonal structure. However, if we consider the eight-site model similar to that assumed for Ti displacement in $BaTiO_3$, then it becomes clear that Ca can displace along the equivalent directions [113], [1-13], [-113], [-1-13], or [11-3], [1-1-3], [-11-3], [-1-1-3]. The activation barrier for Ca moving between these equivalent states has been evaluated to be less than 3 meV. Therefore, thermal and spatial averaging among these states allows the preservation of the overall tetragonal symmetry detected from X-ray diffractions. It should be noted that the estimated displacement of Ca is approximately 0.1 Å (Fig. 10(d)), which is larger than the 0.05 Å shift of Ti in the tetragonal structure of $BaTiO_3$ (see Ref. 1 and Fig. 10(a)).

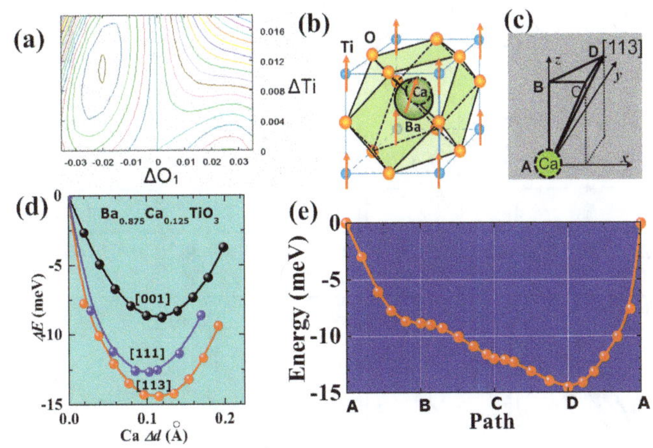


**Figure 10.** (a) Two-dimensional contour map of potential energy of $BaTiO_3$ as a function of Ti and $O_1$ displacement along the [001] direction of the polar *c*-axis. (b) Schematic of Ca off-centering in the bulky Ba sites of the perovskite structure, in which the atomic shifts are shown by the arrows. (c) Direction of Ca-shift for the first principles calculations of $Ba_{7/8}Ca_{1/8}TiO_3$. (d) Change of potential energy of $Ba_{7/8}Ca_{1/8}TiO_3$ along the [001], [111], and [113] directions. (e) Change of potential energy of $Ba_{7/8}Ca_{1/8}TiO_3$ along the path shown in (c).

## 2.5. Polarization and strain responses

For many technical applications, understanding the physical properties of a ceramics sample is of great importance. Figure 11 shows the variation of polarization, bipolar-, and unipolar-field-induced strains with the Ca substitution in the $(Ba_{1-x}Ca_x)TiO_3$ ceramics, which were measured at room temperature. One interesting finding is that the saturation polarization is nearly insensitive to the Ca substitution within the limit of solid solution[8] as shown in Fig. 11(a). This finding is predictable when considering the composition dependence of the tetragonality in the $(Ba_{1-x}Ca_x)TiO_3$ system. As shown in Fig. 9(b), the tetragonality has an approximate value of 1.01 within the solid solution limit. As mentioned above, the tetragonality of the ferroelectric perovskite oxides is predicted to be proportional to its spontaneous polarization by the theoretical calculations.[20, 21]

For the strain response under an electric field, with the exception of $BaTiO_3$ ($x$=0), $(Ba_{1-x}Ca_x)TiO_3$ ceramics show nearly the same level of strain response under the same electric field. For examples, the strain was observed to be approximately 0.08% for a unipolar field of 40 kV/cm, which corresponds to a level of piezoelectric response of 200 pm/V. The observation of the large strain response in $BaTiO_3$ is not surprising because its *T-O* phase transition is located at a temperature close to room temperature. Around the phase transition, a large response of physical properties generally occurs.

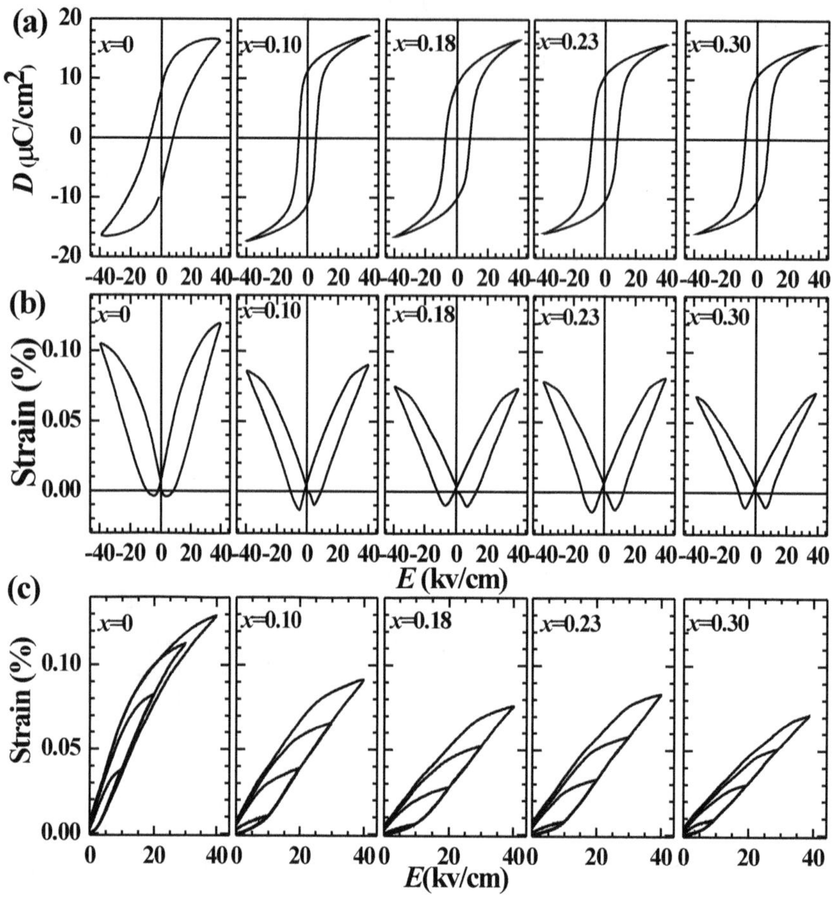


**Figure 11.** (a) Polarization, (b) bipolar-field and (c) unipolar-field strains under electric field in the $(Ba_{1-x}Ca_x)TiO_3$ ceramics.

## 3. Effects of Ca-substitution in the $Ba(Ti,Zr)O_3$ solid solution

To confirm the effects of Ca off-centering in Ba-based perovskite oxides, we also performed investigations on the system of $Ba(Ti,Zr)O_3$ solid solutions, which have been intensively studied since the mid-1950s.[25, 26] The most amazing finding in this system is that it demonstrates very large piezoelectric response, comparable to that of industrial PZT. A high electromechanical coupling factor of 74% and large piezoelectric coefficients of 340 pC/N under a high field were observed in this system.[27, 28] Another interesting thing in this system is that the *O*- or *R*-phase can be tuned to room temperature through controlling Zr substitution, which is of great significance for ferroelectric phase modification. However, as shown in the phase diagram reported by Kell and Hellicar (Fig. 12),[25] the problem of this system is that the Curie point decreases with the increase in the Zr concentration. For Zr concentrations larger than 20 mol%, the Curie point is reduced to a temperature lower than that at room temperature, leading to the disappearance of ferroelectricity in the crystal at room temperature. Here, we show that the Ca off-centering effects mentioned above can also be used to increase the Curie point of $Ba(Ti,Zr)O_3$ and enhance its electromechanical coupling effects through tuning the ferroelectric phase boundaries to room temperature.[9]

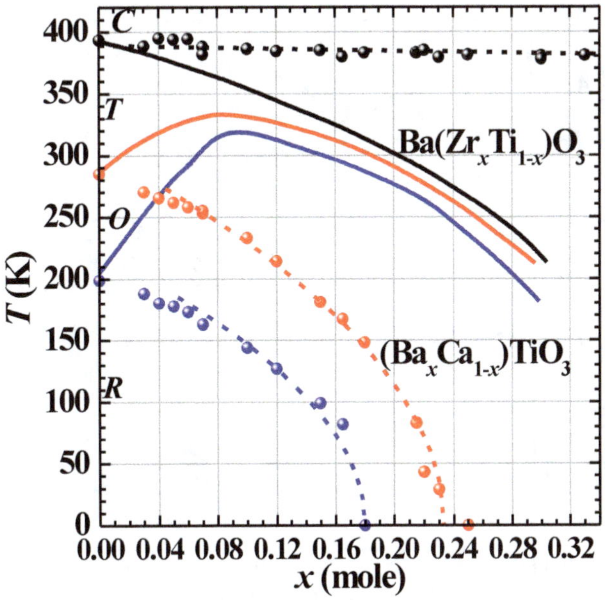


**Figure 12.** Phase diagram of $Ba(Ti_{1-x}Zr_x)O_3$ solid solutions proposed by Kell and Hellicar.[25] For comparison, the phase diagram of $(Ba_{1-x}Ca_x)TiO_3$ is also shown (solid circles).

### 3.1. Sample preparation

In our study, we selected a composition with a Zr concentration of 10 mol%, at which the three successive phases tend to approach each other as shown in Fig. 12. We prepared the $(Ba_{1-x}Ca_x)(Ti_{0.9}Zr_{0.1})O_3$ (BCTZO) ceramics by a solid-state reaction approach. Mixtures of $BaCO_3$, $CaCO_3$, $ZrO_2$, and $TiO_2$ were calcined at 1823 K for 3 h. The calcined powders were ground, pressed, and sintered at 1823 K for 5 h. The ceramic pellets were then electroplated with silver for electrical measurements.

### 3.2. Phase formation and structure transformation at room temperature

At the sintering temperature of 1823 K, a single phase of BCTZO was found to be formed within the composition range of $x \leq 0.18$ beyond which a non-ferroelectric phase with $CaTiO_3$-type orthorhombic structure occurs and coexist with the $BaTiO_3$-type ferroelectric phase. The phase equilibria of $(1-x)Ba(Zr_{0.1}Ti_{0.9})O_3$-$xCaTiO_3$ are very similar to those of $(1-x)BaTiO_3$-$xCaTiO_3$ reported by DeVries and Roy[15] as shown in Fig. 4(a). However, the solid solution limit of $(1-x)Ba(Zr_{0.1}Ti_{0.9})O_3$-$xCaTiO_3$ is approximately half that of $(1-x)BaTiO_3$-$xCaTiO_3$. This fact indicates that the substitution of Zr for Ti in $BaTiO_3$ will reduce the substitution amount of Ca for Ba.

At room temperature, BCTZO with $x$=0 has a ferroelectric rhombohedral structure as shown in phase diagram of Fig. 12. When Ba is substituted with Ca, the structure of BCTZO at room temperature was found to transform from $R$-phase to $O$-phase, and finally to $T$-phase with the increase in Ca concentration. Figure 13 shows the change of lattice parameters with Ca concentration for the BCTZO system. Similar to the unit cell of $(Ba_{1-x}Ca_x)TiO_3$ (Fig. 9(a)), the unit cell of BCTZO shrinks with the substitution of the smaller Ca for the bulky Ba, and its volume is reduced from 65.21 Å$^3$ for $x$=0 to 63.91 Å$^3$ for $x$=0.18.The ferroelectric lattice distortion

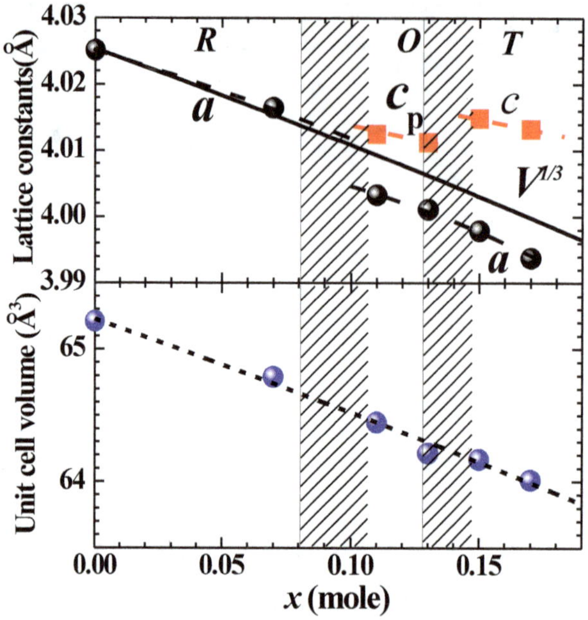


**Figure 13.** Change of the lattice parameters with composition at room temperature for the $(Ba_{1-x}Ca_x)(Ti_{0.9}Zr_{0.1})O_3$ system.

at room temperature is very small in the BCTZO solid solution. The distortion angles $\alpha$ in the *R*-phase and $\beta$ of the monoclinic unit cell in the *O*-phase have deviations of only 0.01° and 0.1° from a right angle, respectively, while the tetragonality *c*/*a* has a value of 1.005 for $x \geq 0.15$ in the *T*-phase.

### 3.3. Phase evolution with temperature

To understand the phase evolution in the BCTZO system, we have measured the temperature variation of dielectric permittivity for different Ca concentrations. The results are summarized in Fig. 14. As reported in many researches, three successive phase transitions are not easy to distinguish for BCTZO with *x*=0 because of the approach of the phase transition. However, these *C-T*, *T-O*, and *O-R* phase transitions as observed in $BaTiO_3$ are clearly demonstrated based on the temperature differentiations of dielectric permittivity shown in Fig. 14. Here, we used the peak of the temperature differentiation of dielectric permittivity to determine the transition temperature of *T-O* and *O-R* phase transitions.

The phase diagrams as functions of Ca concentration and unit cell volume are shown in Fig. 15(a) and (b), respectively. For comparison, a phase diagram of $(Ba_{1-x}Ca_x)TiO_3$ is also shown in the figure. There are several similarities between the BCTZO and $(Ba_{1-x}Ca_x)TiO_3$ systems: (a) Substitution of a smaller Ca for the bulky Ba shifts the *T-O* and *O-R* phase transitions to lower temperatures; in other words, Ca substitution results in the ferroelectric instability of the *O*- and *R*-phases in both systems. (B) By contrast, Ca substitution enhances the ferroelectric stability of the *T*-phase. (c) The chemical-pressure-induced shrink of the unit cell does not reduce the Curie point and weakens the ferroelectricity of both systems. These similarities between BCTZO and $(Ba_{1-x}Ca_x)TiO_3$ systems indicate that the Ca off-centering effects play a critical role in tuning the polarization states in these two systems.

However, there are also some differences between BCTZO and $(Ba_{1-x}Ca_x)TiO_3$ systems: (a) In $(Ba_{1-x}Ca_x)TiO_3$, the Curie point shows a slight decrease with the increase of Ca concentration, but it is increased in the BCTZO system. The Curie point is increased from 363 K for $x$=0 to 376 K for $x$=0.1, after which it seems to reach saturation with further substitution in BCZTO. (b) In $(Ba_{1-x}Ca_x)TiO_3$, the *O*- and *R*-phases completely disappear for Ca-substitution amount of $x$ > 0.233, while in BCTZO, the disappearance of the *O*- and *R*-phases does not occur within the solid solution limit and the *R*-phase is still the ground state as occurs in pure $BaTiO_3$. These facts suggest that the contribution of Ca off-centering displacement to the whole spontaneous polarization in the BCTZO system may be greater than that in the $(Ba_{1-x}Ca_x)TiO_3$ system. This interpretation seems to be reasonable. Since $BaZrO_3$ is not ferroelectric even at zero Kelvin, and substitution of Zr for Ti reduces the ferroelectricity of $Ba(Ti,Zr)O_3$, in contrast to the large Ti displacement in the oxygen octahedron, the same level of Zr displacement is not expected to exist in the oxygen octahedron in $Ba(Ti,Zr)O_3$. Actually, this has been predicted from recent first-principles calculations, which indicates that Zr displacement is extremely small and has a value of about one-sixth of the Ti displacement at the lowest temperature in $Ba(Ti,Zr)O_3$.[29, 30] In contrast, as shown in Fig. 10. Ca displacement is predicted to have a value of two times the Ti displacement from first principles calculations. Therefore, in the BCTZO system, the polarization due to Ca-displacement is able to effectively compensate for the reduction of polarization from B-site atomic displacement due to the substitution of Zr for Ti, leading to the enhancement of ferroelectricity in the BCTZO system.

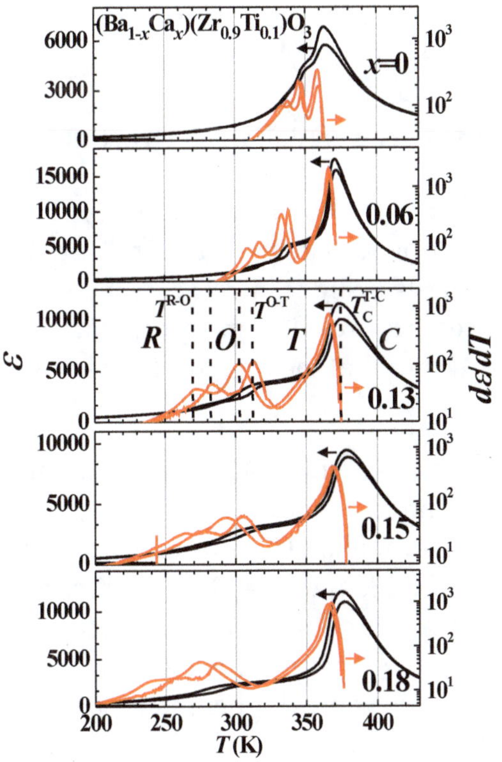


**Figure 14.** Temperature dependence of dielectric permittivity and its temperature differentiation in $(Ba_{1-x}Ca_x)(Ti_{0.9}Zr_{0.1})O_3$ ceramics.

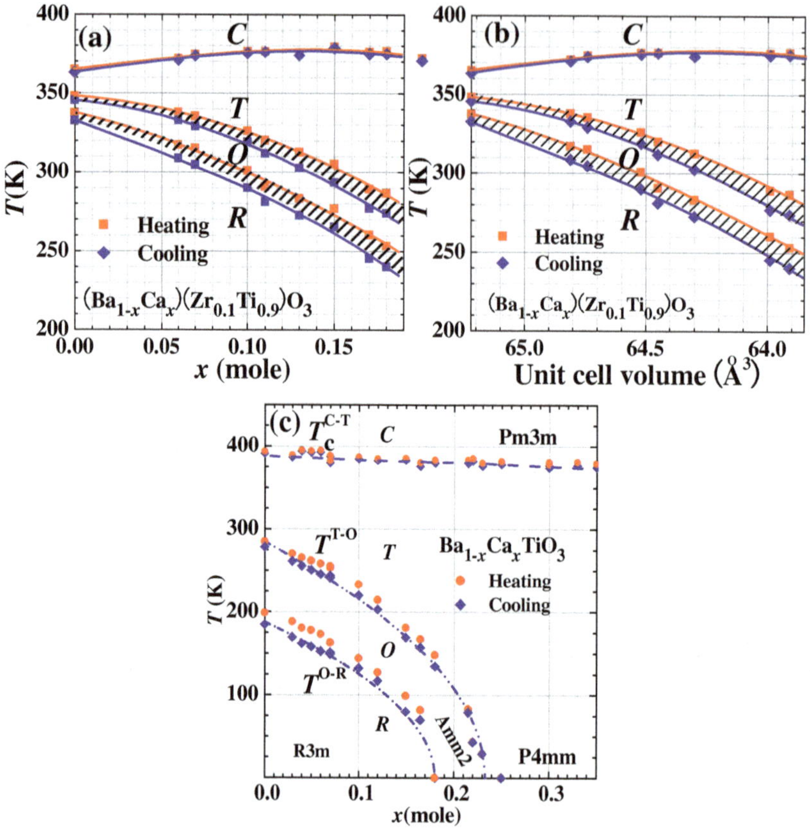


**Figure 15.** Phase diagrams of $(Ba_{1-x}Ca_x)(Ti_{0.9}Zr_{0.1})O_3$ solid solutions as functions of (a) composition and (b) unit cell volume. For comparison, the phase diagram of $(Ba_{1-x}Ca_x)TiO_3$ is also shown in (c).

### 3.4. Polarization and strain responses under an electric field

A *D-E* hysteresis loop of BCTZO is shown in Fig. 16(a). The remanent polarization was observed to have a value of approximately 10 μC/cm$^2$ for the ceramics samples at room temperature. It seems that there is a slight increase in the saturation polarization as the Ca concentration initially increases. This result is in agreement with the variation of Curie point with Ca concentration. It also seems that the coercive field becomes larger with the increase in Ca concentration.

On the other hand, the great enhancement of strain responses under an electric field is clearly observed for Ca-substituted ceramics as shown in Fig. 16(b) and (c). For example, the electric-field-induced strain for $x$=0 has a value of 0.054 % at $E$=10 kV/cm, for the same unipolar field, while it reaches a large value of 0.086% for $x$=0.10, which corresponds to an effective piezoelectric response of 860 pm/V. Figure 16(d) demonstrates the variation of this effective piezoelectric response around room temperature with Ca composition. It is clear that a large effective piezoelectric response with values higher than 800 pm/V has been observed in a wide composition range from $x$=0.06 to $x$=0.16 in the BCTZO ceramics, which is much larger than that obtained in the commercial PZT ceramics. Such extremely large electric-field-induced strain may be of great interest for the development of lead-free piezoelectric ceramics. As shown in Fig. 16(d), large electromechanical coupling effects occur around the *R-O* and *O-T*

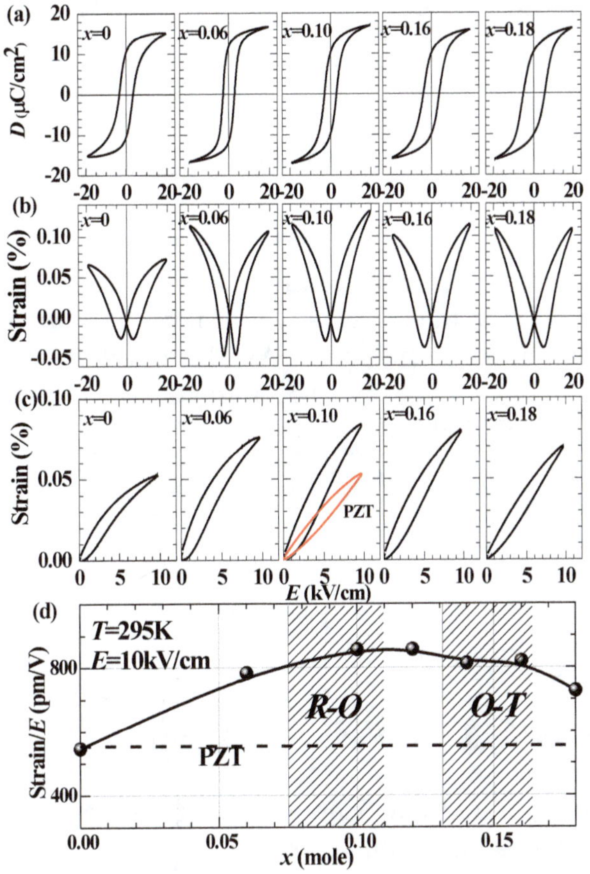


**Figure 16.** (a) Polarization, (b) bipolar-field, and (c) unipolar-field strain under electric field measured at $T$=295 K for $(Ba_{1-x}Ca_x)(Ti_{0.9}Zr_{0.1})O_3$ ceramics. For comparison, the response of commercially used PZT is also shown in (c). (d) Change of the strain at $E$=10kV/cm with composition.

phase boundaries. This indicates that the polymorphic phase transitions play a critical role in the large piezoelectric response in the BCTZO solid solution.

## 4. Summary

Ca off-centering was demonstrated to play a critical role in stabilizing the ferroelectric phase and tuning the polarization states in a $(Ba_{1-x}Ca_x)(Ti_{1-y}Zr_y)O_3$ system. Two typical cases, ($(Ba_{1-x}Ca_x)TiO_3$ with $y$=0 and $(Ba_{1-x}Ca_x)(Ti_{0.9}Zr_{0.1})O_3$ with $y$=0.1), were studied. In both cases, atomic displacement due to Ca off-centering in the bulky Ba sites in the $ABO_3$ perovskite structure provides an approach to compensate for the reduction of ferroelectricity due to chemical pressure, leading to the maintenance of a nearly constant Curie point in the $(Ba_{1-x}Ca_x)TiO_3$ system and an increase in the Curie point in the $(Ba_{1-x}Ca_x)(Ti_{0.9}Zr_{0.1})O_3$ system. The Ca off-centering effects are commonly observed to lead to the shift of the $R$–$O$ and $O$–$T$ phase transitions toward lower temperatures and the ferroelectric stability of the $T$-phase, resulting in the occurrence of quantum phase transitions with interesting physics phenomena at low temperatures in the $(Ba_{1-x}Ca_x)TiO_3$ system and remarkable enhancement of electromechanical coupling effects around room temperature in the $(Ba_{1-x}Ca_x)(Ti_{0.9}Zr_{0.1})O_3$ system over a large composition range.

## Acknowledgements

We thank Prof. Shin-ya Koshihara & Dr. T. Shimizu of the Tokyo Institute of Technology, Mr. T. Kosugi & Prof. S. Tsuneyuki of the University of Tokyo, and Mr. Y. Kamai of the Shizuoka University for their collaboration in this work. We also thank the support from the Collaborative Research Project of the Materials and Laboratory, Tokyo Institute of Technology, and KAKENHI (15H02292 and 26620190).

## Author details

Desheng Fu[1,2*] and Mitsuru Itoh[3]

*Address all correspondence to: fu.tokusho@shizuoka.ac.jp

1 Department of Engineering, Graduate School of Integrated Science & Technology, Shizuoka University, 3-5-1 Johoku, Hamamatsu, Japan

2 Department of Optoelectronics and Nanostructure Science, Graduate School of Science and Technology, 3-5-1 Johoku, Naka-ku, Hamamatsu, Japan

3 Materials and Structures Laboratory, Tokyo Institute of Technology, 4259 Nagatsuta, Yokohama, Japan

## References

[1] Y. Shiozaki, E. Nakamura and T. Mitsui (eds). Ferroelectrics and related substances (Landolt-Bornstein, New Series, Group III, vol. 36, Pt. A1). Berlin: Springer; 2001.

[2] M. Zgonik, P. Bernasconi, M. Duelli, R. Schlesser, P. Günter, M. H. Garrett, D. Rytz, Y. Zhu and X. Wu. Dielectric, elastic, piezoelectric, electro-optic, and elasto-optic tensors of $BaTiO_3$ crystals. Phys. Rev. B 1994;50:5941. DOI: 10.1103/PhysRevB.50.5941.

[3] S.-E. Park and T. R. Shrout. Ultrahigh strain and piezoelectric behavior in relaxor based ferroelectric single crystals. J. Appl. Phys. 1997;82:1804. DOI: 10.1063/1.365983.

[4] R. Tazaki, D. Fu, M. Itoh, M. Daimon and S. Koshihara. Lattice distortion under an electric field in $BaTiO_3$ piezoelectric single crystal. J. Phys. Condens. Matter. 2009;21:215903. DOI: 10.1088/0953-8984/21/21/215903.

[5] J. Rodel, W. Jo, K. T. P. Seifert, E. M. Anton, T. Granzow and D. Damjanovic. Perspective on the development of lead-free piezoceramics. J. Am. Ceram. Soc. 2009;92:1153. DOI: 10.1111/j.1551-2916.2009.03061.x.

[6] D. Fu, M. Itoh and S. Koshihara. Crystal growth and piezoelectricity of $BaTiO_3$-$CaTiO_3$ solid solution. Appl. Phys. Lett. 2008;93:012904. DOI: 10.1063/1.2956400.

[7] D. Fu, M. Itoh, S. Koshihara, T. Kosugi and S. Tsuneyuki. Anomalous phase diagram of ferroelectric $(Ba,Ca)TiO_3$ single crystals with giant electromechanical response. Phys. Rev. Lett. 2008;100:227601. DOI: 10.1103/PhysRevLett.100.227601.

[8] D. Fu, M. Itoh and S. Koshihara. Invariant lattice strain and polarization in $BaTiO_3$–$CaTiO_3$ ferroelectric alloys. J. Phys. Condens. Matter. 2010;22:052204. DOI: 10.1088/0953-8984/22/5/052204.

[9] D. Fu, Y. Kamai, N. Sakamoto, N. Wakiya, H. Suzuki and M. Itoh. Phase diagram and piezoelectric response of $(Ba_{1-x}Ca_x)(Zr_{0.1}Ti_{0.9})O_3$ solid solution. J. Phys. Condens. Matter. 2013;25:425901. DOI: 10.1088/0953-8984/25/42/425901.

[10] A. von Hippel. Ferroelectricity, domain structure, and phase transitions of barium titanate. Rev. Mod. Phys. 1950;22:221. DOI: 10.1103/RevModPhys.22.221.

[11] W. Cochran. Crystal stability and the theory of ferroelectricity. Adv. Phys. 1960;9:387. DOI: 10.1080/00018736000101229.

[12] T. Mitsui and W. B. Westphal. Dielectric and x-ray studies of $Ca_xBa_{1-x}TiO_3$ and $Ca_xSr_{1-x}TiO_3$. Phys. Rev.1961;124:1354. DOI: 10.1103/PhysRev.124.1354.

[13] R. D. Shannon. Revised effective ionic radii and systematic studies of interatomic distances in halides and chalcogenides. Acta Crystallogr. Sect. A Cryst. Phys. Diffr. Theor. Gen. Crystallogr. 1976;32:751. DOI: 10.1107/S0567739476001551.

[14] T. Shimizu, D. Fu, H. Taniguchi, T. Taniyama and M. Itoh. Origin of the dielectric response in $Ba_{0.767}Ca_{0.233}TiO_3$. Appl. Phys. Lett. 2012;100:102908. DOI: 10.1063/1.3693524.

[15] R. C. DeVries and R. Roy. Phase Equilibria in the system $BaTiO_3$-$CaTiO_3$. J. Am. Ceram. Soc. 1955;38:142. DOI: 10.1111/j.1151-2916.1955.tb14918.x.

[16] Ch. Kuper, R. Pankrath and H. Hesse. Growth and dielectric properties of congruently melting $Ba_{1-x}Ca_xTiO_3$ crystals. Appl. Phys. A Mater. Sci. Process. 1997;65:301. DOI: 10.1007/s003390050583.

[17] T. Schneider, H. Beck and E. Stoll. Quantum effects in an n-component vector model for structural phase transitions. Phys. Rev. B 1976;13:1123. DOI: 10.1103/PhysRevB.13.1123.

[18] U. T. Hochli and L. A. Boatner. Quantum ferroelectricity in $K_{1-x}Na_xTaO_3$ and $KTa_{1-y}Nb_yO_3$. Phys. Rev. B 1979;20:266. DOI: 10.1103/PhysRevB.20.266.

[19] J. Iniguez and D. Vanderbilt. First-principles study of the temperature-pressure

phase diagram of $BaTiO_3$. Phys. Rev. Lett. 2002;89:115503. DOI: 10.1103/PhysRevLett.89.115503.

[20] J. B. Neaton, C. L. Hsueh and K. M. Rabe. Enhanced polarization in strained $BaTiO_3$ from first principles. arXiv:condmat/0204511;2002.

[21] S. Tinte, K. M. Rabe, and D. Vanderbilt. Anomalous enhancement of tetragonality in $PbTiO_3$ induced by negative pressure. Phys. Rev. B 2003;68:144105. DOI: 10.1103/PhysRevB.68.144105.

[22] M. Malinowski, K. Lukaszewicz and S. Asbrink. The influence of high hydrostatic pressure on lattice parameters of a single crystal of $BaTiO_3$. J. Appl. Crystallogr. 1986;19:7. DOI: 10.1107/S0021889886090088.

[23] T. Ishidate, S. Abe, H. Takahashi and N. Mori. Phase Diagram of $BaTiO_3$. Phys. Rev. Lett. 1997;78:2397. DOI: 10.1103/PhysRevLett.78.2397.

[24] T. Kosugi, S. Tsuneyuki, D. Fu., M. Itoh, S. Koshihara. First-principles calculation of $Ba_{1-x}Ca_xTiO_3$ single crystal. Meeting abstract of the physical society of Japan 62(issue 2, part 4): 988. Hokkaido University. Sept. 22, 2007.

[25] R. C. Kell and N. J. Hellicar. Structural transitions in barium titanate-zirconate transducer materials. Acta Acustica united with Acustica. 1956;6:235.

[26] B. Jaffe, W. R. Cook, Jr. and H. Jaffe. Piezoelectric ceramics. London: Academic; 1967.

[27] P. W. Rehrig, S.-E. Park, S. Trolier-McKinstry, G. L. Messing, B. Jones and T. R. Shrout. Piezoelectric properties of zirconium-doped barium titanate single crystals grown by templated grain growth. J. Appl. Phys. 1999;86:1657. DOI: 10.1063/1.370943.

[28] Z. Yu, C. Ang, R. Guo and A. S. Bhalla. Piezoelectric and strain properties of $Ba(Ti_{1-x}Zr_x)O_3$ ceramics. J. Appl. Phys. 2002;92:1489. DOI: 10.1063/1.1487435.

[29] C. Laulhé, A. Pasture, F. Hippert and J. Kreisel. Random local strain effects in homovalent-substituted relaxor ferroelectrics: A first-principles study of $BaTi_{0.74}Zr_{0.26}O_3$. Phys. Rev. B 2010;82:13210. DOI: 10.1103/PhysRevB.82.132102.

[30] A. R. Akbarzadeh, S. Prosandeev, E. J. Walter, A. Al-Barakaty, and L. Bellaiche. Finite-temperature properties of $Ba(Zr,Ti)O_3$ relaxors from first principles. Phys. Rev. Lett. 2012;108:257601. DOI: 10.1103/PhysRevLett.108.257601.

9

# New Properties and Applications of Polyvinylidene-Based Ferroelectric Polymer

J.L. Wang, X.J. Meng and J.H. Chu

**Abstract**

There are different kinds of novel properties and applications of polyvinylidene difluoride (PVDF)-based ferroelectric polymer films. Several issues associated with the structure, properties, and applications of PVDF-based ferroelectric polymer films are discussed. The main achievements of the research include high electric tunability of relaxor ferroelectric Langmuir–Blodgett (LB) terpolymer films, the creep process of the domain switching in poly(vinylidene fluoride-trifluoroethylene) ferroelectric thin films, transition from relaxor to ferroelectric-like phase in poly(vinylidene fluoride-trifluoroethylene -chlorofluoroethylene) terpolymer ultrathin films, abnormal polarization switching of relaxor terpolymer films at low temperatures, huge electrocaloric effect in LB ferroelectric polymer thin films, self-polarization in ultrathin LB polymer films, enhanced dielectric and ferroelectric properties in artificial polymer multilayers, and transition of polarization switching from extrinsic to intrinsic in ultrathin PVDF homopolymer films.

In addition, the recent progresses of PVDF-based ferroelectric polymer films are connected with potential applications.

**Keywords:** Poly(vinylidene fluoride), Langmuir–Blodgett technique, Ferroelectric polymer, Ferroelectric relaxor

## 1. Introduction

Polyvinylidene difluoride (PVDF), consisting of $(–CF2–CH2–)_n$, with a carbon chain and hydrogen and fluorine atom on the two sides of carbon, respectively, is not a new synthesized material, which has been found in 50 years ago. In the beginning, PVDF was studied for its high dielectric permittivity and for the diversity of crystalline-phase types. About 10 or 15 years later, piezoelectricity and ferroelectricity properties were found in PVDF materials [1–6].

As it is known, PVDF was found in possession of four or more crystalline-phase types. It consists of the $\alpha$-phase, $\beta$-phase, $\gamma$-phase, and $\sigma$-phase. Phase types are related to various molecular configurations. In these different phase types, the $\beta$-phase is of polar form and the $\alpha$-phase is of antipolar form; the molecular configuration is shown in Fig. 1. The antipolar form $\alpha$-phase PVDF can be transformed to the $\beta$-phase polar form by rapid cooling from the melting or stretching along the carbon chain [1,2,5].

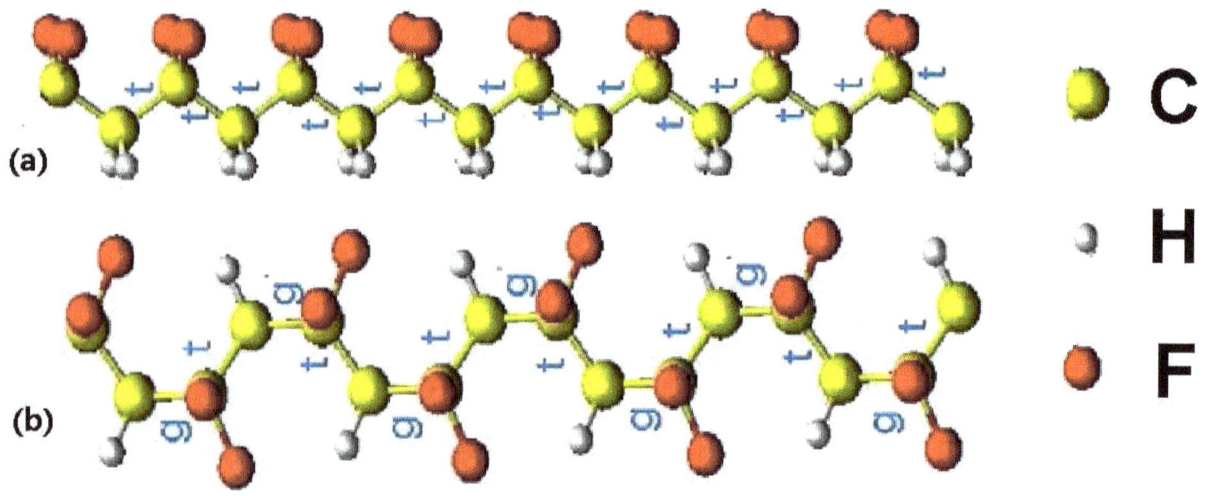


**Figure 1.** The typical molecular configuration of PVDF (a) $\alpha$-phase, (b) $\beta$-phase

The other two types of PVDF phases derive from the $\beta$-phase and $\alpha$-phase. Recently, it has been proved that the $\sigma$-phase PVDF is also a polar form. When the poly(trifluoroethylene) is added into the PVDF, a new type of copolymer P(VDF-TrFE), consisting of $-((-CF2-CH2)_x-(-CF2-CHF-)_{1-x})_n-$ chains, is achieved. The copolymers with 50 % or less trifluoroethylene (TrFE) are ferroelectric, though with reduced polarization and transition temperatures, because some of the hydrogen atoms are replaced with fluorine, reducing the net dipole moment polymer chains [5].The $\beta$-phase P(VDF-TrFE) with ferroelectricity can be obtained easily, so the copolymer has been studied widely and popularly.

The terpolymer derived from the P(VDF-TrFE) is the ferroelectric relaxor polymer P(VDF-TrFE-CFE), which adds chlorofluoroethylene into the P(VDF-TrFE) copolymer. PVDF-based polymer films possess many special properties: dielectric, ferroelectric, piezoelectric, pyroelectric properties and so on. Based on these properties, these films can be used as transducers, ferroelectric memory, gate of transistor, and uncooled infrared sensor. In the following paragraph, some special properties and potential applications of PVDF-based polymer films will be introduced.

## 2. Method

There are several methods for preparing the PVDF-based films, for instance, spin coating, wire bar method, and Langmuir–Blodgett (LB) technique [7–9]. The mostly used film preparing method is spin coating, which is widely used for preparing the films with thickness larger than 30 nm. It is very difficult to get the ultrathin PVDF-based films based on spin coating and other normal methods. The LB method can be used to prepare the ultrathin two-dimensional ferroelectrics. In this chapter, the films of PVDF-based films mostly fabricated using LB technology. In 1995, S. P. Palto, L. M. Blinov, and V. M. Fridkin et al. began studies of ultrathin ferroelectric LB films of P(VDF-TrFE) copolymers with trifluoroethylene, P(VDF-TrFE) [9]. In 2007, J.L. Wang et al. prepared the ultrathin P(VDF-TrFE-CFE) terpolymer LB films [8]. Recently, we achieved the ultrathin PVDF homopolymer films with good ferroelectric performance. The detail of preparing the PVDF-based polymer films by LB technique is described as follows. The typical characteristic of the LB method is one monolayer (ML) at a time by repeatedly dipping a substrate into a liquid subphase coated with a monolayer of the desired polymer. The PVDF-based polymer should be firstly dissolved in dimethylformamide to form a dilute solution 0.01–0.05 wt%. The liquid subphase is ultrapure water with 18 MΩ/m. After the PVDF-based polymer solution is dropped into the water, about 30 min later, the PVDF-based polymer molecule chain will be floating on the surface of the water (as shown in Fig. 2a). In this case, the films are not uniform. The next step is pressing the bar of the LB technique until the surface pressure is up to 5 mN/m or less (as described in Fig. 2b). In this processing, if the surface is too large, the film on the water surface will collapse. The final step is shown in Fig. 2c, dipping the substrate horizontally on the surface of the water covered with monolayer PVDF molecule. Then raising the substrate slowly, one monolayer of PVDF ultrathin film is achieved. Repeating this process, the different thicknesses of PVDF-based film will be produced.

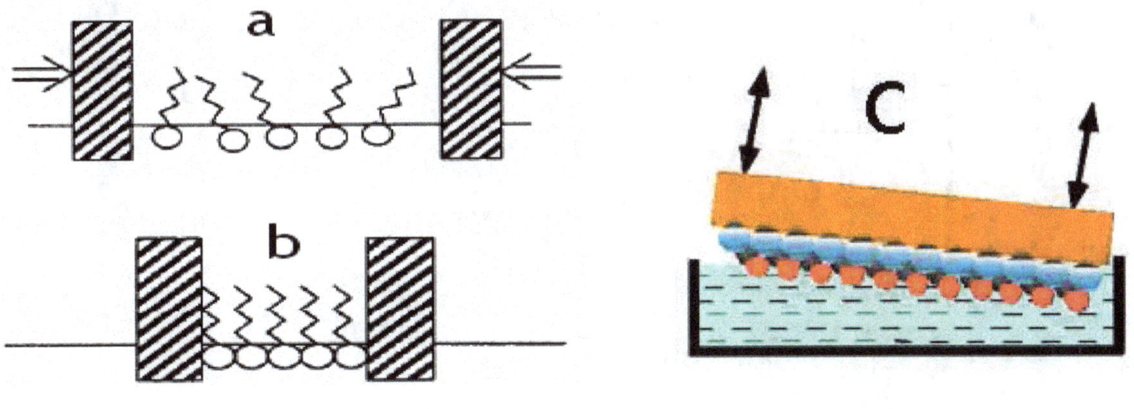


**Figure 2.** The horizontal transfer of LB technique: (a) and (b) the processing of the LB films; (c) transferring the PVDF-based film to the substrate

PVDF-based polymer films possess many special properties, for example, dielectric, ferroelectric, piezoelectric, and pyroelectric properties and so on. Based on these properties, the films can be used for preparing the transducers, ferroelectric memory, gate of transistor, and uncooled infrared sensor. In the following paragraph, some special properties and applications will be introduced.

## 3. Dielectric tunability properties of PVDF-based polymer

The dielectric tunability of the PVDF has been studied by Lu et al. in 2008; the huge tunability can be reached in the P(VDF-TrFE) copolymer films, but in the P(VDF-TrFE-CFE) relaxor terpolymer films, the tunability is lower than the copolymer's [10]. For the traditional way of the relaxor ferroelectrics, the electric tunabilities are very large. Since the film thickness is controlled on a molecule level, ferroelectric LB polymer films have shown some exceptional properties, such as excellent crystallinity, two-dimensional ferroelectricity, surprising giant breakdown voltage, etc.; it is expected that terpolymer films derived from the LB technology can provide a special microstructure to study the origin of the excellent properties of relaxor ferroelectric terpolymers. High-quality ultrathin films of both ferroelectric P(VDF-TrFE) and relaxor ferroelectric P(VDF-TrFE-CFE) have been successfully fabricated by using the LB technique.

The P(VDF-TrFE-CFE) shows a typical relaxor from temperature dependences of the dielectric constant and dielectric loss versus frequency, as it is shown in Fig. 3.

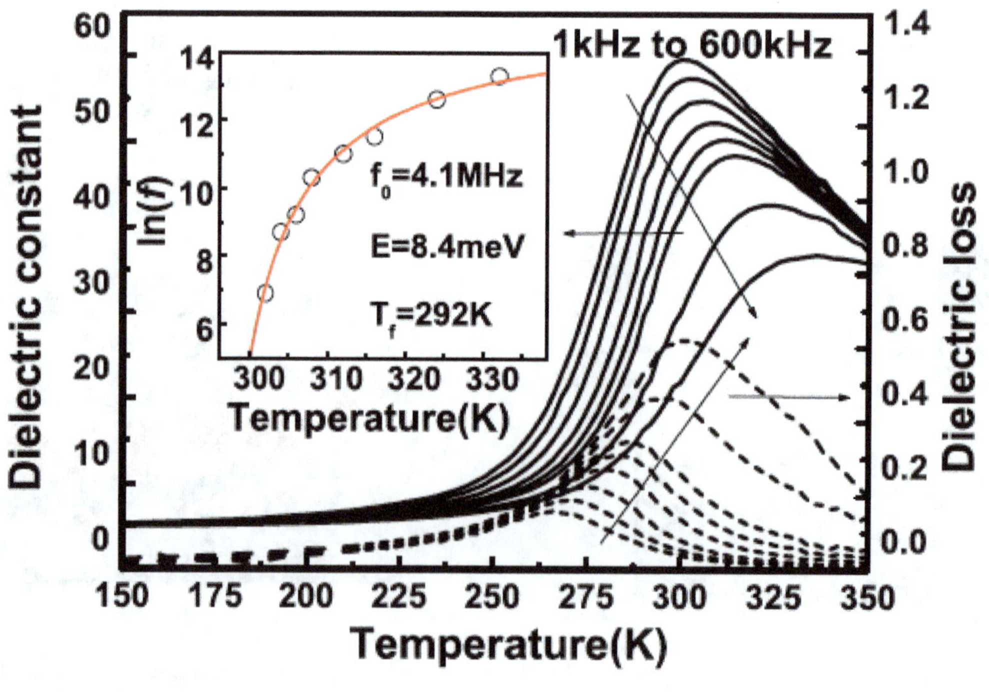


**Figure 3.** The dielectric constant and dielectric loss of P(VDF-TrFE-CFE) LB films as a function of temperature. The inset is the peak temperature of dielectric constant $T_m$ as a function of ln(*f*) and the parameters fitted with the Vogel–Fulcher relation

A tunability of 80 % at 240 MV/m was obtained in P(VDF-TrFE-CFE) terpolymer LB films (shown in Fig. 4), which is due to the highly ordered molecules and the high breakdown electric filed.

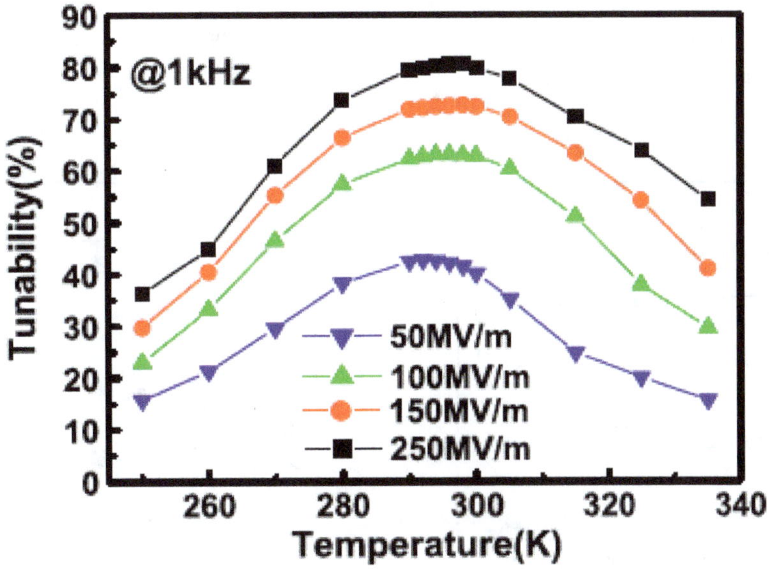


**Figure 4.** The tunability as a function of temperature for P(VDF-TrFE-CFE) LB films measured at 1 kHz, under the dc electric fields of 50, 100, 150, and 250 MV/m, respectively

What are the reasons for the large tunability in our terpolymer LB films? In our opinion, they should be associated with the special microstructure of the terpolymer LB films. It is known to us that LB polymers demonstrate some exceptional features such as good crystallinity and highly planar ordered and close parallel packing of the molecules.

## 4. Huge electrocaloric effect in LB ferroelectric polymer thin films

Recently, the huge electrocaloric effect (ECE) resulting from changes in the entropy and temperature of a material under an applied electric field has attracted the attention of researchers to ferroelectric materials [11,12]. The ECE occurs in both ferroelectric and paraelectric phases and is found to be larger in the paraelectric phase just above the ferroelectric–paraelectric phase transition [13].

In a working cycle-based ECE, the working material contacts the load and absorbs entropy from it. Then the material is isolated from the load and an electric field is applied. With an increase in the electric field, the polarization and temperature of the working material increase under adiabatic condition. The material is then placed in thermal contact with the heat sink and transfers the entropy absorbed from the load to the heat sink. Then the material is isolated from the heat sink. As the applied field is reduced, the temperature of the material decreases back to the temperature of the cooling load. Thus, the larger the ECE of the working material, the better the efficiency of cooling.

We present a detailed investigation of the ECE of P(VDF-TrFE) and P(VDF-TrFE-CFE) films grown by the LB technique on technologically desirable aluminized flexible polyimide substrates. Our results clearly show a large ECE, as the variations of adiabatic temperature $\Delta T$ of P(VDF-TrFE) films (at 390 K under 300 $MVm^{-1}$) and P(VDF-TrFE-CFE) films (at 350 K under 350$MVm^{-1}$) are 21 K (shown in Fig. 5).

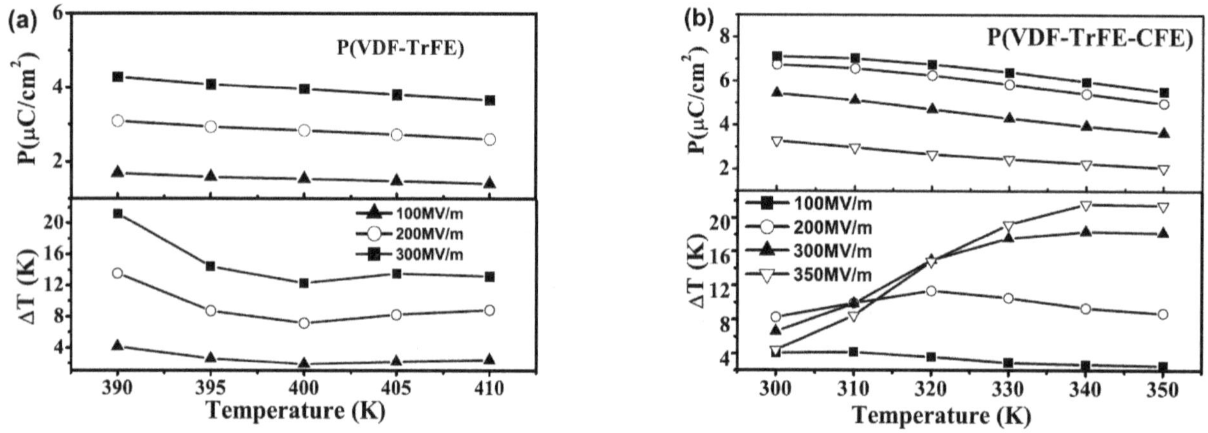


**Figure 5.** Polarizations from *P-E* loops and adiabatic temperature change $\Delta T$ versus temperature in different electric fields for (a) P(VDF-TrFE) films and(b) P(VDF-TrFE-CFE) films

Especially for P(VDF-TrFE-CFE) terpolymer films, the peak of $\Delta T$ is close to room temperature. Considering their practical availability, ferroelectric PVDF-based polymers are potentially applicable for refrigeration by ECE. Professor J. F. Scott has pointed out that PVDF-based polymers are the most promising commercial material due to their large temperature cooling per volt and to their scalability in size and shape [14].

## 5. Ferroelectric-like phase transition in P(VDF-TrFE-CFE) terpolymer ultrathin films

P(VDF-TrFE-CFE) terpolymer has attracted considerable attention for its fruitful properties and related potential application. Investigations have shown that the all-trans (*TTTT*) polar conformations have been converted into shorter all-trans conformations or polar nano-domains by the induced defect modifications [15]. Further studies verify that there coexist various molecular conformations in the terpolymer, such as trans-gauche (*TGTG),* and less polar *TTTG* conformations, and the *TTTG* is the dominant one responsible for the relaxor nature. Klein et al. found that the crystallization conditions can also impact the microstructures and the ferroelectric properties [16]. We have observed that the electric fields can change the length of the conformation.

The temperature dependence of the complex permittivity for the terpolymer films measured at 1 kHz in the heating process is shown in Fig. 6. For the film with 10 nm, the temperature of permittivity maximum $T_m$ is at 308 K, which is consistent with the bulk terpolymer or thick films. Note that $T_m$ increases with the decreasing thickness, and it is up to 360 K for the film with one transferred layer, which is even higher than the Curie temperature $T_c$ of the copolymers with the same composition.

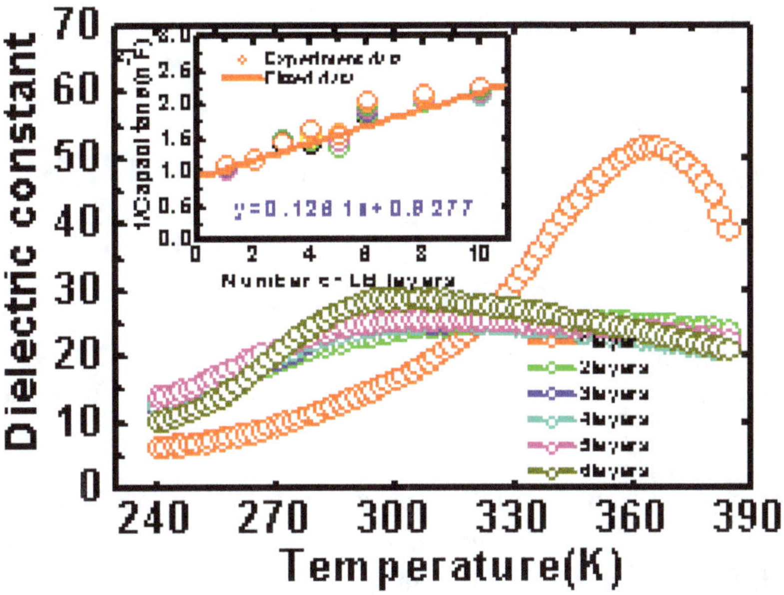


**Figure 6.** The dielectric constant of different thickness terpolymer films as a function of temperature of at 1 k Hz

The coercive voltage is 1.2 V, which is also in accord with the polarization versus electric field (*P-E*) loop and *C-V* curve. In addition, the temperature dependence of the *C-V* curves for the terpolymer film with one monolayer is measured and shown in Fig. 7. The butterfly-shape feature disappears as the temperature is higher than 360 K, indicating a transition from ferroelectric-like to paraelectric phase. This is in accord with the temperature dependence of the capacitance.

A hypothesis can be deduced from the above results that all-trans-like molecular conformations form in the ultrathin terpolymer films. As mentioned earlier, the dominant conformation in P(VDF-TrFE-CFE) terpolymer is the less polar *TTTG* conformation. It has been reported that the polar phase or ferroelectric-like phase can be induced by controlling the kinetic factors of the annealing process of terpolymer because the crystal structure is determined by both thermodynamics and kinetics. As it is known, the large PVDF dipoles can induce mirror charges on conducting substrates, thus forming an extrinsic vertical electric field on the substrate [17]. Here, it is considered that the terpolymer dipoles are also large enough to induce mirror charges on the conducting Al electrodes, which then can lead to a vertical electric field. Under such high vertical electric field, the less polar conformations are converted into higher polar conformations.

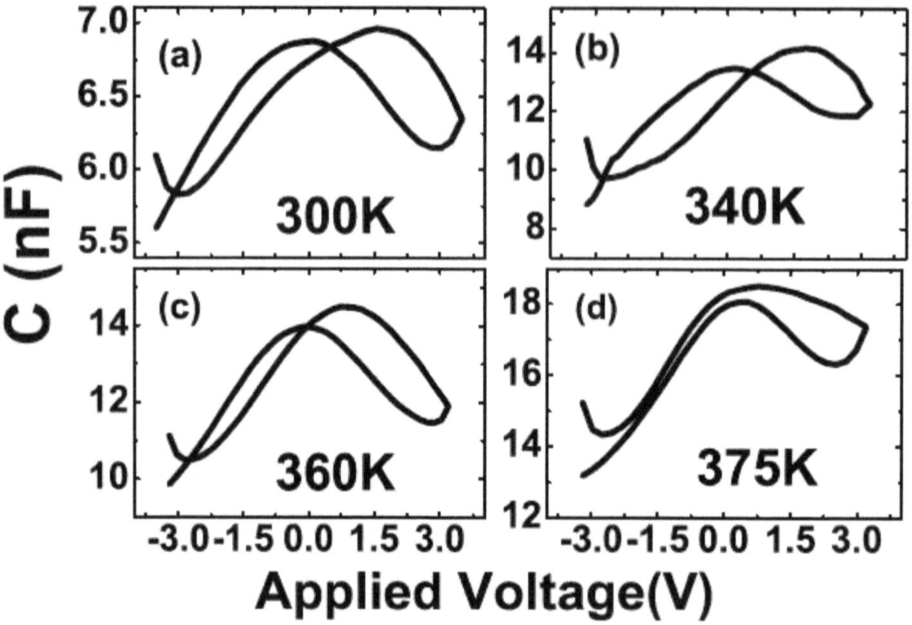


**Figure 7.** The temperature dependence of *C-V* curve at 10 kHz for one transferred layer terpolymer film. (a) 300 K, (b) 340 K, (c) 360 K, and (d) 375 K

The ferroelectric-like phase transition is observed in the P(VDF-TrFE-CFE) terpolymer films as the thickness is lower than 3 nm. The ferroelectric-like features are considered to result from the induced electric field due to the mirror charges in the electrodes.

## 6. Abnormal polarization switching of relaxor terpolymer films at low temperature

The temperature dependences of the dielectric and ferroelectric properties of terpolymer films produced using the LB method were systematically investigated, with an emphasis on the nature of the ferroelectricity at low temperatures [18].

The change of remanent polarization ($P_r$) with temperature is not sharp, and a broad peak at around 265 K is seen for $\partial P_r/\partial T$ (Fig. 8(a)).

A broad peak at ~270 K is displayed in the plot of $\partial D/\partial E_{E=0}$ versus T (Fig. 8(b)). The coercive field ($E_c$) increases slowly with decreasing temperature to ~265 K, below which it increases rapidly (Fig. 8(c)). This verifies that a ferroelectric phase transition occurs at ~270 K in the P(VDF-TrFE-CFE) films. The $P_r$ value of the P(VDF-TrFE-CFE) film is smaller than that of the P(VDF-TrFE) film, and the magnitude of the $P_r$ of the P(VDF-TrFE-CFE) film is still smaller than that of the film's saturation polarization ($P_s$) (Fig. 8(d)).

This suggests that some less-polar molecular conformations (TTTG') still affect the polarization switching. A deviation from Merz's law was observed in the relationship between the coercive

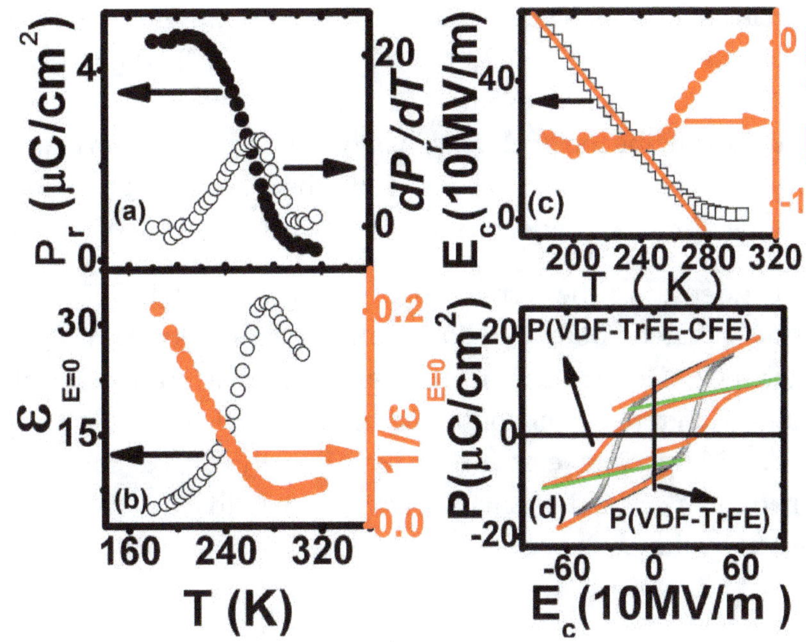


**Figure 8.** Temperature dependence of (a) $P_r$ (left), $\partial Pr/\partial T$ (right); (b) $E_c$ (left), $1\partial D/\partial E_{E=0}$ (right); and (c) $E_c$ derived from *P-E* loop at zero $P_r$ (left), $\partial E_C/\partial T$(right). (d) *P-E* loops of P(VDF-TrFE-CFE) and P(VDF-TrFE) films at 210 K and 100 Hz

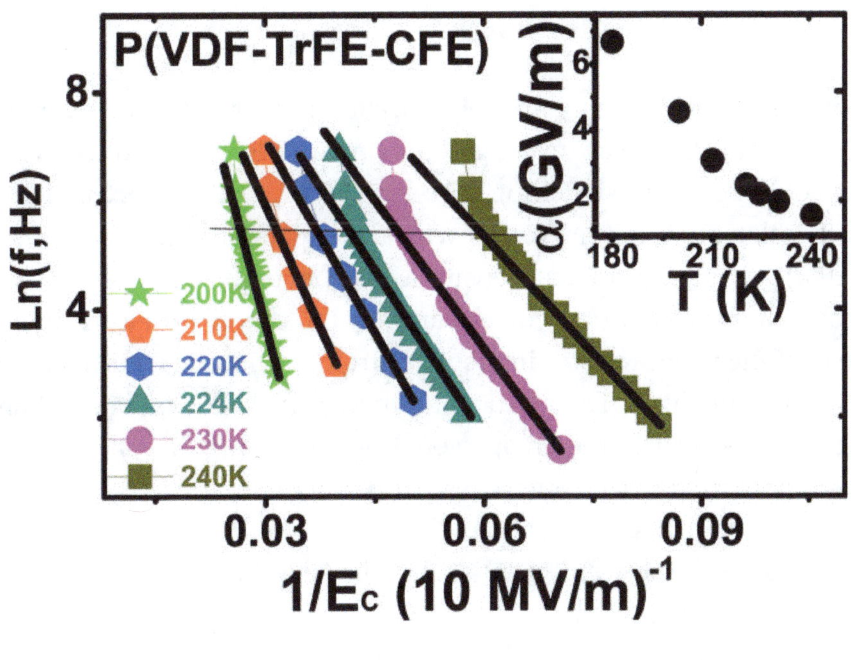


**Figure 9.** Relationship between $1/E_c$ and ln*f* for relaxor P(VDF-TrFE-CFE) terpolymer films

field and the frequency (Fig. 9). The deviation from Merz's law at high frequency further evidences the presence of *TTTG'* conformations in the ferroelectric state.

The relaxor P(VDF-TrFE-CFE) terpolymer, the CFE monomer is introduced into the ferroelectric P(VDF-TrFE) copolymer as a defect, leading to all-trans polar conformations being converted into less-polar conformations, i.e., *TTTG'*, which are considered to be nanometer-sized all-trans conformations. The dynamics of the *TTTG'* conformations of P(VDF-TrFE-CFE) films under electric fields or on cooling demonstrate a behavior different from that of the polar nanoregions (PNRs) of the Pb(Mg1/3Nb2/3)O3 (PMN) system [19]. The weak intermolecular interactions impede the development of some *TTTG′* conformations into all-trans ones along the direction perpendicular to the molecular chains, so these *TTTG′* conformations can largely retain their dynamics. This may be responsible for both the relaxor nature of the P(VDF-TrFE-CFE) terpolymer and the difference between its $P_r$ and *Ps* values, even in its ferroelectric state.

The temperature dependences of the ferroelectricity of P(VDF-TrFE-CFE) terpolymer films were systemically investigated. Both the polarization current ($\partial P_r/\partial T$) and the dielectric response derived from the *P-E* loop at zero field, suggesting that a ferroelectric phase transition occurs at 270 K. Distinct differences were observed in the P(VDF-TrFE-CFE) terpolymer films compared with perovskite relaxors, e.g., a broad peak in the $\partial P_r/\partial T$ curve, deviation from Merz's law at high frequency, and a smaller activation field [20]. These differences are considered to be caused by the existence of the less-polar *TTTG* conformation in the ferroelectric state.

## 7. The creep process of the domain switching in P(VDF-TrFE) ferroelectric thin films

The polarization switching behavior in poly(vinylidene fluoride-trifluoroethylene) P(VDF-TrFE) (70/30 mol%) thin films was investigated using a pulse transient current method [21]. The dependence of the domain switching current on the coercive electric field was derived. The current in the plateau region increases with the capacitor areas, whereas $t_w$ is basically constant (Fig. 10(a)), while the current density is independent of the capacitor areas (Fig. 10 (b)). This indicates that the charging current around the coercive field is limited by domain switching instead of the series resistor in the measurement circuit. Thus, it can reflect directly the speed of domain switching. The relationship between switching current and the electric field across the film fits well to the creep model with $\mu$=0.5 (Fig. 11). The exact dynamical exponent $\mu$ was found to be 0.5011, and resultant parameters $d_i$=2 and n=1, respectively. (An interface is characterized by its dimension $d_i$ ($d_i$=1 for a line or 2 for a surface) and can move in n transverse directions.) The result implies that the two-dimensional domain walls propagate along one transverse direction.

Considering the derived parameters $d_i$=2 and n=1 in the present study, a model was proposed for the polarization switching process in a crystalline lamella of the P(VDF-TrFE). Firstly, 180 $^{o}$ rotation of dipolar appears along a single-chain molecule; secondly, intermolecular expansion of chain rotations along external applied electric field with the switched molecular chains as the center because of the minimization of the depolarization energy; thirdly, domain walls with $d_i$=2 appeared at both sides of the switched dipolar plane. The domain wall, assumed to have a shape like a thin slab, propagates slowly, corresponding to n=1, till the completion of the domain switching in the lamellae

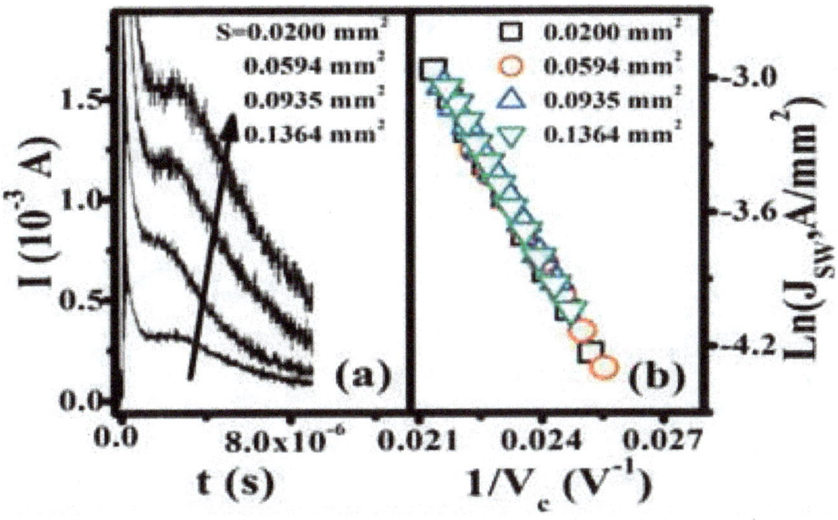


**Figure 10.** The variation of transient current with different capacitor areas under $V_a$ =40 V, $R_L$=1 KΩ. (b) $J_{sw}$ dependence of $V_c$ with different capacitor areas for P(VDF-TrFE) film

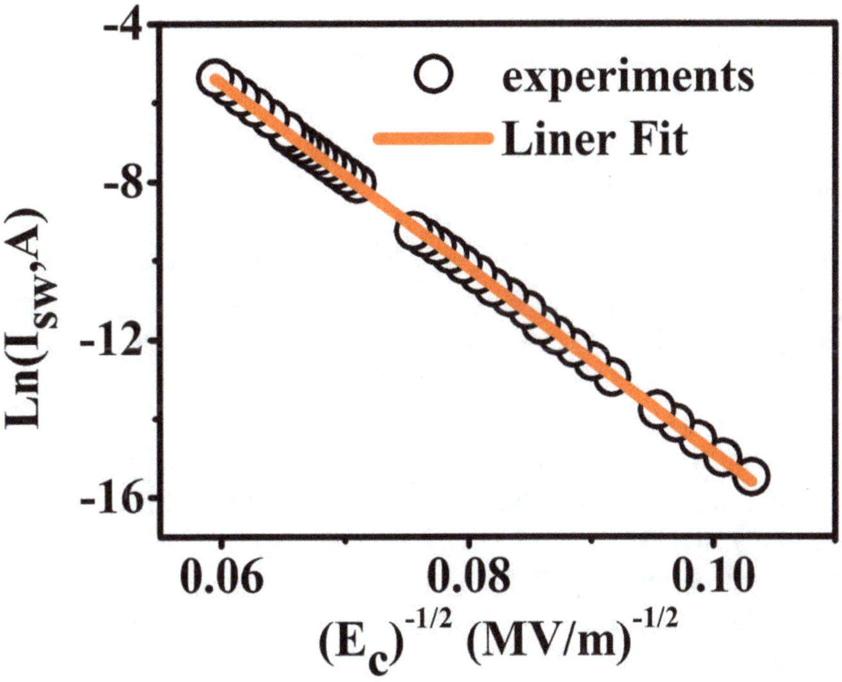


**Figure 11.** Dependence of domain switching current on the reciprocal of the square root of $E_c$.

## 8. Self-polarization in ultrathin LB polymer films

Ultrathin copolymer films of P(VDF-TrFE) were deposited on Al-coated polyimide substrates, by the LB method. A top Al electrode was evaporated onto the polymer film to form an Al/polymer/Al structured infrared detector. The pyroelectric voltage response of the detector under various polarizing processes was characterized. The detector with only one transferred polymer layer exhibited a preferential polarization direction. This was considered to result from the self-polarization of the ultrathin polymer film [22]. It was due to the preferred alignment of the dipoles on the Al substrates. This process can be applied for designing stable fast-response infrared detectors.

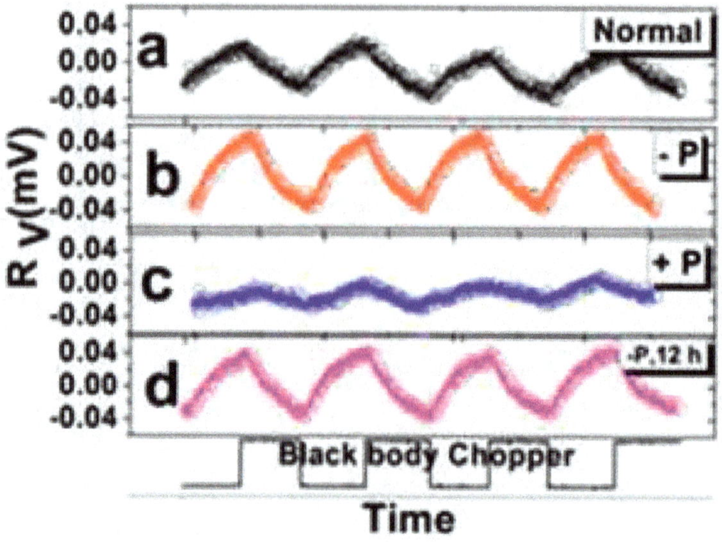


**Figure 12.** Pyroelectric voltage of the device under infrared radiation (a) before poling, (b) after poling at -1 V, (c) after poling at +1 V, and (d) 12 h after poling at -1 V

The fresh unpolarized device shows an appreciable pyroelectric voltage response, suggesting a preferential polarization. Upon polarizing the device at -1 V, the voltage response increases by a factor of 2, compared with the fresh device. Upon polarizing at +1 V, the voltage response decreases, in comparison with the fresh device. The applied 1 V is higher than the coercive electric field of the P(VDF-TrFE) ultrathin film with only 1 ML, reported in our previous investigation. Thus, the pyroelectric voltage responses under different polarizing directions should exhibit a 180° phase difference, but no such phase difference is observed in Fig. 12. This may be due to the back switching of domains, after removal of the positive poling voltage. It also suggests that the preferential polarization of the fresh device is aligned from the bottom electrode to the surface of the P(VDF-TrFE) film. The unpolarized detector exhibited a preferential voltage response.

This was considered to result from the self-polarization of the ultrathin P(VDF-TrFE) polymer film, due to the preferred alignment of the dipoles on the Al substrates. This result can be used to fabricate fast-response room temperature infrared detectors.

## 9. Electronic transport property in ferroelectric polymer films

The leakage current mechanism of ferroelectric copolymer of P(VDF-TrFE) prepared by LB was investigated in the temperature range from 100 K to 350 K. The electron as the dominant injected carrier was observed in the ferroelectric copolymer films. The transport mechanisms in copolymer strongly depend on the temperature and applied voltage. From 100 K to 200 K, Schottky emission dominates the conduction. With the increase of temperature, the Frenkel–Poole emission instead of the Schottky emission conducts the carrier transport. When the temperature gets to 260 K, the leakage current becomes independent of temperature, and the space charge limited current conduction was observed [23].

The P(VDF-TrFE) film shows a saturated hysteresis loop with a remanent polarization ($P_r$) of ~6.8 μC/cm$^2$ and saturation polarization ($P_s$) of ~11 μC/cm$^2$, respectively, which indicates good ferroelectricity. It was previously reported that the electron affinity of β-P(VDF-TrFE) was about 4 eV, based on a density functional theory study [24]. The work function value of Au and Al metals are 5.1 eV and 4.1 eV [25], respectively.

The conduction through the lowest unoccupied molecular level (LUMO) of P(VDF-TrFE) and the leakage current is controlled either by the interface energy barrier that exists between the Fermi level of the metal and the LUMO level of the polymer or by the bulk-controlled mechanics such as Frenkel–Poole emission and space-charge-limited current (SCLC) conduction [26,27]. The schematic diagram of the band structure of the P(VDF-TrFE) and the work functions of Au and Al are presented in the figure. The temperature dependence of the I-V behaviors from 100 K to 350 K was measured, and the temperature dependence of the current density under voltage 5 V (about 70 MV/m) is presented in Fig. 13. It can be seen that the current density increased with the temperature, increasing from 100 K to 260 K, but it is nearly independent of temperature as the temperature is higher than 260 K.

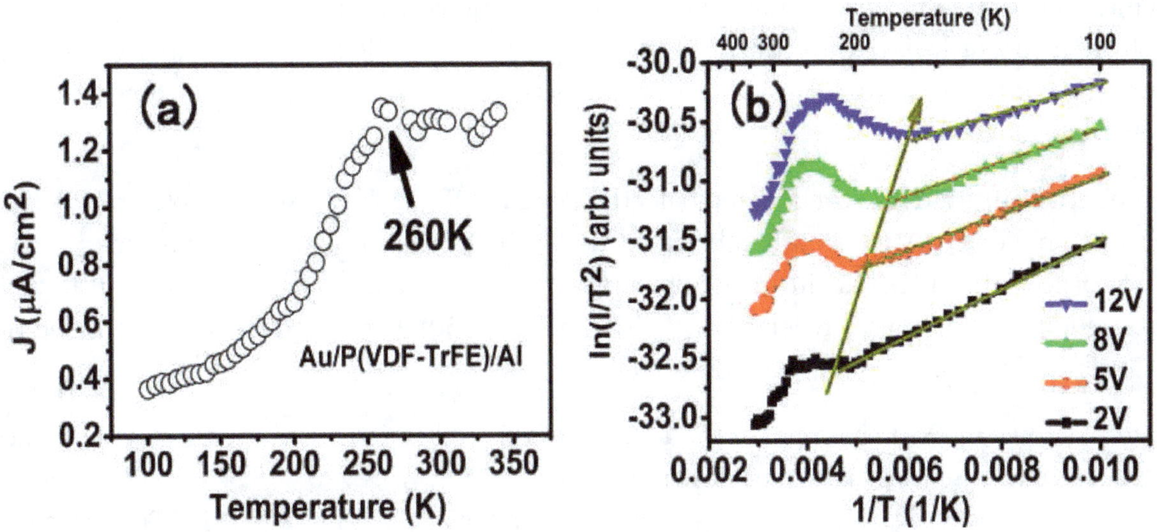


**Figure 13.** Temperature dependence of the current for P(VDF–TrFE) film under 5 V. (b) ln(I/T$^2$) vs. 1/T plots at different voltages. The direction of the arrow indicates a reducing temperature for increased voltage

The electric conduction of the ferroelectric P(VDF-TrFE) copolymer films has been comprehensively investigated. It is found that the electrons are the dominant injected carriers in the P(VDF-TrFE) films, and the charge injection occurs either at the polymer/electrode interface or in the bulk polymer films. Various transport mechanisms are observed in the P(VDF-TrFE) films, which are influenced by both temperature and applied voltage.

Schottky emission and Frenkel–Poole emission are found to be the dominant transport mechanism in the temperature range from 100 K to 200 K and the range from 200 K to 260 K, respectively. Space-charge-limited current conduction is the main transport mechanism as the temperature is higher than 260 K.

## 10. Enhanced electric properties in the artificial polymer multilayers

Multilayers consisting of alternative ferroelectric P(VDF-TrFE) copolymer and relaxor P(VDF-TrFE-CFE) terpolymer with different periodicities in thickness were prepared. A superlattice-like structure is shown in the polymer multilayer as the periodic thickness is lower than a critical value. The dielectric constant of the multilayer with a small periodic thickness is two times higher than that of the P(VDF-TrFE) copolymer over a temperature range between 300 K and 350 K. The multilayer also shows a good ferroelectricity in the same temperature range. The enhanced electrical properties of the multilayers are due to the long-range ferroelectric coupling [28,29].

Organic ferroelectric polymers have recently attracted much attention for their potential applications in flexible electronic devices, such as display, solar cell, information storage, and so on [30–32]. However, compared with its inorganic counterpart, organic ferroelectric polymer has some drawbacks, e.g., the lower dielectric constant and electric polarization, which is an obstacle for their practical applications. It is well known that many artificial superlattices (SL) and multilayers (ML) based on perovskite ferroelectrics show some amazing properties [33–35].

High-quality ultrathin films of both ferroelectric P(VDF-TrFE) and relaxor ferroelectric P(VDF-TrFE-CFE) have been successfully fabricated using the LB technique, which provide precise control of the film thickness in molecular scale. In the present study, periodic multilayers composed of alternating ferroelectric P(VDF-TrFE, 70/30) layer and relaxor ferroelectric P(VDF-TrFE-CFE, 56.2/36.3/7.5) layer were fabricated.

Figure 14 shows the XRD patterns of the multilayers with different periodicities. Two diffraction peaks were observed in the multilayer structure of $(15/15)_2$, i.e., $2\theta = 18.57$ and 19.97, which are assigned to the typical (110,200) reflection of the P(VDF-TrFE-CFE) and (110,200) reflection of P(VDF-TrFE), respectively. Compared with the pure P(VDF-TrFE-CFE) and P(VDF-TrFE), there is no shift in the (110,200) peaks for the $(15/15)_2$ multilayer, suggesting that both the terpolymer and copolymer components in the multilayer keep their original phase structure. When the periodic thickness is decreased to five transfer layers, a shift is observed in both the diffraction peaks. Simultaneously, the intensity of the diffraction peak associated with the pure terpolymers becomes weaker.

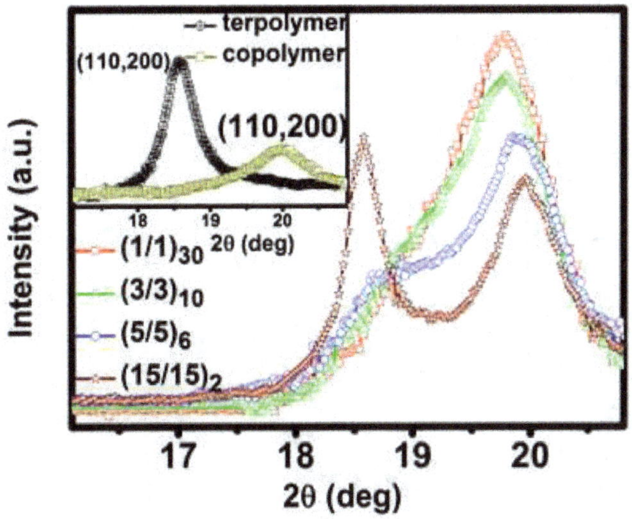


**Figure 14.** The XRD patterns of the multilayers with various periodicities. The inset shows the XRD patterns of copolymer and terpolymer

The temperature dependence of the dielectric constant of the multilayer structures on the heating process is shown in Fig. 15. Dielectric constant calculated with a series capacitor model of the individual copolymer and terpolymer is also shown in Fig. 15. It can be seen that the temperature dependence of the dielectric constant for the multilayer with the transfer number m larger than 10 is analogous to that of the series model, which shows a platform between 300 K and 360 K. This suggests that these multilayers with thick interlayer are just a combination of individual copolymer and terpolymer components.

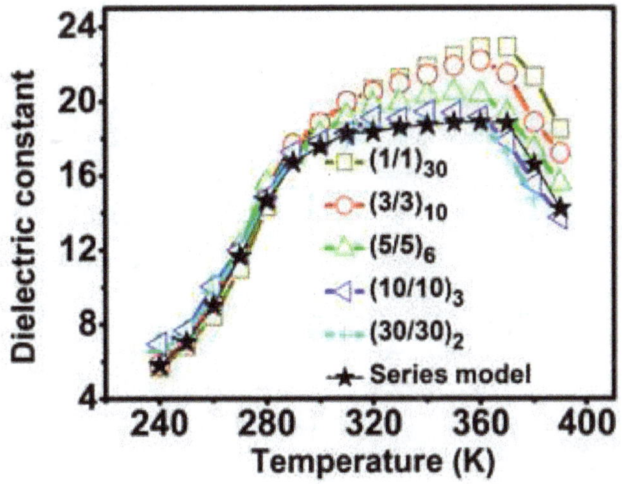


**Figure 15.** The dielectric constant of multilayers with different periodicities and the calculated dielectric constant using series model as a function of temperature

The *P-E* hysteresis loops for the multilayers as well as the pure copolymer and terpolymer films are shown in Fig. 16. The $(1/1)_{30}$ multilayer structure displays a good ferroelectricity with a remanent polarization ($P_r$) of ~ 5.2 μC/cm², which is just a little smaller than that of pure P(VDF-TrFE) thin films. The temperature dependence of the $P_r$ of the multilayers with various periodic thicknesses is summarized in the inset of Fig. 3.

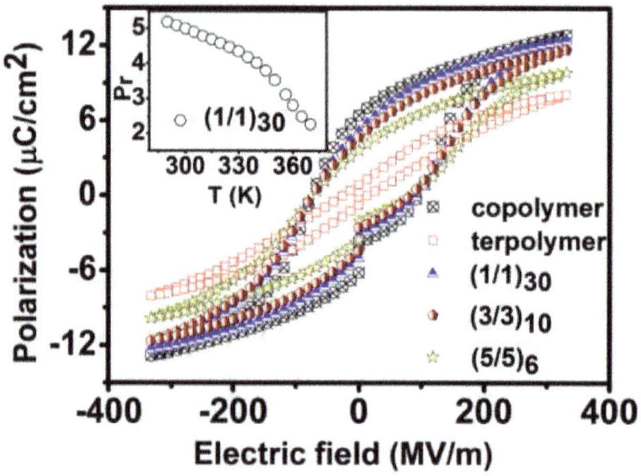


**Figure 16.** *P-E* hysteresis loops of the multilayers with different periodicities, pure copolymer and terpolymer films at 1 kHz. The inset shows $P_r$ of (1/1)30 multilayer as a function of temperature

The piezoelectric properties of the copolymer P(VDF-TrFE) and terpolymer P(VDF-TrFE-CFE) multilayers are presented in Fig. 17.

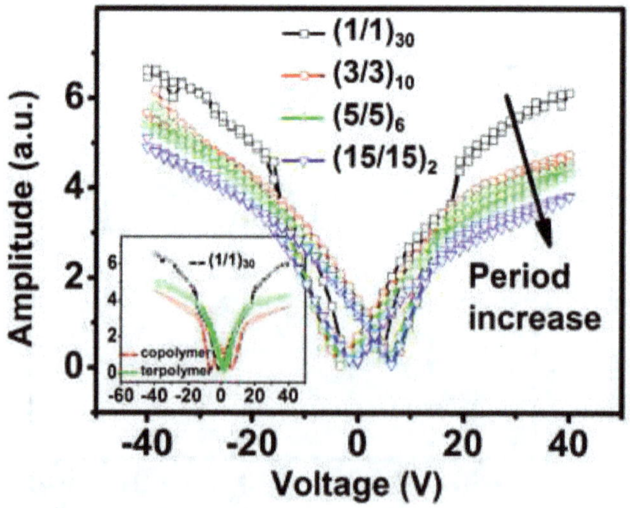


**Figure 17.** The piezoelectric response curves of multilayers with different periodicities. The inset compares the piezoelectric response curves of the (1/1)30 multilayer, P(VDF-TrFE) and P(VDF-TrFE-CFE) homogenous films

The piezoelectricity of the samples was measured by piezoresponse force microscopy (PFM). To increase the degree of accuracy and allow meaningful comparison of the piezoresponses of the samples, measurements were made at ten different locations for all samples and then averaged. The piezoelectric coefficient $d_{33}$ for various multilayer structures are listed in Fig. 19. The $d_{33}$ of the pure copolymer and terpolymer are 41 pC/N and 52 pC/N, respectively, which is close to the reported value for the copolymer (49 pC/N) and equal to that observed for the terpolymer. Interestingly, an enhanced piezoelectric response of 81.6 pC/N is observed for the $(1/1)_{30}$ multilayers, which is nearly 57 % larger than the pure terpolymer. As the periodic thickness increases, the piezoelectric response of the multilayers decreases dramatically.

In summary, the multilayers composed of alternating P(VDF-TrFE) copolymer and P(VDF-TrFE-CFE) terpolymer layers have been prepared, and their crystal structure and dielectric and ferroelectric properties have been studied. The multilayers with a periodic thickness of ~3 nm shows a superlattice-like crystal structure, high dielectric constant, good ferroelectricity, and piezoelectricity over a wide temperature range from 300 K to 350 K. The long-range ferroelectric coupling is considered to be dominant for multilayers with a smaller periodic interlayer.

## 11. Polarization switching properties of PVDF homopolymer films

PVDF homopolymer thin films have been prepared by the Langmuir–Blodgett technique, and their electrical properties have comprehensively been studied [36]. The PVDF homopolymer films show better ferroelectricity with higher polarization and higher breakdown electric field than that of P(VDF-TrFE) copolymer films.

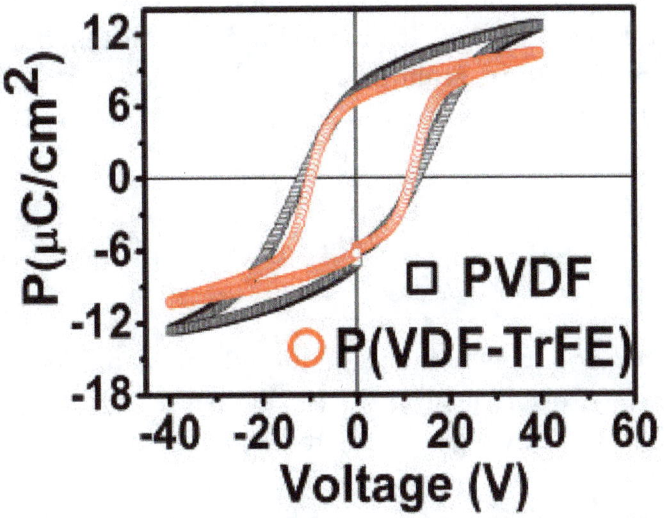


**Figure 18.** *P-E* loops of PVDF and P(VDF-TrFE) copolymer LB films

The phase image of the piezoresponse shows a polarization switching in the PVDF homopolymer films, suggesting a typical ferroelectric feature. The ferroelectric *P-E* loops of the PVDF homopolymer and P(VDF-TrFE) films are shown in Fig. 18. The well *P-E* hysteresis loop of the PVDF films also confirms its ferroelectricity. It can also be seen that both the $P_r$ and $E_c$ of the PVDF LB films are higher than that of P(VDF-TrFE) LB films. In addition, the breakdown voltage of the PVDF LB films is found to be above 50 V, whereas the breakdown voltage of the P(VDF-TrFE) LB films is only 40 V.

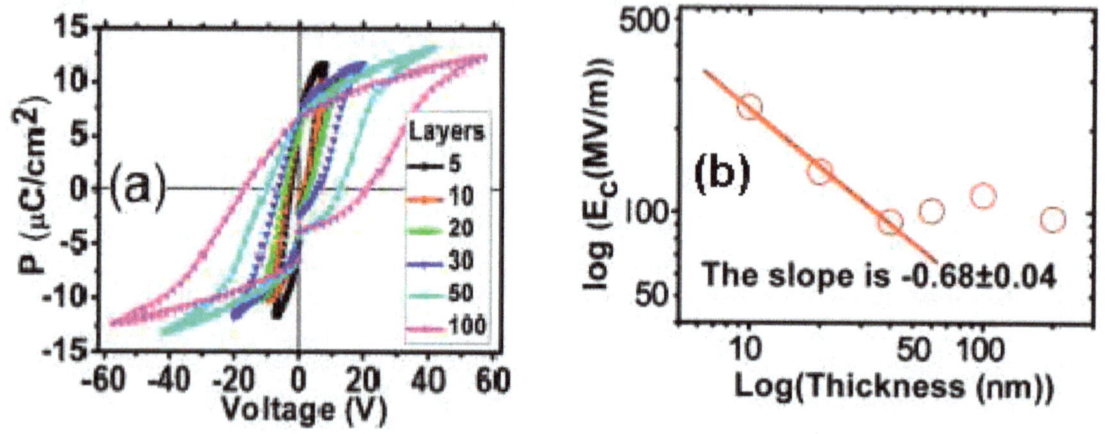


**Figure 19.** (a) The P-E loops of the PVDF homopolymer LB films with 10 to 200 transferred layers. (b) Log $E_c$ vs.log thickness of the PVDF homopolymer LB films

In this study, the film thickness dependence of the $E_c$ of the PVDF homopolymer LB films is investigated. Figure 19(a) shows the *P-E* loops of the PVDF homopolymer LB films with various transferred layers. Note that the $E_c$ is basically constant between 200 nm and 44 nm but dramatically increases with the thickness decreasing in the range from 45 nm to 11 nm (Fig. 19(b)). Thus, the switching in the thicker thickness range is attributed to be extrinsic. Further investigation on the thickness dependence of the $E_c$ of our PVDF homopolymer films in the thinner thickness range, i.e., from 45 nm to 11 nm, reveals that the relationship basically complies with the semi-empirical scaling law of the Janovec–Kay–Dunn (JKD) scaling, suggesting that the switching is neither extrinsic nor intrinsic in this range.

## 12. Conclusion

PVDF-based ferroelectric polymers have been studied for many years, and widely used in many electronic devices, for example, transducers, actuators, switches, and infrared sensors. Nonetheless, there are lots of novel properties that need to be explored. In this chapter, the LB method has been used for preparing ultrathin films of PVDF-based films. In addition, many special characteristics of these films have been generalized. These special properties include high tunability, huge electrocaloric effect, polarization switching, self-polarization, and enhanced electric properties in the artificial polymer multilayers. Besides these properties and potential applications, ferroelectric polymers possess many other advantages for applications. The advantages of this type of ferroelectric polymers include low cost, ease and flexibility of fabrication in different kinds of thin film forms, and resistance to degradation caused by strain. PVDF-based polymers are also more readily altered to conform to complex device requirements imposed by the environment, size, shape, physical flexibility, reliability, durability, and other constraints. PVDF-based films can be easily patterned for integrated electronic applications.

## Author details

J.L. Wang*, X.J. Meng and J.H. Chu

*Address all correspondence to: jlwang@mail.sitp.ac.cn

National Laboratory for Infrared Physics, Shanghai Institute of Technical Physics, Chinese Academy of Sciences, Shanghai, China

## References

[1] Kohji Tashiro, Masamichi Kobayashi, Hiroyuki Tadokoro, Eiichi Fukada. Calculation of elastic and piezoelectric constants of polymer crystals by a point charge model: application to poly(vinylidene fluoride) form I. Macromolecules 13 691–698 (1980).

[2] Yasuhiro Takahashi, Hiroyuki Tadokoro. Crystal structure of form III of poly(vinylidene fluoride). Macromolecules 13 (5), 1317–1318 (1980).

[3] G. T. Davis, J. E. McKinney, M. G. Broadhurst, and S. C. Roth. Electric-field-induced phase changes in poly(vinylidene fluoride). J. Appl. Phys. 49, 4998–5003 (1978).

[4] K. Tashiro. Ferroelectric Polymers, H. S. Nalwa, Ed. (New York: Dekker) pp. 63–181 (1995).

[5] J. F. Legrand. Structure and ferroelectric properties of P(VDF-TrFE) copolymers. Ferroelectrics 91, 303–317 (1989). T. Furukawa. Ferroelectric properties of vinylidene fluoride copolymers. Phase Transit. 18, 143–211 (1989).

[6] A. J. Lovinger. Ferroelectric polymers. Science 220, 1115–1121 (1983).

[7] A. Bune, et al. Novel switching phenomena in ferroelectric LB films. Appl. Phys. Lett. 67, 3975–3977 (1995).

[8] J. L. Wang, X. J. Meng, S. Z. Yuan, J. Yang, J. L. Sun, H. S. Xu, and J. H. Chu. High electric tunability of relaxor ferroelectric LB terpolymer films. Appl. Phys. Lett. 93, 192905 (2008).

[9] S. Palto, et al. Ferroelectric Langmuir-Blodgett films. Ferroelectr. Lett. 19, 65–68 (1995).

[10] S. G. Lu, B. Neese, B. J. Chu, Y. Wang, and Q. M. Zhang, Large electric tunability in poly(vinylidene fluoride-trifluoroethylene) based polymers. Appl. Phys. Lett. 93, 042905 (2008).

[11] P. F. Liu, J. L. Wang, X. J. Meng, J. Yang, B. Dkhil, and J. H. Chu. Huge electro caloric effect in LB ferroelectric polymer thin films. New J. Phys. 12, 023035 (2010).

[12] A. S. Mischenko, Q. Zhang, J. F. Scott, R. W. Whatmore, and N. D. Mathur. Giant electrocaloric effect in thin-film PbZr0.95Ti0.05O3. Science 311, 1270 (2006).

[13] F. Jona and G. Shirane. Ferroelectric Crystals (New York: Dover) p 134 (1993).

[14] J. F. Scott. Electrocaloric materials. Annu. Rev. Mater. Res. 41, 229 (2011).

[15] J. L. Wang, S. Z. Yuan, L. Tian, X. J. Meng, J. L. Sun, and J. H. Chu. Transition from relaxor to ferroelectric-like phase in P(VDF-TrFE-CFE) terpolymer ultrathin films. Appl. Phys. Lett. 98, 052906 (2011).

[16] R. J. Klein, J. Runt, and Q. M. Zhang. Influence of crystallization conditions on the microstructure and electromechanical properties of poly(vinylidene fluoride–trifluoroethylene–chlorofluoroethylene) terpolymers. Macromolecules 36, 7220 (2003).

[17] C. G. Duan, W. N. Mei, W. G. Yin, J. Liu, J. R. Hardy, S. Ducharme, and P. A. Dowben. Simulations of ferroelectric polymer film polarization: the role of dipole interactions. Phys. Rev. B 69, 235106 (2004).

[18] B. B. Tian, X. L. Zhao, B. L. Liu, J. L. Wang, L. Han, J. L. Sun, X. J. Meng, and J. H. Chu. Abnormal polarization switching of relaxor terpolymer films at low temperatures. Appl. Phys. Lett. 102, 072906 (2013).

[19] D. Fu, H. Taniguchi, M. Itoh, S. Koshihara, N. Yamamoto, and S. Mori, Relaxor Pb(Mg1/3 Nb2/3)O3: a ferroelectric with multiple inhomogeneities. Phys. Rev. Lett. 103, 207601 (2009).

[20] W. J. Merz. Domain formation and domain wall motions in ferroelectric BaTiO3 single crystals. Phys. Rev. 95, 690 (1954).

[21] B. B. Tian, Z. H. Chen, A. Q. Jiang, X. L. Zhao, B. L. Liu, J. L. Wang, L. Han, Sh. Sun, J. L. Sun, X. J. Meng, and J. H. Chu. The creep process of the domain switching in P(VDF-TrFE) ferroelectric thin films. Appl. Phys. Lett. 103, 042909 (2013).

[22] X. L. Zhao, B. B. Tian, B. L. Liu, J. L. Wang, L. Han, J. L. Sun, X. J. Meng, and J. H. Chu. Self-polarization in ultrathin LB polymer films. Thin Solid Films 551, 171–173 (2014).

[23] X. L. Zhao, J. L. Wang, B.B. Tian, B. L. Liu, Y. H. Zou, X. D. Wang, S. Sun, J. L. Sun, X. J. Meng, J. H. Chu. Temperature dependence of electronic transport property in ferroelectric polymer films. Appl. Surf. Sci. 316, 497–500 (2014).

[24] E. V. Paramonova, V. S. Bystrov, Y. Dekhtyar, A. Katashev, N. Polyaka, A. V. Bystrova, A. V. Sapronova, V. M. Fridkin, H. Kliem, A. L. Kholkin. Computational studies

of PVDF and P(VDF-TrFE) nanofilms polarization during phase transition revealed by emission spectroscopy. Math. Bioinform. 6, 273 (2011).

[25] H. B. Michaelson. The work function of the elements and its periodicity. J. Appl. Phys. 48, 4729 (1977).

[26] S. M. Sze, K. K. Ng. Physics of Semiconductor Devices, 3rd ed. (Hoboken: John Wiley & Sons) (2007).

[27] A. Rose. Space-charge-limited currents in solids. Phys. Rev. 97, 1538 (1955).

[28] X. L. Zhao, J. L. Wang, B. L. Liu, B. B. Tian, Y. H. Zou, S. Sun, J. L. Sun, X. J. Meng, and J. H. Chu. Enhanced dielectric and ferroelectric properties in the artificial polymer multilayers. Appl. Phys. Lett. 104, 082903 (2014).

[29] X. L. Zhao, J. L. Wang, B. B. Tian, B. L. Liu, X. D. Wang, S. Sun, Y. H. Zou, T. Lin, J. L. Sun, X. J. Meng, and J. H. Chu. Enhanced piezoelectric response in the artificial ferroelectric polymer multilayers Appl. Phys. Lett. 105, 222907 (2014).

[30] C. R. McNeill, K. Asadi, B. Watts, P. W. M. Blom, and D. M. de Leeuw. Structure of phase-separated ferroelectric/semiconducting polymer blends for organic non-volatile memories. Small 6, 508 (2010).

[31] S. R. Forrest, The path to ubiquitous and low-cost organic electronic appliances on plastic. Nature 428, 911 (2004).

[32] T. J. Reece, S. Ducharme, A. V. Sorokin, and M. Poulsen. Nonvolatile memory element based on a ferroelectric polymer Langmuir–Blodgett film. Appl. Phys. Lett. 82, 142 (2003).

[33] J. Shen and Y. Ma. Long-range coupling interactions in ferroelectric superlattices. Phys. Rev. B 61, 14279 (2000).

[34] A. Sarkar and S. B. Krupanidhi. Ferroelectric interaction and polarization studies in BaTiO3 SrTiO3 superlattice. J. Appl. Phys. 101, 104113 (2007).

[35] S. J. Callori, J. Gabel, D. Su, J. Sinsheimer, M. V. Fernandez-Serra, and M. Dawber, Ferroelectric PbTiO3 /SrRuO3 superlattices with broken inversion symmetry. Phys. Rev. Lett. 109, 067601 (2012).

[36] J. L. Wang, B. L. Liu, X. L. Zhao, B. B. Tian, Y. H. Zou, S. Sun, H. Shen, J. L. Sun, X. J. Meng, and J. H. Chu. Transition of the polarization switching from extrinsic to intrinsic in the ultrathin PVDF homopolymer films. Appl. Phys. Lett. 104, 182907 (2014).

# Permissions

All chapters in this book were first published in FMSC, by InTech Open; hereby published with permission under the Creative Commons Attribution License or equivalent. Every chapter published in this book has been scrutinized by our experts. Their significance has been extensively debated. The topics covered herein carry significant findings which will fuel the growth of the discipline. They may even be implemented as practical applications or may be referred to as a beginning point for another development.

The contributors of this book come from diverse backgrounds, making this book a truly international effort. This book will bring forth new frontiers with its revolutionizing research information and detailed analysis of the nascent developments around the world.

We would like to thank all the contributing authors for lending their expertise to make the book truly unique. They have played a crucial role in the development of this book. Without their invaluable contributions this book wouldn't have been possible. They have made vital efforts to compile up to date information on the varied aspects of this subject to make this book a valuable addition to the collection of many professionals and students.

This book was conceptualized with the vision of imparting up-to-date information and advanced data in this field. To ensure the same, a matchless editorial board was set up. Every individual on the board went through rigorous rounds of assessment to prove their worth. After which they invested a large part of their time researching and compiling the most relevant data for our readers.

The editorial board has been involved in producing this book since its inception. They have spent rigorous hours researching and exploring the diverse topics which have resulted in the successful publishing of this book. They have passed on their knowledge of decades through this book. To expedite this challenging task, the publisher supported the team at every step. A small team of assistant editors was also appointed to further simplify the editing procedure and attain best results for the readers.

Apart from the editorial board, the designing team has also invested a significant amount of their time in understanding the subject and creating the most relevant covers. They scrutinized every image to scout for the most suitable representation of the subject and create an appropriate cover for the book.

The publishing team has been an ardent support to the editorial, designing and production team. Their endless efforts to recruit the best for this project, has resulted in the accomplishment of this book. They are a veteran in the field of academics and their pool of knowledge is as vast as their experience in printing. Their expertise and guidance has proved useful at every step. Their uncompromising quality standards have made this book an exceptional effort. Their encouragement from time to time has been an inspiration for everyone.

The publisher and the editorial board hope that this book will prove to be a valuable piece of knowledge for researchers, students, practitioners and scholars across the globe.

# List of Contributors

**Aimé Peláiz-Barranco and Yuslín González Abreu**
Facultad de Física - Instituto de Ciencia y Tecnología de Materiales, Universidad de La Habana. San Lázaro y L, Vedado. La Habana, Cuba

**José de los Santos Guerra**
Grupo de Ferroelétricos e Materiais Multifuncionais, Instituto de Física, Universidade Federal de Uberlândia. Uberlândia - M G, Brazil

**Jinfei Wang**
Functional Materials Research Laboratory, College for Materials Science and Engineering, Tongji University, Caoan, Shanghai, China
Dwo Chemical (China) Investment, LTD. Pudong district, Shanghai, China

**Tongqing Yang**
Functional Materials Research Laboratory, College for Materials Science and Engineering, Tongji University, Caoan, Shanghai, China

**Pierre Saint-Grégoire**
University of Nîmes, Department of Sciences and Arts, Nimes cedex, France

**Takeo Sasaki**
Tokyo University of Science,Tokyo, Japan

**S. Mukherjee**
Department of Physics and Center for Nanoscience and Nanotechnology, National Sun Yat-Sen University, Kaohsiung, Taiwan
UGC-DAE Consortium for Scientific Research, Mumbai Centre, Bhabha Atomic Research Centre, Trombay, Mumbai, India

**T. H. Kao, H. C. Wu, K. Devi Chandrasekhar and H. D. Yang**
Department of Physics and Center for Nanoscience and Nanotechnology, National Sun Yat-Sen University, Kaohsiung, Taiwan

**Vladimir Koval**
Institute of Materials Research, Slovak Academy of Sciences, Kosice, Slovakia

**Giuseppe Viola**
Department of Applied Science and Technology, Institute of Materials Physics and Engineering, Torino, Italy
School of Engineering and Materials Science, Queen Mary University of London, London, UK

**Yongqiang Tan**
School of Physics, Shandong University, China

**Guilhermina F. Teixeira, Rafael A. Ciola and Maria A. Zaghete**
Department of Biochemistry and Chemistry Technology, Chemistry Institute, São Paulo State University - UNESP, Araraquara, Brazil

**Walter K. Sakamoto**
Department of Physics and Chemistry, School of Engineering, São Paulo State University, UNESP, Ilha Solteira, Brazil

**Takeshi Shigenari**
Department of Applied Physics and Chemistry, The University of Electro-Communications, Chofu-shi, Tokyo, Japan

**Toshio Ogawa**
Department of Electrical and Electronic Engineering, Shizuoka Institute of Science and Technology, Toyosawa, Fukuroi, Shizuoka, Japan

**Desheng Fu**
Department of Engineering, Graduate School of Integrated Science & Technology, Shizuoka University, 3-5-1 Johoku, Hamamatsu, Japan
Department of Optoelectronics and Nanostructure Science, Graduate School of Science and Technology, 3-5-1 Johoku, Naka-ku, Hamamatsu, Japan

**Mitsuru Itoh**
Materials and Structures Laboratory, Tokyo Institute of Technology, 4259 Nagatsuta, Yokohama, Japan

**J.L. Wang, X.J. Meng and J.H. Chu**
National Laboratory for Infrared Physics, Shanghai Institute of Technical Physics, Chinese Academy of Sciences, Shanghai, China

# Index